ALGEBRA

Larry C. Grove
Department of Mathematics
The University of Arizona

DOVER PUBLICATIONS, INC.
Mineola, New York

Bibliographical Note

This Dover edition, first published in 2004, is an unabridged republication of the work originally published in the "Pure and Applied Mathematics" series by Academic Press, New York, in 1983. An errata list has been added on pp. xv and xvi in the present edition.

Library of Congress Cataloging-in-Publication Data

Grove, Larry C.
Algebra / Larry C. Grove.
p. cm.
Originally published: New York : Academic Press, 1983. (Pure and applied mathematics ; 110).
Includes index.
ISBN 0-486-43947-X (pbk.)
1. Algebra, Abstract. I. Title. II. Pure and applied mathematics (Academic Press) ; 110.

QA162.G76 2004
512'.02—dc22

2004056232

Manufactured in the United States of America
Dover Publications, Inc., 31 East 2nd Street, Mineola, N.Y. 11501

Contents

Preface

It is fairly standard at present for first-year graduate students in mathematics in the United States to take a course in abstract algebra. Most, but not all, of them have previously taken an undergraduate algebra course, but the content and substance of that course vary widely. Thus the first graduate course usually begins from first principles but proceeds at a faster pace.

This book is intended as a textbook for that first graduate course. It is based on several years of classroom experience. Any claim to novelty must be on pedagogical grounds. I have attempted to find and use presentations and proofs that are accessible to students, and to provide a reasonable number of concrete examples, which seem to me necessary in order to breathe life into abstract concepts.

My own practice in teaching has been to treat the material in Chapters I–V as the basic course, and to include material from Chapter VI as time permits. There are in Chapters I–V, however, several sections that can be omitted with little consequence for later chapters; examples include the sections on generators and relations, on norms and traces, and on tensor products. The selection of "further topics" in Chapter VI is naturally somewhat arbitrary. Everyone, myself included, will find unfortunate omissions, and *further* further topics will no doubt be inserted by many who use the book. The topics in Chapter VI are more or less independent of one another, but they tend to draw freely on the first five chapters.

There are two types of exercises. Some are sprinkled throughout the text; these are usually straightforward and are intended to clarify the concepts as they appear. The results of those exercises are often assumed in the following textual material. The other exercises are at the ends of the chapters. They vary widely in difficulty, and are only rarely referred to later. Of course, not all of the exercises are new, and I am indebted to a wide variety of sources.

My debts to earlier textbooks will be clear to those familiar with the sources, but particular mention should be made of the works of Artin [1–4], Van der Waerden [37], Jacobson [17], Zariski–Samuel [41], and Curtis–Reiner [8]. I have followed Kaplansky's elegant version of the Fundamental Theorem of Galois theory.

I have learned more than I can reasonably acknowledge from my colleagues, past and present. I hope they know who they are and accept my gratitude. The same applies to a large number of students, who have suffered through several preliminary versions and who have prompted many improvements. I must single out Kwang Shang Wang and Javier Gomez Calderon, who ferreted out large numbers of mistakes, misprints, and obscurities by means of several careful rereadings.

Finally, my best thanks go to Helen for all the typing and all the rest.

List of Symbols

$\leq$	Subgroup of
$\lhd$	Normal subgroup of
$\prod$	Product
$\times$	Cartesian product, direct product
$\oplus$	Direct sum
$\vee$	Join
$\bigotimes_R$	Tensor product over ring R
$\mathbb{A}$	Field of algebraic numbers
$a \mid b$	a is a divisor of b; b is a multiple of a
$a \sim b$	a and b are associates
$[A, B]$	The group generated by commutators $[x, y]$, $x \in A$ and $y \in B$
ACC	Ascending chain condition
A_n	Alternating group on n letters
Aut G	The automorphism group of group G
$\mathbb{C}$	The field of complex numbers
$\mathrm{cf}(G)$	Class functions on group G
$C_G(x)$	Centralizer of element x in group G
C_{ij}	Column operation; add column i to column j
C_{ijr}	Column operation; add r times column i to column j
$\mathrm{cl}(x)$	Conjugacy class containing group element x
$C(M) = C_R(M)$	Centralizer in ring R of module M
DCC	Descending chain condition
$\deg f(x)$, $\deg f(X)$	Degree of polynomial
$\deg T$	Degree of representation
D_m	Dihedral group of order $2m$
E_{ij}	Matrix unit, ij-entry is 1 and others are 0
$\mathrm{End}(A)$	Endomorphism ring of abelian group A
$F(a)$	Simple field extension with primitive element a
FHT	Fundamental homomorphism theorem (for groups)
FHTM	Fundamental homomorphism theorem (for modules)

FHTR	Fundamental homomorphism theorem (for rings)
$\mathscr{F}H$	Fixed field of subgroup H
F_q	Galois field with q elements
F_R	Field of fractions of integral domain R
$F(S)$	Extension field of F generated by set S
$f'(x)$	Derivative of polynomial $f(x)$
ϕ_∞	Absolute value function on $\mathbb{C}$
ϕ_p	p-adic valuation on $\mathbb{Q}$
G'	Derived group (commutator subgroup) of group G
GCD	Greatest common divisor
G_f	Galois group of polynomial $f(x)$
$GF(q)$	Galois field with q elements
$G^{(k)}$	Subgroup in the derived series of group G
$G(K:F)$	Galois group of field K over subfield F
$\mathscr{G}L$	Subgroup of Galois group fixing elements of intermediate field L
$GL(n, F)$, $GL(n, q)$, $GL(V)$	General linear group
$\mathbb{H}$	Hamilton's ring of quaternions
$\text{Hom}_R(M, N)$	R-homomorphisms from module M to module N
$\sqrt{I}$	Radical of ideal I
$I(G)$	Inner automorphism group of group G
$\text{Im}(f)$	Image of mapping f
$\text{Int}_S(R)$	Integral closure of subring R in ring S
$(I:R)$	Quotient of ideal I in ring R
$\text{Irr}(G)$	Set of irreducible characters of group G
$J(R)$	Jacobson radical of ring R
$\ker f$	Kernel of homomorphism f
$\ker \chi$	Kernel of character χ
LCM	Least common multiple
$L_k(G)$	Subgroup in descending central series of group G
$m_a(x) = m_{a,F}(x)$	Minimal polynomial over field F of algebraic element a
$M_n(R)$	$n \times n$ matrices over ring R
$M[r]$	Submodule of M annihilated by ring element r
$N = N_{K/F}$	Norm function from field K to subfield F
$N_G(A)$	Normalizer in group G of subset A
$\text{Orb}_G(s)$	Orbit of element s under action of group G
$\text{Perm}(S)$	Group of permutations of set S
PID	Principal ideal domain
$\text{PSL}(n, F)$, $\text{PSL}(n, q)$, $\text{PSL}(V)$	Projective special linear group
$\mathbb{Q}$	Field of rational numbers
Q_2	Quaternion group, order 8
Q_m	Generalized quaternion group, order $4m$
$\mathbb{R}$	Field of real numbers
R^*	Nonzero elements in ring R
RG	Group algebra of group G over commutative ring R
R_{ij}	Row operation; add row i to row j
R_{ijr}	Row operation; add r times row i to row j
R_m	Ring of algebraic integers in quadratic field $\mathbb{Q}(\sqrt{m})$

$R\langle S\rangle$	R-module generated by set S
$R[x]$	Ring of polynomials over ring R in indeterminate x
$R(x)$	Field of rational functions over integral domain R in indeterminate x
$R[x_1,\ldots,x_n] = R[X]$	Ring of polynomials over ring R in indeterminates $x_1,\ldots,x_n$
$R(x_1,\ldots,x_n) = R(X)$	Field of rational functions over integral domain R in indeterminates $x_1,\ldots,x_n$
(S)	Ideal generated by subset S of a ring
$\langle S\rangle$	Subgroup generated by subset S of a group
(s)	Principal ideal generated by ring element s
$\mathrm{SL}(n, F)$, $\mathrm{SL}(n, q)$, $\mathrm{SL}(V)$	Special linear group
S_n	Symmetric group on n letters
$\langle S \mid R\rangle$	Group with generating set S subject to set R of relations
$\mathrm{Stab}_G(s)$	Stabilizer of element s under action of group G
$\sigma_i = \sigma_i(x_1,\ldots,x_n)$	ith symmetric polynomial in indeterminates $x_1,\ldots,x_n$
$\mathrm{TD}(K:F)$	Transcendence degree of field K over subfield F
$\mathrm{Tr} = \mathrm{Tr}_{K/F}$	Trace function from field K to subfield F
UFD	Unique factorization domain
$U(R)$	Group of units in ring R with 1
$V_a(f)$	Variation of a sequence of polynomials at element a of an ordered field
V_T	Module over polynomial ring determined by linear transformation T of vector space V
$\lvert x\rvert$	Order of group element x
$[x, y]$	Commutator of group elements x and y
$\mathbb{Z}$	Ring of integers
$\mathscr{Z} = \mathscr{Z}_{G,S}$	Cycle index of permutation group G on set S
$Z(G)$	Center of group G
$Z_i = Z_i(G)$	Subgroup in ascending central series of group G
$\mathbb{Z}_n$	Ring of integers mod n
$\mathbb{Z}_{p^\infty}$	Divisible abelian p-group; p-primary component of $\mathbb{Q}/\mathbb{Z}$

Introduction

The conventions and notation of elementary set theory are assumed to be familiar to the reader. If $\{S_\alpha : \alpha \in A\}$ is any family of sets, indexed by a set A, we shall write $\prod\{S_\alpha : \alpha \in A\}$, or simply $\prod_\alpha S_\alpha$, for their Cartesian product. Thus $\prod\{S_\alpha : \alpha \in A\}$ is the set of all functions $f: A \to \bigcup\{S_\alpha : \alpha \in A\}$ for which $f(\alpha) \in S_\alpha$, all $\alpha \in A$. If the family $\{S_\alpha\}$ is finite, say $\{S_1, \ldots, S_n\}$, or countable, say $\{S_1, S_2, \ldots\}$, we may write $S_1 \times S_2 \times \cdots \times S_n$, or $S_1 \times S_2 \times \cdots$, respectively, for the Cartesian product. In those cases the elements of the Cartesian product are conveniently represented as ordered n-tuples $(x_1, x_2, \ldots, x_n)$, or sequences $(x_1, x_2, \ldots)$, respectively, where $x_i \in S_i$ for each i. If S and T are sets we write $S \backslash T$ for the relative complement of T in S, i.e., $S \backslash T = \{x \in S : x \notin T\}$.

The cardinality of any set S will be denoted by $|S|$.

A *binary operation* on a set S is a function from the Cartesian product $S \times S$ to the set S. For our purposes a binary operation will often be called *multiplication*, with notation $(x, y) \mapsto xy$, or *addition*, with notation $(x, y) \mapsto x + y$. A binary operation (say multiplication) on a set S is called *associative* if $x(yz) = (xy)z$ for all $x, y, z \in S$.

We shall have occasion to use *Zorn's Lemma*, an equivalent of the set-theoretic *Axiom of Choice*. A brief discussion, with an example of an application, appears in an appendix.

It is assumed that the reader is conversant with the material of a first course in linear algebra, including standard matrix operations and basic facts concerning vector spaces and linear transformations. The existence of a basis and dimension for a vector space are proved in the appendix.

We shall denote the set of integers by $\mathbb{Z}$, the rational numbers by $\mathbb{Q}$, the real numbers by $\mathbb{R}$, and the complex numbers by $\mathbb{C}$. Frequent use will be made of the division algorithm in $\mathbb{Z}$. Also, familiarity with Euler's totient function ϕ will be required on occasion. Details can be found in any book on elementary number theory or in almost any undergraduate abstract algebra book.

Errata

page	line	
7	-16	The Corollary should follow the proof of Proposition 1.7.
11	18	... G to G/K.
20	-5	Proposition 2.1.
29	6	$p_i f = f_i$ (*italic* f)
30	-14	Exercise 4.2.
39	9	"*and*" should *not* be italicized.
40	13	$P = < x >$
41	-8	3(c) above
51	-12	R/I (*italic* I)
61	-14	$a,\, b \in R^*$,
66	7	... no solutions in integers since $m \leq -15$
69	12	... is irreducible and nonconstant.
69	-3	... is irreducible and nonconstant (and hence ...
70	6	The k should be italicized.
70	16	The a should be italicized.
72	-3	The m should be italicized.
77	3	Omit Exercise 24; it is Proposition 5.17.
78	16	... is a nonzero prime ideal ...
95	-12	Exercise I.2.4
99	9	$1 \leq i \leq k$.
113	-7	The "if" part of the conclusion makes implicit use of Proposition 4.1.
115	-2	$\sum_{k=1}^{r}$
116	3-4	with s As down ... $=$ $(\det A)^s$ and $\mathrm{trace}(t_a) =$ $s{\cdot}\mathrm{trace}\, A$, ...
116	7	Replace the script F by a script G.

120	5	Suppose $F \subseteq E \subseteq K$, $F \subseteq L \subseteq K$, with L a finite Galois extension of F. Show that the join $E \vee L$ is finite and Galois over E, ...
121	16	If p is a prime ...
121	-10	$x^4 + x^3 + x^2 + x + 1 = \ldots$
123	12	Assume that char(F) is not 2.
162	2	The diagram needs to be modified. Delete the arrow labeled f_1; reverse the top vertical arrow labeled i, and relabel it as f_1.
163	17	Italicize the b in *(ra)*b.
164	17	Italicize the m in (m).
175	-10	$\ldots r + I = r(e + I)$ for all ...
175	-9	left ideal I
177	15	J is a right ideal
178	-12	$1 \leq k \in \mathbb{Z}$
186	12	$I_j \cong \Delta_{n_j}^{(j)}$ ... division ring $\Delta^{(j)}$.
187	1	... with the ring $\Delta_{n_j}^{(j)}$ of ...
187	2	division ring $\Delta^{(j)}$ by ...
192	-5	... are isomorphic $\mathbb{C}$-algebras.
193	11	... e is a left relative unit ...
193	12	modular ideal N.
220	15	If R is integrally closed, $[K\colon F]$ is finite, and $a \in S$, ...
257	-2	... of changes in sign (ignoring 0's) ...
271	-18	if $\tilde{T}$ is a representation
271	-17	composition $\tilde{T}\eta$
282	-1	$R_{-39} = \{a + b\sqrt{-39}/2\colon a, b \in \mathbb{Z},\ a \equiv b (\mathrm{mod}\ 2)\}$
294	-19	Elementary divisors, 139, 144
298	18	Simple ring should refer to p. 51 rather than p. 50.

ALGEBRA

Chapter I | Groups

1. GROUPS, SUBGROUPS, AND HOMOMORPHISMS

A nonempty set with an associative binary operation is called a *semigroup*, and a semigroup S having an *identity element* 1 such that $1x = x1 = x$ for all $x \in S$ is called a *monoid*. Most of the algebraic systems discussed herein will be semigroups or monoids, but almost always with further requirements imposed, so the semigroup or monoid aspect will seldom be explicitly emphasized.

One trivial consequence of the definition of a monoid deserves mention.

Proposition 1.1. The identity element of a monoid S is unique.

Proof. Suppose 1 and e are identities in S. Then $1 = 1e = e$.

A *group* is a set G with an associative binary operation (usually called multiplication) and an identity element 1 satisfying the further requirement that for each $x \in G$ there is an *inverse* element $y \in G$ such that $xy = yx = 1$.

Proposition 1.2. If G is a group and $x \in G$, then x has a unique inverse element.

Proof. Let y and z be inverses for x. Then

$$y = y1 = y(xz) = (yx)z = 1z = z.$$

The unique inverse for $x \in G$ is denoted by x^{-1}. Note that $(x^{-1})^{-1} = x$.

Proposition 1.3. If G is a group and $x, y \in G$, then $(xy)^{-1} = y^{-1}x^{-1}$.

Proof

$$(xy)(y^{-1}x^{-1}) = ((xy)y^{-1})x^{-1} = (x(yy^{-1}))x^{-1} = (x1)x^{-1} = xx^{-1} = 1,$$

and similarly $(y^{-1}x^{-1})(xy) = 1$.

As Coxeter [7] has pointed out, the "reversal of order" in Proposition 1.3 becomes clear when we think of the operations of putting on our shoes and socks.

If the binary operation of a group G is written as addition, then the identity element is commonly denoted by 0 rather than 1, and the inverse of x by $-x$ rather than x^{-1}. It is customary to use additive notation only if $x + y = y + x$ for all $x, y \in G$.

In general, a group G (multiplicative again) is called *abelian* (or *commutative*) if $xy = yx$ for all $x, y \in G$.

We write $x^0 = 1, x^1 = x, x^2 = xx$, and in general $x^n = x^{n-1}x$ for $1 \leq n \in \mathbb{Z}$. Define $x^{-n} = (x^{-1})^n$, again for $1 \leq n \in \mathbb{Z}$. It is easy to verify by induction that the usual laws of exponents hold in any group, viz.,

$$x^m x^n = x^{m+n} \qquad \text{and} \qquad (x^m)^n = x^{mn}$$

for all $x \in G$, all $m, n \in \mathbb{Z}$. The additive analog of x^n is nx, so the additive analogs of the laws of exponents are $mx + nx = (m + n)x$ and $n(mx) = (mn)x$.

Exercise 1.1. Verify the laws of exponents for groups.

EXAMPLES

1. Let $G = \{1, -1\} \subseteq \mathbb{R}$, with multiplication as usual. Then G is a group.
2. Let $G = \mathbb{Z}, \mathbb{Q}, \mathbb{R}$, or $\mathbb{C}$, with the usual binary operation of addition. Then G is a group.
3. Let $G = \mathbb{Q}\backslash\{0\}$, the set of nonzero rational numbers, under multiplication. Then G is a group. Similarly this holds for $\mathbb{R}\backslash\{0\}$ and $\mathbb{C}\backslash\{0\}$, but *not* for $\mathbb{Z}\backslash\{0\}$. (Why?)
4. Let S be a nonempty set. A *permutation* of S (sometimes called a *bijection* of S) is a 1–1 function ϕ from S onto S. Let G be the set of all permutations of S. If $\phi, \theta \in G$, we define $\phi\theta$ to be their *composition* product, i.e., $\phi\theta(s) = \phi(\theta(s))$ for all $s \in S$. Composition is a binary operation on G (verify), and it is associative, for if $\phi, \theta, \sigma \in G$ and $s \in S$, then

$$(\phi(\theta\sigma))(s) = \phi(\theta\sigma(s)) = \phi[\theta(\sigma(s))],$$

and

$$((\phi\theta)\sigma)(s) = \phi\theta(\sigma(s)) = \phi[\theta(\sigma(s))].$$

G has an identity element, the permutation $1 = 1_S$ defined by $1(s) = s$, all $s \in S$, and each $\phi \in G$ has an inverse ϕ^{-1} defined by $\phi^{-1}(s_1) = s_2$ if and only if $\phi(s_2) = s_1$ (there are a few details to be verified). Thus G is a group; we write $G = \text{Perm}(S)$. This example is of considerable importance and will be pursued much further.

5. As a special case of the preceding example take $S = \{1, 2, 3, \ldots, n\}$. The group G of all permutations of S is called the *symmetric group* on n letters and is denoted by $G = S_n$. If $\phi \in S_n$, it is convenient to display the function ϕ explicitly in the form

$$\phi = \begin{pmatrix} 1 & 2 & \cdots & n \\ \phi(1) & \phi(2) & \cdots & \phi(n) \end{pmatrix}.$$

For example, if $n = 3$, then $\phi = \left(\begin{smallmatrix} 1 & 2 & 3 \\ 2 & 3 & 1 \end{smallmatrix}\right)$ is the permutation that maps 1 to 2, 2 to 3, and 3 to 1. The notation makes it quite simple to carry out explicit computations of the composition product. Suppose, for example, that $n = 3$ and $\phi = \left(\begin{smallmatrix} 1 & 2 & 3 \\ 2 & 3 & 1 \end{smallmatrix}\right)$, $\theta = \left(\begin{smallmatrix} 1 & 2 & 3 \\ 3 & 2 & 1 \end{smallmatrix}\right)$. Note from the definition of $\phi\theta$ in Example 4 that θ acts first and ϕ second. Thus θ maps 1 to 3 and ϕ then maps 3 to 1, and so the composite $\phi\theta$ maps 1 to 1. Similarly, $\phi\theta$ maps 2 to 3 and maps 3 to 2. Thus

$$\phi\theta = \begin{pmatrix} 1 & 2 & 3 \\ 2 & 3 & 1 \end{pmatrix}\begin{pmatrix} 1 & 2 & 3 \\ 3 & 2 & 1 \end{pmatrix} = \begin{pmatrix} 1 & 2 & 3 \\ 1 & 3 & 2 \end{pmatrix}.$$

Observe that

$$\theta\phi = \begin{pmatrix} 1 & 2 & 3 \\ 3 & 2 & 1 \end{pmatrix}\begin{pmatrix} 1 & 2 & 3 \\ 2 & 3 & 1 \end{pmatrix} = \begin{pmatrix} 1 & 2 & 3 \\ 2 & 1 & 3 \end{pmatrix} \neq \phi\theta,$$

so S_3 is *not* an abelian group. It is easy to see that S_n is, likewise, not abelian for any $n > 3$, although S_1 and S_2 are abelian.

6. Let T be an equilateral triangle in the plane with center O. Let D_3 denote the set of *symmetries* of T, i.e., distance-preserving functions from the plane onto itself that carry T onto T (as a set of points). The elements of D_3 are called *congruences* of the triangle T in plane geometry. With composition as the binary operation, D_3 is a group. Let us list its elements explicitly. There is, of course, the identity function 1, with $1(x) = x$ for all x in the plane. There are two counterclockwise rotations, ϕ_1 and ϕ_2, about O as center through angles of 120° and 240°, respectively, and three mirror reflections $\theta_1, \theta_2, \theta_3$ across the three lines passing through the vertices of T and through O (see Fig. 1).

It is edifying to cut a cardboard triangle, label the vertices, and determine composition products explicitly. The result is the "multiplication table" (Fig. 2) for D_3.

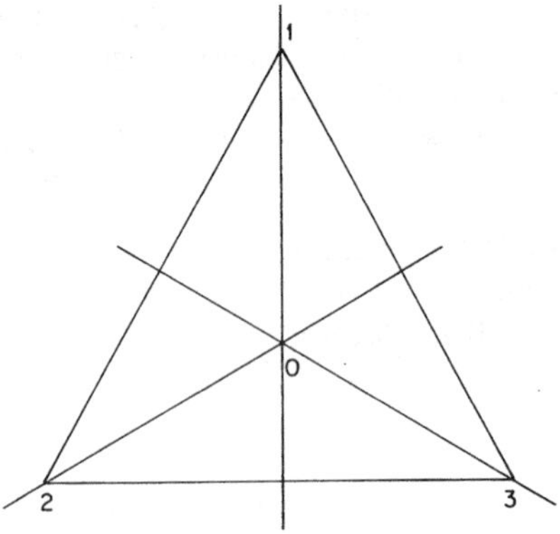

Figure 1

$\cdot$	1	ϕ_1	ϕ_2	θ_1	θ_2	θ_3
1	1	ϕ_1	ϕ_2	θ_1	θ_2	θ_3
ϕ_1	ϕ_1	ϕ_2	1	θ_3	θ_1	θ_2
ϕ_2	ϕ_2	1	ϕ_1	θ_2	θ_3	θ_1
θ_1	θ_1	θ_2	θ_3	1	ϕ_1	ϕ_2
θ_2	θ_2	θ_3	θ_1	ϕ_2	1	ϕ_1
θ_3	θ_3	θ_1	θ_2	ϕ_1	ϕ_2	1

Figure 2

A routine inspection of the table shows that each element has an inverse, and also (if enough time is spent) that the operation is associative. Associativity is also clear from the fact that each element of D_3 is a permutation of the points of the plane. Thus D_3 is a group.

If we let $S = \{1, 2, 3\}$ be the set of vertices of T, then each element of D_3 gives rise to a permutation of S, i.e., to an element of the symmetric group S_3. For example, $\phi_1 \mapsto \binom{1\ 2\ 3}{2\ 3\ 1}$, $\theta_1 \mapsto \binom{1\ 2\ 3}{1\ 3\ 2}$, etc. The result is a 1–1 correspondence between the group D_3 of symmetries of T and the symmetric group S_3. It is instructive to label the elements of S_3 accordingly [e.g., $\alpha_1 = \binom{1\ 2\ 3}{2\ 3\ 1}$, $\beta_1 = \binom{1\ 2\ 3}{1\ 3\ 2}$, etc.], to write out the multiplication table for S_3 and to compare with the table above.

7. This time let T be a square in the plane, with center O, and let D_4 be its set (in fact group) of symmetries. There are four rotations (one of them the identity, through 0°) and four reflections (see Fig. 3). The multiplication table should be computed.

Again each element of D_4 gives rise to a permutation of the set $S = \{1, 2, 3, 4\}$ of vertices of T, i.e., to an element of S_4. For example, the rotation ϕ_1 through 90° counterclockwise about O gives the permutation $\alpha_1 = \binom{1\ 2\ 3\ 4}{2\ 3\ 4\ 1}$. Note in this case, however, that not all elements of S_4 occur. For example, $\binom{1\ 2\ 3\ 4}{2\ 1\ 3\ 4}$ is not the result of any symmetry of the square.

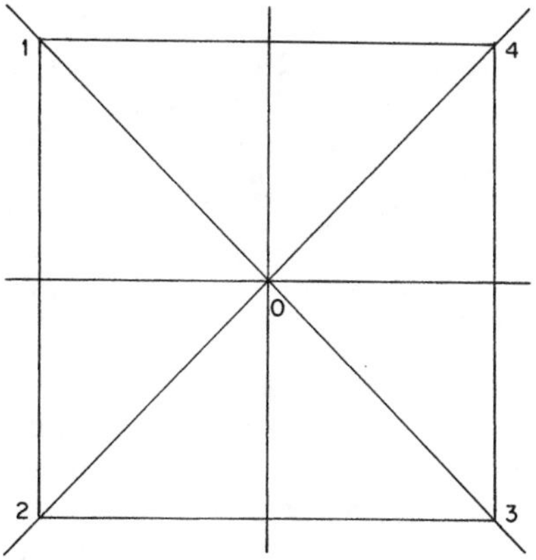

Figure 3

8. The quaternion group Q_2 consists of 8 matrices ± 1, $\pm i$, $\pm j$, $\pm k$ under multiplication, where

$$i = \begin{bmatrix} 0 & -1 & 0 & 0 \\ 1 & 0 & 0 & 0 \\ 0 & 0 & 0 & -1 \\ 0 & 0 & 1 & 0 \end{bmatrix}, \quad j = \begin{bmatrix} 0 & 0 & -1 & 0 \\ 0 & 0 & 0 & 1 \\ 1 & 0 & 0 & 0 \\ 0 & -1 & 0 & 0 \end{bmatrix}, \quad k = \begin{bmatrix} 0 & 0 & 0 & -1 \\ 0 & 0 & -1 & 0 \\ 0 & 1 & 0 & 0 \\ 1 & 0 & 0 & 0 \end{bmatrix}$$

and 1 denotes the 4×4 identity matrix. It is easy to verify that $i^2 = j^2 = k^2 = -1$ and that $ij = k$. All other products can be determined from those. For example, since $ijk = k^2 = -1$ we have $i^2jk = -jk = -i$, and hence $jk = i$. The chief advantage of presenting Q_2 as a set of matrices is that the associative law is automatically satisfied.

9. Klein's 4-group K consists of four 2×2 matrices:

$$1 = \begin{bmatrix} 1 & 0 \\ 0 & 1 \end{bmatrix}, \quad a = \begin{bmatrix} 1 & 0 \\ 0 & -1 \end{bmatrix}, \quad b = \begin{bmatrix} -1 & 0 \\ 0 & 1 \end{bmatrix}, \quad \text{and} \quad c = \begin{bmatrix} -1 & 0 \\ 0 & -1 \end{bmatrix}.$$

Its multiplication table is Fig. 4.

$\cdot$	1	a	b	c
1	1	a	b	c
a	a	1	c	b
b	b	c	1	a
c	c	b	a	1

Figure 4

10. Let T be a regular tetrahedron and let G be the set of all rotations of three-dimensional space that carry T to itself (as a set of points), i.e., all the *rotational symmetries* of T. Thus G consists of the identity 1, rotations through angles of 180° about each of three axes joining midpoints of opposite edges, and rotations through 120° and 240° about each of four axes joining vertices with centers of opposite faces. Thus $|G| = 12$.

Exercise 1.2. Let G be the set of 12 rotational symmetries of a regular tetrahedron.

(1) Verify that G is a group and write out its multiplication table.

(2) Each element of G gives rise to a permutation of the set of vertices of the tetrahedron, numbered 1, 2, 3, and 4. List the resulting permutations in S_4.

(3) Each element of G also gives rise to a permutation of the set of 6 edges of the tetrahedron. List the resulting permutations in S_6.

Exercise 1.3. Describe the groups of rotational symmetries of a cube (there are 24) and of a regular dodecahedron (there are 60). It will be helpful to have cardboard models.

Many more examples will appear as we continue. It will be convenient at this point to introduce some concepts, some terminology, and some elementary consequences of the definitions.

The cardinality $|G|$ of a group G is called its *order*. If G is not finite we usually say simply that G has *infinite order*. An easy counting argument shows that the symmetric group S_n has order $n!$.

A subset H of a group G is called a *subgroup* of G if the binary operation on G restricts to a binary operation on H under which H is itself a group. In that case the identity element of H must be the original identity 1 of G. (Why?) We write $H \leq G$ or $G \geq H$ to indicate that H is a subgroup of G. Referring to the additive groups $\mathbb{Z}$, $\mathbb{Q}$, and $\mathbb{R}$ we have, for example, $\mathbb{Z} \leq \mathbb{Q}$, $\mathbb{Q} \leq \mathbb{R}$, and $\mathbb{Z} \leq \mathbb{R}$.

Proposition 1.4. If H is a nonempty subset of a group G, then $H \leq G$ if and only if $xy^{-1} \in H$ for all $x, y \in H$.

Proof. $\Rightarrow$: Obvious. $\Leftarrow$: Choose $x \in H$ and take $y = x$. Then $xy^{-1} = xx^{-1} = 1 \in H$. Next take $x = 1$ and any $y \in H$ to see that $1y^{-1} = y^{-1} \in H$. Thus $x(y^{-1})^{-1} = xy \in H$ whenever $x, y \in H$, so the multiplication on G restricts to a binary operation on H, which is associative since the original operation on G is associative. Thus H is a group and so $H \leq G$.

Exercise 1.4. If G is a *finite* group and $\varnothing \neq H \subseteq G$, show that H is a subgroup of G if and only if $xy \in H$ whenever $x \in H$, $y \in H$.

Proposition 1.5. If $\{H_\alpha\}$ is any collection of subgroups of a group G, then $\bigcap_\alpha H_\alpha \leq G$.

Proof. Apply the criterion in Proposition 1.4.

If G is a group and S is any subset of G, then by Proposition 1.5 we see that $\bigcap\{H : S \subseteq H \leq G\}$ is a subgroup of G. It is the smallest subgroup of G that contains S; smallest in the sense that it is contained in every subgroup containing S. We write $\langle S\rangle$ for that subgroup and call it the subgroup *generated by S*.

There is a useful alternative description of $\langle S\rangle$ if $S \neq \varnothing$. Let $S^{-1} = \{x^{-1} : x \in S\}$. Choose elements $x_1, x_2, \ldots, x_k \in S \cup S^{-1}$ for any k, $1 \leq k \in \mathbb{Z}$, and form the product $x_1 x_2 \cdots x_k$. The collection of all such elements is a subgroup of G (by Proposition 1.4) that contains S and is contained in every subgroup containing S, hence must be $\langle S\rangle$.

A group G that is generated by a single element, $G = \langle x\rangle$, is called a *cyclic* group. For example, the additive group $\mathbb{Z}$ of integers is a cyclic group, since $\mathbb{Z} = \langle 1\rangle = \langle -1\rangle$. A rotation of the plane about a point through angle $2\pi/n$ generates a cyclic group of order n for each $n \in \mathbb{Z}$, $n \geq 1$.

If G is a group and $x \in G$, then we define the *order* of x, written $|x|$, to be $|\langle x\rangle|$, the order of the cyclic subgroup generated by x. Thus either $|x|$ is infinite or $|x| \in \mathbb{Z}$, in which case $|x|$ is the least positive integer n for which $x^n = 1$.

Proposition 1.6. Suppose x is an element of finite order n in a group G, and suppose $x^m = 1$, $0 < m \in \mathbb{Z}$. Then $n \mid m$.

Proof. Write $m = nq + r$ with $q, r \in \mathbb{Z}$, $0 \leq r < n$. Then $1 = x^m = x^{nq+r} = (x^n)^q x^r = x^r$, so $r = 0$.

Corollary. If $G = \langle x\rangle$ is cyclic of finite order n and $k \mid n$, $0 < k \in \mathbb{Z}$, then $\langle x^{n/k}\rangle$ is the unique subgroup of order k in G.

Proof. Clearly $x^{n/k}$ has order k. If x^s has order k, then $x^{sk} = 1$, so $n \mid sk$, say $rn = sk$. But then $x^s = (x^{n/k})^r \in \langle x^{n/k}\rangle$.

Exercise 1.5. If x and y are commuting elements (i.e., $xy = yx$) in a group G, show that $|xy|$ divides $\mathrm{LCM}(|x|, |y|)$; equality holds if $\langle x\rangle \cap \langle y\rangle = 1$.

Proposition 1.7. A subgroup of a cyclic group is cyclic.

Proof. Say $H \leq G = \langle x\rangle$. If $H = 1$ it is cyclic, so suppose $H \neq 1$. Choose $x^m \in H$ with $0 < m \in \mathbb{Z}$ and m minimal. If $x^k \in H$ write $k = mq + r$, with q, $r \in \mathbb{Z}$ and $0 \leq r < m$. Then $x^r = x^{k-mq} = x^k(x^m)^{-q} \in H$, so $r = 0$ by the minimality of m, and hence $x^k = (x^m)^q \in H$. Thus $H = \langle x^m\rangle$ is cyclic.

Exercise 1.6. (1) Suppose $G = \langle x\rangle$ is infinite. (a) If $m \neq k$ in $\mathbb{Z}$, show that $x^m \neq x^k$. (b) Show that $G = \langle x\rangle = \langle x^{-1}\rangle$, but that $G \neq \langle x^k\rangle$ if $k \neq 1, -1$.

(2) Suppose $G = \langle x\rangle$ is finite of order n. (a) If $m, k \in \mathbb{Z}$, show that $x^m = x^k$ if and only if $m \equiv k (\mathrm{mod}\, n)$, i.e., $n \mid m - k$. (b) Show that $G = \langle x^m\rangle$

if and only if $(m, n) = 1$, i.e., m and n are relatively prime. Thus the number of different generators for G is $\phi(n)$, ϕ being Euler's totient function.

If $H \leq G$ and $x, y \in G$ we say that x and y are *congruent* mod H, and write $x \equiv y (\text{mod } H)$, if $y^{-1}x \in H$. It is easily checked that congruence mod H is an equivalence relation on G, so G is partitioned into equivalence classes. Note that $x \equiv y (\text{mod } H)$ if and only if $y^{-1}x = h \in H$, or $x = yh$ for some $h \in H$. Thus the equivalence class containing y is $\{yh: h \in H\}$, which we write as yH and call the *left coset* of H containing y. Note that $xH = yH$ if and only if $x \equiv y (\text{mod } H)$. The number (possibly infinite) of distinct left cosets of H in G is called the *index* of H in G and is denoted by $[G:H]$.

Theorem 1.8 (Lagrange's Theorem). If G is a finite group and $H \leq G$, then $|H|$ is a divisor of $|G|$. In fact $|G| = [G:H]|H|$.

Proof. The mapping $h \mapsto xh$ is a 1–1 correspondence between H and the left coset xH so $|H| = |xH|$ for all $x \in G$. Since G is the disjoint union of $[G:H]$ left cosets, each with $|H|$ elements, the theorem follows.

A *homomorphism* f from a group G to a group H is a function $f: G \to H$ such that $f(xy) = f(x)f(y)$ for all $x, y \in G$. If f is 1–1 it is called a *monomorphism*; if it is onto it is called an *epimorphism*. If f is both 1–1 and onto it is called an *isomorphism*. In that case f^{-1} is also an isomorphism, from H to G, and we say that G and H are *isomorphic*. When G and H are isomorphic we write $G \cong H$.

A homomorphism from G to G is called an *endomorphism* of G, and an isomorphism of G with itself is called an *automorphism* of G. We shall write $\text{Aut}(G)$ for the set of all automorphisms of a group G.

If $f: G \to H$ is a homomorphism, then the *kernel* of f is defined as $\ker f = \{x \in G: f(x) = 1 \in H\}$.

Proposition 1.9. If $f: G \to H$ is a homomorphism, then $\ker f \leq G$, and f is a monomorphism if and only if $\ker f = 1$.

Proof. Since $f(1) = f(1 \cdot 1) = f(1)f(1)$, we may multiply by $f(1)^{-1}$ to see that $f(1) = 1$. Thus

$$1 = f(1) = f(xx^{-1}) = f(x)f(x^{-1}),$$

so $f(x^{-1}) = f(x)^{-1}$ for all $x \in G$. If $x, y \in \ker f$, then

$$f(xy^{-1}) = f(x)f(y^{-1}) = f(x)f(y)^{-1} = 1 \cdot 1 = 1,$$

so $xy^{-1} \in \ker f$, and thus $\ker f \leq G$. For $x, y \in G$ we have $f(x) = f(y)$ if and only if $1 = f(x)f(y)^{-1} = f(xy^{-1})$, i.e., if and only if $xy^{-1} \in \ker f$. If $\ker f = 1$, then $xy^{-1} = 1$, or $x = y$, so f is 1–1. The converse is clear since $f(1) = 1$.

EXAMPLES

1. Take $G = \mathbb{R}$, the additive group of real numbers, and $H = \{r \in \mathbb{R}: r > 0\}$, with ordinary multiplication. Define $f: G \to H$ by setting $f(r) = e^r$. Then f is an isomorphism and the inverse isomorphism is the natural logarithm function.

2. Let G be the group of symmetries of an equilateral triangle, with notation as in Example 6, p. 3, and let $H = \{\pm 1\}$ under multiplication. Define $f(\phi) = 1$ for each rotation ϕ and $f(\theta) = -1$ for each reflection θ. Then f is a homomorphism. (Verify.)

3. If G is an abelian group and $n \in \mathbb{Z}$, then the function $f: G \to G$ defined by $f(x) = x^n$ for all $x \in G$ is an endomorphism of G since $(xy)^n = x^n y^n$ for all x, $y \in G$.

4. Suppose $\langle x \rangle$ and $\langle y \rangle$ are cyclic groups with $|x| = |y|$. Then the function $f: \langle x \rangle \to \langle y \rangle$, given by $f(x^k) = y^k$, is well defined. That is clear if $|x|$ is infinite, whereas if $|x| = |y| = n$ and if $x^m = x^k$, then $x^{m-k} = 1$, so $n \mid m - k$, and thus also $y^{m-k} = 1$, or $y^m = y^k$. It is easy to see that f is a homomorphism. If $x^m \in \ker f$, then $y^m = 1$, in which case $x^m = 1$ since $|x| = |y|$, so f is 1–1. It is clearly onto, so it is in fact an isomorphism. We have established that any two cyclic groups of the same order are isomorphic.

5. If $G = S_3$ and H is cyclic of order 6, then G and H are *not* isomorphic since H is abelian and G is not. In general, in order to establish nonisomorphism it is necessary to exhibit some group-theoretical property that one of the groups has and the other does not have.

6. The gist of the remarks in Examples 6 and 7, pp. 3–4, is that the group D_3 of symmetries of the triangle is isomorphic with the symmetric group S_3, and the group D_4 of symmetries of the square is isomorphic with a subgroup of the symmetric group S_4.

Exercise 1.7. Show that D_4 is not isomorphic with the quaternion group Q_2.

Proposition 1.10. If G is a group, then Aut G is a group with composition as multiplication.

Proof. Since Aut G is a subset of Perm(G) we may apply Proposition 1.4. If $f, g \in \text{Aut } G$ and x, $y \in G$, then

$$\begin{aligned}(fg^{-1})(xy) &= f(g^{-1}(xy)) = f(g^{-1}(x)g^{-1}(y)) \\ &= f(g^{-1}(x))f(g^{-1}(y)) = (fg^{-1})(x)(fg^{-1})(y),\end{aligned}$$

so fg^{-1} is a homomorphism. Also, $fg^{-1} \in \text{Perm}(G)$, so $fg^{-1} \in \text{Aut } G$ and Aut G is a group.

If $f:G \to H$ is a homomorphism, set $K = \ker f$. If $x \in G$ and $y \in K$ note that

$$f(x^{-1}yx) = f(x^{-1})f(y)f(x) = f(x)^{-1} \cdot 1 \cdot f(x) = 1,$$

so $x^{-1}yx \in K$. If we write $x^{-1}Kx$ for $\{x^{-1}yx: y \in K\}$ we have observed that $x^{-1}Kx \subseteq K$ for all $x \in G$.

Subgroups with the property just described are called *normal* subgroups. In general, then, if $H \le G$ we say that H is *normal* in G if $x^{-1}Hx \subseteq H$ for all $x \in G$. Note that then $H \subseteq xHx^{-1} \subseteq H$, so in fact $x^{-1}Hx = H$ for all $x \in G$. We write $H \lhd G$, or $G \rhd H$, if H is normal in G.

Observe that if G is abelian, then every subgroup is normal. If $G = S_3$, then the subgroup $K = \langle\left(\begin{smallmatrix}1&2&3\\2&3&1\end{smallmatrix}\right)\rangle$ is normal in G, but the subgroup $H = \langle\left(\begin{smallmatrix}1&2&3\\2&1&3\end{smallmatrix}\right)\rangle$ is not, since $\left(\begin{smallmatrix}1&2&3\\1&3&2\end{smallmatrix}\right)\left(\begin{smallmatrix}1&2&3\\2&1&3\end{smallmatrix}\right)\left(\begin{smallmatrix}1&2&3\\1&3&2\end{smallmatrix}\right) = \left(\begin{smallmatrix}1&2&3\\3&2&1\end{smallmatrix}\right) \notin H$.

For any group G define the *center* of G to be

$$Z(G) = \{x \in G: xy = yx, \text{ all } y \in G\}.$$

It is easy to verify (Do so!) that $Z(G) \lhd G$.

If we had defined congruence of x and y in G modulo a subgroup H to mean that $xy^{-1} \in H$, then the equivalence classes would have been *right cosets* Hx. If $H \le G$, then $xH \mapsto Hx^{-1}$ is a 1–1 correspondence between the sets of left and right cosets of H in G.

Exercise 1.8. If $H \le G$ show that $H \lhd G$ if and only if every left coset of H is also a right coset.

If $H \lhd G$ write G/H for the set of all cosets of H in G. Note that $|G/H| = [G:H]$, and that $|G/H| = |G|/|H|$ by Lagrange's Theorem if G is finite. If xH, $yH \in G/H$, define a product $(xH)(yH) = xyH$. The product is well defined since $H \lhd G$, for if $xH = uH$ and $yH = vH$, then $xyH = xHy = uHy = uyH = uvH$. It is a routine matter to verify that G/H, with the binary operation just defined, is itself a group with $1H = H$ as its identity element. It is called the *quotient group*, or *factor group*, of G modulo H.

Define a map $\eta: G \to G/H$ by setting $\eta(x) = xH$. Then

$$\eta(xy) = xyH = xHyH = \eta(x)\eta(y),$$

so η is a homomorphism, in fact it is clearly an epimorphism. We call η the *canonical quotient map* from G to G/H. Note that $x \in \ker \eta$ if and only if $xH = H$, i.e. $\ker \eta = H$.

When G is written additively and $H \lhd G$ we shall still write G/H for the quotient group of G modulo H, but the cosets of H in G are usually written in the form $x + H$ rather than xH.

For an important example take $G = \mathbb{Z}$ and let $H = n\mathbb{Z} = \{nk: k \in \mathbb{Z}\}$ for some $n \in \mathbb{Z}$, $n > 0$. Then, if $x, y \in \mathbb{Z}$, we have $x \equiv y(\text{mod } H)$ if and only if $x - y \in H$, i.e., if and only if $n \mid x - y$, which is the usual definition of

$x \equiv y(\text{mod } n)$ from elementary number theory. Thus the cosets of $H = n\mathbb{Z}$ in $\mathbb{Z}$ are the classes of integers that are congruent mod n. By the division algorithm in $\mathbb{Z}$ every integer is congruent to one of $0, 1, 2, \ldots, n - 1$, all of which are incongruent. If we write $\bar{k}$ for the coset $k + n\mathbb{Z}$, then $\mathbb{Z}/n\mathbb{Z} = \{\bar{0}, \bar{1}, \bar{2}, \ldots, \overline{n-1}\}$, a group of order n. This group is called the group of *integers mod n* and will be denoted by $\mathbb{Z}_n$. Note that $\mathbb{Z}_n$ is cyclic, since $\mathbb{Z}_n = \langle \bar{1} \rangle$.

Theorem 1.11 (The Fundamental Homomorphism Theorem). Suppose $f: G \to H$ is a homomorphism from G *onto* H, and that $K = \ker f$. Then $K \lhd G$ and $G/K \cong H$.

Proof. We have already observed that $K \lhd G$. If $xK = yK$, then $y^{-1}x \in K$, so $1 = f(y^{-1}x) = f(y)^{-1}f(x)$, and hence $f(x) = f(y)$. Thus if we set $\bar{f}(xK) = f(x)$, then $\bar{f}$ is a well-defined map from G/K onto H. Furthermore,

$$\bar{f}(xKyK) = \bar{f}(xyK) = f(xy) = f(x)f(y) = \bar{f}(xK)\bar{f}(yK),$$

so $\bar{f}$ is a homomorphism. Finally,

$$\ker \bar{f} = \{xK : f(x) = 1\} = \{xK : x \in K\} = K,$$

the identity element of G/K, so $\bar{f}$ is an isomorphism by Proposition 1.9.

Let us examine Theorem 1.11 and its proof a bit more closely. Note that $xK = \eta(x)$, where η is the canonical quotient map from G to G/K, Thus we have defined $\bar{f}(xK) = \bar{f}(\eta(x)) = f(x)$, so that f is the composition product of η and $\bar{f}$. That information is captured conveniently in the following diagram.

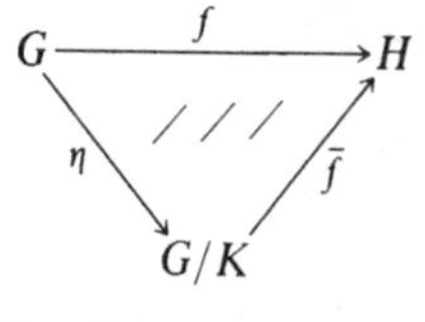

The shading indicates that the diagram *commutes*, which means simply that $f = \bar{f}\eta$.

Even if $f: G \to H$ is not an epimorphism, its image, which we denote by $\text{Im}(f)$, is a subgroup of H, as is easily seen by Proposition 1.4. Of course, f may be viewed as a homomorphism from G onto $\text{Im}(f)$, and the Fundamental Homomorphism Theorem (FHT) gains the slightly more general form $G/\ker f \cong \text{Im}(f) \leq H$. Diagramatically we have

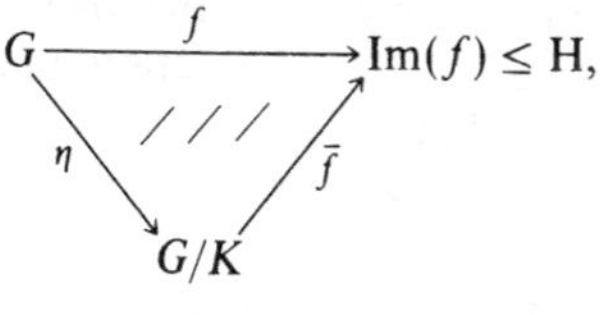

where $K = \ker f$.

Proposition 1.12. Suppose $f: G \to H$ is an epimorphism, with $\ker f = K$. Then $L \leftrightarrow f^{-1}(L) = \{x \in G: f(x) \in L\}$ is a 1–1 correspondence between the set of all subgroups L of H and the set of all subgroups of G that contain K. Furthermore, $L \lhd H$ if and only if $f^{-1}(L) \lhd G$.

Proof. If $L \le H$, then $f^{-1}(L) \le G$ by Proposition 1.4, and it is clear that $K \le f^{-1}(L)$. Note that $f(f^{-1}(L)) = L$, and hence $L \mapsto f^{-1}(L)$ is 1–1. If $K \le M \le G$, set $L = f(M)$. Then $L \le H$ and it is clear that $M \le f^{-1}(L)$. If $x \in f^{-1}(L)$, then $f(x) \in L = f(M)$, so $f(x) = f(y)$ for some $y \in M$. It follows that $y^{-1}x \in \ker f = K \le M$, so $x \in yM = M$, and hence $f^{-1}(L) = M$. Thus $L \mapsto f^{-1}(L)$ is onto. The proof of the statement concerning normality is left as an (easy) exercise.

Corollary. If $K \lhd G$, then all subgroups of G/K have the form M/K, where $K \le M \le G$, and $M/K \lhd G/K$ if and only if $M \lhd G$.

Theorem 1.13 (The Freshman Theorem). Suppose G is a group, $H \lhd G$, $K \lhd H$, and $K \lhd G$. Then $H/K \lhd G/K$, and $G/H \cong (G/K)/(H/K)$.

Proof. If $xK = yK$ in G/K, then $xKH = yKH = xH = yH$ since $K \le H$. Thus $f: G/K \to G/H$, with $f(xK) = xH$, is well defined. Also, $f(xKyK) = f(xyK) = xyH = xHyH = f(xK)f(yK)$, so f is a homomorphism from G/K onto G/H. Since $\ker f = \{xK : xH = H\} = \{xK : x \in H\} = H/K$, we see that $H/K \lhd G/K$ and by the FHT that $(G/K)/(H/K) \cong G/H$.

2. PERMUTATION GROUPS

Suppose that S is a set, G is a group, and $\phi: G \to \text{Perm}(S)$ is a homomorphism from G into the group of all permutations of S. Then we say that G is a *permutation group on* S, or simply that G *acts on* S. If ϕ is a monomorphism we say that G acts *faithfully* on S. It is customary to suppress mention of ϕ when there is only one permutation action under consideration, and if $x \in G$, $s \in S$, we simply write xs rather than $\phi(x)(s)$ for the image of s under the permutation $\phi(x)$. As a result we have a mapping $(x, s) \mapsto xs$ from $G \times S$ to S, which satisfies the following requirements:

(a) $(xy)s = x(ys)$, and
(b) $1s = s$

for all $x, y \in G$, and all $s \in S$. Conversely, any mapping from $G \times S$ to S that satisfies (a) and (b) serves to define a permutation action of G on S if we set $\phi(x)(s) = xs$ for all $x \in G$ and all $s \in S$.

If G is a group acting on a set S and $s \in S$ we define the *stabilizer* of s in G to be

$$\text{Stab}_G(s) = \{x \in G : xs = s\}$$

and we define the *orbit* of s under G to be

$$\mathrm{Orb}_G(s) = \{xs : x \in G\}.$$

If we define $s \sim t$ in S to mean that $s = xt$ for some $x \in G$, then $\sim$ is an equivalence relation and the orbits are its equivalence classes. Thus S is a disjoint union of orbits.

If $\mathrm{Orb}_G(s) = S$ for some (and hence for every) $s \in S$, we say that G is *transitive* on S. This means that given any s and t in S there is some $x \in G$ such that $xs = t$.

Proposition 2.1. If a group G acts on a set S and $s \in S$, then $\mathrm{Stab}_G(s) \leq G$ and $[G : \mathrm{Stab}_G(s)] = |\mathrm{Orb}_G(s)|$.

Proof. Set $H = \mathrm{Stab}_G(s)$. If $x, y \in H$, then $ys = s$, so $s = y^{-1}s$, and hence $(xy^{-1})s = x(y^{-1}s) = xs = s$, so that $xy^{-1} \in H$ and $H \leq G$. We have $xs = ys$ if and only if $y^{-1}x \in H$ if and only if $xH = yH$. Thus $xs \mapsto xH$ is a well-defined 1–1 correspondence between $\mathrm{Orb}_G(s)$ and the set of left cosets of $H = \mathrm{Stab}_G(s)$ in G.

As a first example take $S = G$ and define $\phi(x)\,y = xy$ to be the usual product in G. Then

$$\ker \phi = \{x \in G : xy = y \text{ for all } y \in G\} = 1$$

so G acts faithfully. This faithful representation of G as a permutation group on $S = G$ is called the *left regular representation* of G.

Theorem 2.2 (Cayley's Theorem). If G is any group, then G is isomorphic with a transitive group of permutations acting on a set S (viz., $S = G$).

Proof. Only the transitivity remains to be verified. Given $y, z \in S = G$, take $x = zy^{-1} \in G$. Then $xy = zy^{-1}y = z$.

If G is a group and $x, y \in G$, then the *conjugate* of x by y is defined to be $x^y = y^{-1}xy$. Note that $x^{yz} = (x^y)^z$.

As a second example of a permutation action take $S = G$ again, but this time if $x \in G$ and $y \in S$, then the action of x on y is conjugation of y by x^{-1}, i.e., $\phi(x)y = y^{x^{-1}} = xyx^{-1}$. If $x, y \in G$ and $z \in S$, then

$$\phi(xy)z = z^{(xy)^{-1}} = z^{y^{-1}x^{-1}} = (z^{y^{-1}})^{x^{-1}} = \phi(x)(z^{y^{-1}}) = \phi(x)\phi(y)z,$$

so ϕ is a homomorphism. Note that the kernel of the action is $\{x \in G : y = xyx^{-1}, \text{ all } y \in G\} = Z(G)$, the center of G, so the action is faithful if and only if $Z(G) = 1$. If $y \in G$, then $\mathrm{Orb}_G(y)$ is called the *conjugacy class* of G containing y and is denoted by $\mathrm{cl}(y)$. The stabilizer of $y \in G$ is $\{x \in G : xy = yx\}$, which is commonly called the *centralizer* of y in G and denoted $C_G(y)$. The next proposition is a direct consequence of Proposition 2.1.

Proposition 2.3. If G is a group and $x \in G$, then $|\text{cl}(x)| = [G:C_G(x)]$.

Note that $|\text{cl}(x)| = 1$ if and only if $x^y = x$ for all $y \in G$, which is if and only if $x \in Z(G)$. If G is a finite group, let us choose representatives $x_1, x_2, \ldots, x_k$ of the conjugacy classes having more than one element (if there are any), i.e., the classes *not* in the center of G. Thus G is the disjoint union of $Z(G)$ and the sets $\text{cl}(x_i)$, $1 \le i \le k$, and we have the *class equation* for G:

$$|G| = |Z(G)| + \sum_{i=1}^{k} |\text{cl}(x_i)| = |Z(G)| + \sum_{i=1}^{k} [G:C_G(x_i)].$$

By Lagrange's Theorem (1.8) and Propositions 2.1 and 2.3 each summand on the right-hand side of the class equation is a divisor of $|G|$.

Proposition 2.4. If $p \in \mathbb{Z}$ is a prime, $0 < n \in \mathbb{Z}$, and G is a group with $|G| = p^n$, then $Z(G) \neq 1$.

Proof. In the class equation $|G|$ and all $|\text{cl}(x_i)|$ are divisible by p, so $|Z(G)|$ must be divisible by p and hence must be larger than 1.

Exercise 2.1. If $G/Z(G)$ is cyclic show that $G = Z(G)$ is abelian. Conclude that if p is a prime and $|G| = p^2$, then G is abelian.

Next take S to be the set of all subsets of a group G and define an action of G on S by setting $\phi(x)A = xAx^{-1} = A^{x^{-1}}$ for $x \in G$, $A \in S$ (we agree that $\varnothing^x = \varnothing$). The elements of $\text{Orb}_G(A)$ are called the *G-conjugates* of A; $\text{Stab}_G(A)$ is called the *normalizer* of A in G and is denoted by $N_G(A)$.

Proposition 2.5. If G is a group and $A \subseteq G$, then the number of distinct G-conjugates of A in G is $[G:N_G(A)]$.

Theorem 2.6 (The Isomorphism Theorem). Suppose $H, K \le G$ and $K \le N_G(H)$. Then $KH = HK \le G$, $H \lhd KH$, $K \cap H \lhd K$, and $KH/H \cong K/K \cap H$.

Proof. By KH we mean, of course, $\{xy : x \in K, y \in H\}$. Let us show first that $KH \le G$. Take $x, u \in K$ and $y, v \in H$. Then

$$(xy)(uv)^{-1} = xyv^{-1}u^{-1} = xu^{-1}uyv^{-1}u^{-1} = (xu^{-1})(yv^{-1})^{u^{-1}} \in KH,$$

since $K \le N_G(H)$. Note that $xy = y^{x^{-1}}x$ and $yx = xy^x$, so $HK = KH$. Since $K \le N_G(H)$ it is immediate that $H \lhd KH$. Define $f: K \to KH/H$ by setting $f(x) = xH$. Then f is a homomorphism; it is onto since $xyH = xH$, all $y \in H$. Since

$$\ker f = \{x \in K : xH = H\} = \{x \in K : x \in H\} = K \cap H,$$

the result follows from the FHT.

Exercise 2.2. (1) Suppose, in the context of Theorem 2.6, that HK is finite. We may conclude that $|KH| \cdot |K \cap H| = |K| \cdot |H|$. Suppose more generally that A and B are arbitrary finite subgroups of G and define

$AB = \{ab : a \in A \text{ and } b \in B\}$ (it may well not be a subgroup). Show that $|AB| \cdot |A \cap B| = |A| \cdot |B|$. [*Hint*: The relation $(a_1, b_1) \sim (a_2, b_2)$ if and only if $a_1 b_1 = a_2 b_2$ is an equivalence relation on the Cartesian product $A \times B$. What are the equivalence classes?]

(2) If $C \leq B \leq G$ show that $[G:C] = [G:B][B:C]$. Conclude that if $[G:A]$ and $[G:B]$ are finite, then $[G:A \cap B]$ is finite.

Suppose G is a group and $H \leq G$, and set $S = \{xH : x \in G\}$, the set of all left cosets of H in G. Then G acts naturally on S, with $\phi(x)yH = xyH$, all $x, y \in G$. An element $x \in G$ is in the kernel of the action if and only if $xyH = yH$, or $y^{-1}xy \in H$, or $x \in yHy^{-1}$ for all $y \in G$. Thus the kernel is $K = \bigcap\{H^z : z \in G\}$, and the action is faithful if and only if all the conjugates of H intersect in 1.

If $[G:H]$ is finite, say $[G:H] = n$, then the permutation action of G on S is a homomorphism ϕ from G into Perm(S), which is, of course, isomorphic with the symmetric group S_n and has order $n!$. By the FHT we see that $G/K \cong \text{Im}(\phi)$, and so $[G:K] \mid n!$ by Lagrange's Theorem.

Exercise 2.3. Suppose G is finite, $H \leq G$, $[G:H] = n$, and $|G| \nmid n!$. Show that there is a normal subgroup K of G, $K \neq 1$, such that $K \leq H$.

Theorem 2.7 (Cauchy). Suppose G is a finite group, $p \in \mathbb{Z}$ is a prime, and $p \mid |G|$. Then G has an element x of order p.

Proof (McKay [26]). Let S be the subset of the p-fold Cartesian product $G \times G \times \cdots \times G$ consisting of all $(x_1, x_2, \ldots, x_p)$ for which $x_1 x_2 \ldots x_p = 1$, except that the element $(1, 1, \ldots, 1)$ is excluded. If $(x_1, x_2, \ldots, x_p)$ is to be in S we may choose $x_1, x_2, \ldots, x_{p-1}$ arbitrarily; then $x_p = x_{p-1}^{-1} \ldots x_1^{-1}$ is determined. Thus $|S| = |G|^{p-1} - 1$, so $p \nmid |S|$. Let $C = \langle z \rangle$ be a cyclic group of order p (C has no a priori relationship to G). We may define an action of C on S by specifying that $z(x_1, x_2, \ldots, x_p) = (x_2, x_3, \ldots, x_p, x_1)$ since $x_2 x_3 \ldots x_p x_1 = (x_1 \ldots x_p)^{x_1} = 1$. Since p is prime, each C-orbit in S must have either 1 or p elements by Proposition 2.1. If all orbits had p elements, then $|S|$ would be a multiple of p, so there must be an orbit with one element, which necessarily has the form $(x, x, \ldots, x)$, $x \neq 1$, hence $x^p = 1$.

For an application of Cauchy's Theorem suppose G is a group of order 28. Then G has an element of order 7 and hence has a cyclic subgroup H of order 7. As in the discussion preceding Exercise 2.3, G acts on the set of left cosets of H, and there is consequently a homomorphism $\phi: G \to S_4$. But $|S_4| = 24$ and $|G| = 28 \nmid 24$, so if $K = \ker \phi$, then $K \neq 1$. Since $1 \neq K = \bigcap\{H^x : x \in G\} \leq H$, of prime order, it is clear that $K = H$ and the subgroup of order 7 must be normal.

Exercise 2.4. If $|G| = p^n$, p a prime, show that G has subgroups $G_0, G_1, \ldots, G_n$ with $1 = G_0 \leq G_1 \leq \cdots \leq G_n = G$ such that $[G_i : G_{i-1}] = p$, $1 \leq i \leq n$. [*Hint*: Try induction, choose $G_1 \leq Z(G)$, and consider G/G_1.]

3. THE SYMMETRIC AND ALTERNATING GROUPS

The symmetric group S_n acts on the set $S = \{1, 2, \ldots, n\}$, so if σ is a fixed element in S_n, then the cyclic group $\langle\sigma\rangle$ also acts on S. Write $T_1, T_2, \ldots, T_k$ for the $\langle\sigma\rangle$-orbits in S and define permutations $\sigma_1, \sigma_2, \ldots, \sigma_k$ as follows: σ_i acts as σ does on T_i but acts as the identity on the remainder of S. Clearly $\sigma = \sigma_1\sigma_2 \cdots \sigma_k$, and the permutations $\sigma_1, \sigma_2, \ldots, \sigma_k$ all commute with one another since the orbits $T_1, \ldots, T_k$ are pairwise disjoint. Since T_i is a $\langle\sigma\rangle$-orbit, its elements are permuted cyclically by σ, i.e., if $s \in T_i$ and $|T_i| = n_i$, then s is sent to σs, which is sent to $\sigma^2 s$, etc., until finally $\sigma^{n_i - 1}s$ is sent back to s by σ.

In general a permutation ϕ that permutes a subset T of S cyclically and fixes all elements of $S\backslash T$ is called a *cycle* (a *k-cycle* if $|T| = k$). A cycle ϕ admits the convenient notation

$$\phi = (s\ \phi s\ \phi^2 s \cdots \phi^{k-1}s),$$

each element being mapped by ϕ to the next element to the right except that the last element is mapped back to the first, and elements not mentioned are understood to be fixed.

For example, the permutation $\phi = \binom{1\,2\,3\,4\,5}{3\,4\,5\,1\,2}$ is a cycle in S_5, and it can be written as $\phi = (13524)$, or as $\phi = (35241)$, etc.

If cycles ϕ_1 and ϕ_2 permute the elements of T_1 and T_2, and if $T_1 \cap T_2 = \varnothing$, we say that ϕ_1 and ϕ_2 are *disjoint* cycles. Disjoint cycles clearly commute with one another.

A perusal of the above discussion serves to establish the next proposition.

Proposition 3.1. If $\sigma \in S_n$, then σ can be expressed as a product of disjoint cycles. The expression is unique except for the order of occurrence of the factors, since disjoint cycles commute.

It is customary to suppress 1-cycles (i.e., fixed points) when writing a permutation as a product of cycles. For an example take $\sigma = \binom{1\,2\,3\,4\,5\,6\,7\,8\,9}{2\,7\,3\,5\,8\,9\,1\,6\,4} \in S_9$. The $\langle\sigma\rangle$-orbits are $\{1, 2, 7\}$, $\{3\}$, and $\{4, 5, 6, 8, 9\}$, and σ may be written as $\sigma = (127)(45869)$, the 1-cycle (3) being suppressed.

Note that if σ is a k-cycle, then $|\sigma| = k$. If $\sigma = \sigma_1\sigma_2 \cdots \sigma_m$, disjoint, with σ_i a k_i-cycle, then $|\sigma|$ is the least common multiple of $k_1, k_2, \ldots, k_m$ (see Exercise 1.5). The inverse of a k-cycle is easily obtained by writing the entries in the cycle in reverse order. For example, if $\sigma = (12345)$, then $\sigma^{-1} = (54321)$.

A 2-cycle (ab) is called a *transposition.* Every cycle is easily expressed as a product of transpositions, e.g.,

$$(123 \cdots k) = (1k)(1\ k-1) \cdots (13)(12).$$

Thus by Proposition 3.1 *every* permutation can be written as a product of transpositions. We say that $\sigma \in S_n$ is *even* if it is possible to write σ as a product of an even number of transpositions; otherwise we say that σ is *odd*. Thus a product $\sigma_1\sigma_2$ is even if σ_1 and σ_2 are both even or if they are both odd, but $\sigma_1\sigma_2$ is odd if one of the factors is even and the other is odd. If we set $H = \{\pm 1\}$ under multiplication it follows that the function $f\colon S_n \to H$, defined by

$$f(\sigma) = \begin{cases} 1 & \text{if } \sigma \text{ is even,} \\ -1 & \text{if } \sigma \text{ is odd,} \end{cases}$$

is a homomorphism. The kernel of f is the normal subgroup of S_n consisting of all even permutations; it is called the *alternating group* on n letters and is denoted by A_n.

Proposition 3.2. The alternating group A_n has index 2 in the symmetric group S_n if $n \geq 2$.

Proof (Spitznagel [35]). If we show that the homomorphism f above maps S_n onto H, then $S_n/A_n \cong H$ by the FHT, so $|S_n/A_n| = [S_n : A_n] = |H| = 2$. Thus we need only prove the existence of an odd permutation. If the transposition (12) were even we would be able to write the identity permutation 1 as a product of an *odd* number of transpositions. Suppose we have done so, $1 = (ab)\cdots$, using a minimal number of transpositions and with the smallest possible number of as appearing. At least one more a must appear, since $1a \neq b$, so suppose (ac) is the next one to the right. Note that $(de)(ac) = (ac)(de)$ if (de) and (ac) are disjoint, and $(dc)(ac) = (ad)(cd)$, so we may move the second a to the left and write $1 = (ab)(af)\cdots$, with the same minimality conditions met. But now if $b = f$ we may reduce the number of transpositions by 2, and if $b \neq f$, then $(ab)(af) = (af)(bf)$ and we may reduce the number of as, in both cases a contradiction. Thus (12) is odd and the proof is complete.

Exercise 3.1. If $\sigma \in S_n$ show that σ can not be written once as the product of an even number of transpositions and another time as the product of an odd number.

Let $\sigma = (a_1 a_2 \cdots a_k)$ be a cycle in S_n, let τ be any other element of S_n, and consider the conjugate $\tau\sigma\tau^{-1}$. Write $b_i = \tau(a_i)$, $1 \leq i \leq k$, and agree that $a_{k+1} = a_1$, $b_{k+1} = b_1$. Then

$$\tau\sigma\tau^{-1}(b_i) = \tau\sigma\tau^{-1}(\tau a_i) = \tau\sigma(a_i) = \tau(a_{i+1}) = b_{i+1}, \qquad 1 \leq i \leq k.$$

Furthermore, $\tau\sigma\tau^{-1}(s) = s$ if $s \neq b_1, \ldots, b_k$ (Why?), and so

$$\tau\sigma\tau^{-1} = \tau(a_1 a_2 \cdots a_k)\tau^{-1} = (b_1 b_2 \cdots b_k),$$

or

$$\tau(a_1 a_2 \cdots a_k)\tau^{-1} = (\tau a_1 \tau a_2 \cdots \tau a_k),$$

and the conjugate is another k-cycle, obtained by replacing each a_i by $\tau(a_i)$. For example, if $\sigma = (324)$ and $\tau = (13524)$, then $\tau\sigma\tau^{-1} = (541)$, since $\tau(3) = 5$, $\tau(2) = 4$, and $\tau(4) = 1$.

When $\sigma \in S_n$ is expressed as a product of disjoint cycles suppose there are k_j j-cycles, $1 \le j \le n$ (so that $k_1 + 2k_2 + 3k_3 + \cdots + nk_n = n$). Then we say that σ has *cycle type* $(k_1, k_2, \ldots, k_n)$. For example, in S_9 the permutation $\sigma = \binom{1\,2\,3\,4\,5\,6\,7\,8\,9}{2\,7\,3\,5\,8\,9\,1\,6\,4} = (127)(45869)$ has $k_1 = 1$, $k_3 = 1$, $k_5 = 1$, and all other $k_i = 0$, so its cycle type is $(1,0,1,0,1,0,0,0,0)$.

Proposition 3.3. If $\sigma \in S_n$, then the conjugacy class $\mathrm{cl}(\sigma)$ consists of all $\phi \in S_n$ that have the same cycle type as σ.

Proof. Write $\sigma = \sigma_1\sigma_2 \cdots \sigma_m$ as a product of disjoint cycles (include all 1-cycles throughout this proof). Then, for any $\tau \in S_n$,

$$\tau\sigma\tau^{-1} = \tau\sigma_1\tau^{-1}\,\tau\sigma_2\tau^{-1} \cdots \tau\sigma_m\tau^{-1},$$

so $\tau\sigma\tau^{-1}$ has the same cycle type as σ by the discussion above. Conversely, suppose that $\phi \in S_n$ has the same cycle type as σ. Say

$$\sigma = (a_1a_2\cdots)(b_1b_2\cdots)\cdots,$$
$$\phi = (c_1c_2\cdots)(d_1d_2\cdots)\cdots,$$

with the cycles appearing in order of increasing lengths in both cases. Define $\tau \in S_n$ by setting $\tau(a_i) = c_i$, $\tau(b_i) = d_i$, etc. As above we see that $\tau\sigma\tau^{-1} = \phi$, so $\phi \in \mathrm{cl}(\sigma)$.

Corollary. If $n \ge 3$, then $Z(S_n) = 1$.

Exercise 3.2. (1) Write out the conjugacy classes explicitly in S_3 and S_4.

(2) What are the conjugacy classes in A_4?

(3) Since $|A_4| = 12$, any subgroup of order 6 would be normal. Use (2) to show that A_4 has no subgroup of order 6. Conclude that the converse to Lagrange's Theorem is false.

Proposition 3.4. If $n \ge 5$, then all 3-cycles are conjugate in A_n.

Proof. Let (ijk) be any 3-cycle. Then $(ijk) = \sigma(123)\sigma^{-1}$ for some $\sigma \in S_n$. If $\sigma \in A_n$ we are finished. If not set $\tau = \sigma(45)$. Then $\tau \in A_n$ and

$$\tau(123)\tau^{-1} = \sigma(45)(123)(45)\sigma^{-1} = \sigma(123)\sigma^{-1} = (ijk).$$

Proposition 3.5. If $n \ge 3$, then A_n is generated by 3-cycles.

Proof. If i, j, k, and m are distinct, then $(ij)(ik) = (ikj)$ and $(ij)(km) = (jmk)(ikj)$. The result follows.

A group G is called *simple* if its only normal subgroups are 1 and G. For example, any group of prime order is simple by Lagrange's Theorem. The concept is of central importance in the structure theory of finite groups. The next theorem will be important later when we apply the Galois Theory to the solution of polynomial equations.

Theorem 3.6. If $n \neq 4$, then the alternating group A_n is simple.

Proof. A_1, A_2, and A_3 are simple since their orders are 1, 1, and 3. Suppose $n \geq 5$ and take $H \lhd A_n$, $H \neq 1$. By Propositions 3.4 and 3.5 it will suffice to show that there is a 3-cycle in H in order to conclude that $H = A_n$. Choose a prime p such that $p \mid |H|$, and choose an element $\sigma \in H$ of order p (Cauchy's Theorem, 2.7). Then σ is a product of k disjoint p-cycles for some k. If $p = 3$ and $k = 1$ we are finished. Otherwise there are four cases to consider.

Case 1: $p > 3$. Say $\sigma = (a_1 a_2 \cdots a_p) \cdots$. Then

$$\sigma(a_1 a_2 a_3)\sigma^{-1}(a_1 a_3 a_2) = (a_2 a_3 a_4)(a_1 a_3 a_2) = (a_1 a_4 a_2) \in H.$$

Case 2: $p = 3$ and $k > 1$. Say $\sigma = (a_1 a_2 a_3)(a_4 a_5 a_6) \cdots$. Then

$$\sigma(a_1 a_2 a_4)\sigma^{-1}(a_1 a_4 a_2) = (a_2 a_3 a_5)(a_1 a_4 a_2) = (a_1 a_4 a_3 a_5 a_2) \in H,$$

and we are back in Case 1.

Case 3: $p = 2$, $k = 2m \geq 2$, and there is some letter a_5 fixed by σ. Say $\sigma = (a_1 a_2)(a_3 a_4) \cdots$. Then

$$\sigma(a_1 a_2 a_5)\sigma^{-1}(a_1 a_5 a_2) = (a_2 a_1 a_5)(a_1 a_5 a_2) = (a_1 a_2 a_5) \in H.$$

Case 4: $p = 2$ and $k = 2m \geq 2$. If $\sigma = (a_1 a_2)(a_3 a_4)$, then we are back in Case 3, so suppose $\sigma = (a_1 a_2)(a_3 a_4)(a_5 a_6) \cdots$. Then

$$\sigma(a_1 a_2 a_5)\sigma^{-1}(a_1 a_5 a_2) = (a_2 a_1 a_6)(a_1 a_5 a_2) = (a_1 a_5)(a_2 a_6) \in H.$$

Again we are back in Case 3 and the proof is complete.

Exercise 3.3. Find a normal subgroup of order 4 in A_4. Thus A_4 is *not* simple.

4. THE SYLOW THEOREMS

We have observed (Exercise 3.2.3) that the converse to Lagrange's Theorem fails. The converse does hold, however, for prime power divisors of the order of a finite group by a remarkable theorem due to the Norwegian mathematician Sylow.

If G is a finite group and $p \in \mathbb{Z}$ is a prime, then a *p-Sylow subgroup* of G is a subgroup P such that $|P| = p^k$ is the highest power of p that divides $|G|$ (we write $p^k \parallel |G|$ to indicate that $p^k \mid |G|$ but $p^{k+1} \nmid |G|$).

Theorem 4.1 (The First Sylow Theorem). If G is a finite group and $p \in \mathbb{Z}$ is a prime, then G has a p-Sylow subgroup.

Proof. Induction on $|G|$. The result is trivial if $|G| = 1$ or if $p \nmid |G|$, so assume that $p \mid |G| > 1$ and that the theorem holds for all groups of smaller order. If there exists $H \leq G$, $H \neq G$, such that $p \nmid [G:H]$, then H has a p-Sylow subgroup P which is also p-Sylow in G. Assume then that $p \mid [G:H]$ for all proper subgroups H of G. By the class equation (p. 14)

$$|G| = |Z(G)| + \sum_{i=1}^{k} [G:C_G(x_i)]$$

we see that $p \mid |Z(G)|$. By Cauchy's Theorem (2.7) there is an element $x \in Z(G)$ of order p. Set $K = \langle x \rangle$. Then $K \lhd G$ since $x \in Z(G)$, and if $p^m \parallel |G|$, then $p^{m-1} \parallel |G/K|$. By the induction hypothesis G/K has a p-Sylow subgroup $P_1 = P/K$ of order p^{m-1}, where $P \leq G$. But then $|P| = |P_1| \cdot |K| = p^m$, so P is p-Sylow in G.

As a consequence of Exercise 2.4 and the First Sylow Theorem we see that if p is a prime and p^i divides $|G|$ then G has a subgroup of order p^i.

Exercise 4.1. A group G is called a *p-group* (p a prime) if every element of G has order a power of p. Show that a finite group G is a p-group if and only if $|G|$ is a power of p.

Proposition 4.2. Suppose P is a p-Sylow subgroup of G and that $H \leq G$ is a p-group. Then $H \cap N_G(P) = H \cap P$.

Proof. Set $H_1 = H \cap N_G(P)$. Clearly $H \cap P \leq H_1$. By the Isomorphism Theorem (2.6) we have $H_1P/P \cong H_1/(H_1 \cap P)$, and so $[H_1P:P] = [H_1:H_1 \cap P] = p^m$ for some m, since $H_1 \leq H$, a p-group. Thus $|H_1P| = [H_1P:P] \cdot |P| = p^m \cdot |P|$, and H_1P is a p-group. Thus $H_1P = P$ since P is p-Sylow; hence $H_1 \leq P$, and so $H_1 = H \cap P$.

Theorem 4.3 (The Second Sylow Theorem). Suppose G is a finite group, $p \in \mathbb{Z}$ is a prime, P is a p-Sylow subgroup of G, and $H \leq G$ is a p-group. Then H is contained in some conjugate P^x of P. In particular, all p-Sylow subgroups of G are conjugate with one another.

Proof. Let $\mathcal{S}$ be the set of all G-conjugates of P; then H acts on $\mathcal{S}$ by conjugation. Let $\mathcal{S}_1, \mathcal{S}_2, \ldots, \mathcal{S}_k$ be the H-orbits in $\mathcal{S}$ and choose $P_i \in \mathcal{S}_i$, $1 \leq i \leq k$. Then

$$\mathrm{Stab}_H(P_i) = \{h \in H : P_i^h = P_i\} = H \cap N_G(P_i) = H \cap P_i$$

by Proposition 4.2, so $|\mathcal{S}_i| = [H:H \cap P_i]$ by Proposition 2.1 Note that $|\mathcal{S}| = [G:N_G(P)]$ by Proposition 2.5, so $p \nmid |\mathcal{S}|$. Since $|\mathcal{S}| = \sum_{i=1}^k |\mathcal{S}_i| = \sum_{i=1}^k [H:H \cap P_i]$ and each $[H:H \cap P_i]$ is a power of p we must have $[H:H \cap P_i] = p^0 = 1$ for some i. Thus $H \cap P_i = H$ and $H \leq P_i = P^x$ for some $x \in G$.

Corollary. If a finite group G has a unique p-Sylow subgroup P for some prime p, then $P \lhd G$.

Theorem 4.4 (The Third Sylow Theorem). If G is a finite group and $p \in \mathbb{Z}$ is a prime, then the number of distinct p-Sylow subgroups of G is congruent to 1 modulo p.

Proof. Let P be p-Sylow and again let $\mathscr{S}$ be the set of all G-conjugates of P. Let P act on $\mathscr{S}$ by conjugation, with P-orbits $\mathscr{S}_1, \mathscr{S}_2, \ldots, \mathscr{S}_k$. Note that $\{P\}$ is an orbit with just one element, say $\mathscr{S}_1 = \{P\}$. If $P_i \in \mathscr{S}_i$, $2 \leq i \leq k$, then $P_i \neq P$ and then $\text{Stab}_P(P_i) = P \cap N_G(P_i) = P \cap P_i \neq P$ by Proposition 4.2. Thus $|\mathscr{S}_i| = [P : P \cap P_i]$ is divisible by p if $2 \leq i \leq k$, and so $|\mathscr{S}| = 1 + \sum_{i=2}^{k} |\mathscr{S}_i| \equiv 1 (\text{mod } p)$.

Exercise 4.2. If P is a p-Sylow subgroup of G show that $N_G(N_G(P)) = N_G(P)$.

The Sylow Theorems can be used to show the existence of nontrivial normal subgroups in groups of certain orders, and hence to show that the groups are not simple. A few easy examples of applications of the theorems follow. For the examples let us write S1, S2, and S3 for the first, second, and third Sylow Theorems, respectively.

EXAMPLES

1. Suppose $|G| = 28$. By S3 the number of 7-Sylow subgroups of G is in the list 1, 8, 15, 22, ... But by S2 and Proposition 2.5 the number must be a divisor of 28, and 1 is the only divisor of 28 in the list. Thus there is only one 7-Sylow subgroup, of order 7, and it is normal in G by the corollary to S2.

2. Suppose p and q are primes, with $q < p$, suppose $|G| = pq$, and suppose that $p \not\equiv 1 \pmod q$. By S2, S3, and Proposition 2.5 the number of p-Sylow subgroups of G is $1 + kp$ for some $k \in \mathbb{Z}$ such that $1 + kp \mid pq$, hence $1 + kp \mid q$. Clearly $k = 0$ since $q < p$, and so G has a normal subgroup P of order p. Likewise the number of q-Sylow subgroups has the form $1 + mq$, with $1 + mq \mid p$, so either $1 + mq = 1$ or $1 + mq = p$. But we have assumed that $p \not\equiv 1 \pmod q$, so G also has a normal subgroup Q of order q. Both P and Q are cyclic, say $P = \langle x \rangle$ and $Q = \langle y \rangle$. Then $x^{-1}y^{-1}xy \in P \cap Q = 1$ since P and Q are normal. Thus $xy = yx$ and so $|xy| = |x|\,|y| = pq$ and $G = \langle xy \rangle$ is a cyclic group.

3. Suppose the order of G is $56 = 2^3 \cdot 7$. If G has just one 7-Sylow subgroup it is normal. Otherwise G has 8 different 7-Sylow subgroups, and hence it has $8 \cdot 6 = 48$ elements of order 7. But that leaves only 8 more elements in G, which must constitute a unique (and hence normal) 2-Sylow subgroup. Thus G must have either a normal subgroup of order 7 or of order 8, and cannot be a simple group.

Exercise 4.3. Show that the only simple groups of order less than 36 are those of prime order.

Proposition 4.5. If G is a simple group and $|G| = 60$, then $G \cong A_5$.

Proof. By Exercise 2.3 G can not have subgroups of index 2, 3, or 4. By S3 G must have 6 5-Sylow subgroups, 10 3-Sylow subgroups, and either 5 or 15 2-Sylow subgroups. Suppose there is no subgroup of index 5 in G, so there are 15 2-Sylow subgroups. Let T_1, T_2 be distinct 2-Sylow subgroups, and suppose $T_1 \cap T_2 \neq 1$; hence $|T_1 \cap T_2| = 2$. Set $T = T_1 \cap T_2$ and $F = \langle T_1 \cup T_2 \rangle$. Clearly $T \leq Z(F)$, so $T \lhd F$ and $F \neq G$. But also $|F| > 4$ and $4 \mid |F|$ since $T_1 \leq F$, so $|F| \geq 12$. Consequently $[G:F] \leq 5$, a contradiction. Thus $T_1 \cap T_2 = 1$. But then G has 45 nonidentity elements in 2-Sylow subgroups, 20 in 3-Sylow subgroups, and 24 in 5-Sylow subgroups, for a total of $1 + 45 + 20 + 24 = 90$ elements, again a contradiction. Thus G must have a subgroup of index 5, and hence there is an isomorphism from G to a subgroup H of order 60 in S_5. If $H \neq A_5$, then $H \cap A_5$ would have index 2 and hence would be normal in A_5, contradicting the simplicity of A_5 (see Exercise 12.22). Thus $G \cong A_5$.

5. SOLVABLE GROUPS, NORMAL AND SUBNORMAL SERIES

We know from the FHT that every homomorphic image of a group G is isomorphic with a quotient group of G. Among all possible *abelian* homomorphic images of G we will exhibit one that is in a sense maximal. We take this opportunity to present, by way of example, a notion of *universality* that will prove useful in later sections and chapters.

A pair (U, ε) is called *universal* for a group G (with respect to abelian epimorphic images) if U is an abelian group, $\varepsilon: G \to U$ is an epimorphism, and if given any abelian group A and homomorphism $f: G \to A$, then there is a unique homomorphism $g: U \to A$ for which $f = g\varepsilon$, i.e., the diagram

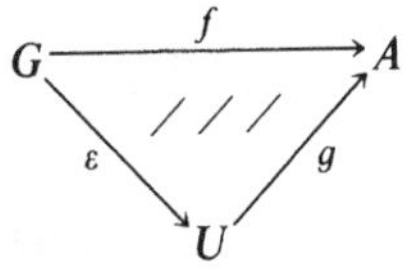

is commutative. We say that f can be "factored through" U.

Proposition 5.1. If a universal pair (U, ε) exists for a group G then U is unique (up to isomorphism).

Proof. Let (U_1, ε_1) be another universal pair for G. We have

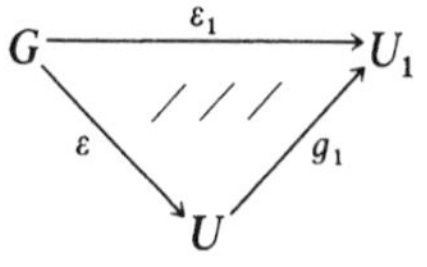

and

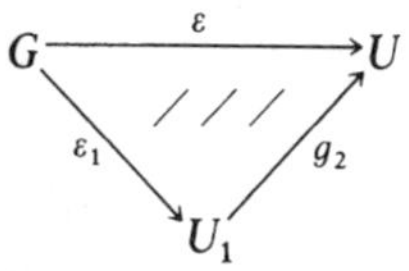

or $\varepsilon_1 = g_1\varepsilon$ and $\varepsilon = g_2\varepsilon_1$. Thus $\varepsilon_1 = g_1 g_2 \varepsilon_1$ and $\varepsilon = g_2 g_1 \varepsilon$. But then we have

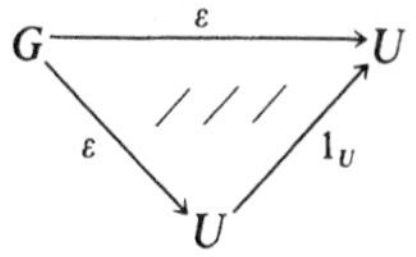

and

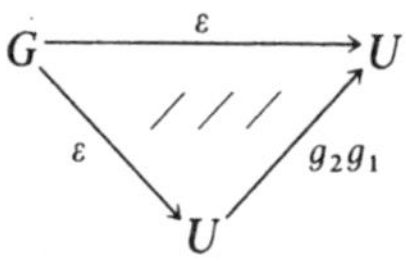

and by the uniqueness in the definition of a universal pair we see that $g_2 g_1 = 1_U$, the identity map on U. Similarly $g_1 g_2$ is the identity map on U_1, and so g_1 and g_2 are inverse isomorphisms.

Exercise 5.1. If (U, ε) is a universal pair for a group G and $h \in \operatorname{Aut}(U)$ show that $(U, h\varepsilon)$ is also universal for G. Conversely if (U, ε_1) is universal for G show that $\varepsilon_1 = h\varepsilon$ for some $h \in \operatorname{Aut}(U)$.

If $x, y \in G$, then their *commutator* is defined to be $[x, y] = x^{-1}y^{-1}xy$. Note that $[x, y] = 1$ if and only if x and y commute. If $f: G \to A$ is any homomorphism from G to an abelian group A, then

$$f([x, y]) = f(x)^{-1}f(y)^{-1}f(x)f(y) = 1$$

and every commutator is in the kernel of f. In particular, if (U, ε) is a universal pair for G, then every commutator $[x, y]$ is in the kernel of ε, since U is abelian.

Exercise 5.2. If $x, y, z \in G$ show that $[x, y]^{-1} = [y, x]$ and $[x, y]^z = [x^z, y^z]$, so inverses and conjugates of commutators are commutators.

For any group G define the *derived group* G' of G to be the subgroup of G generated by all the commutators, i.e.,

$$G' = \langle [x, y] : x, y \in G \rangle.$$

It is clear from Exercise 5.2 that $G' \lhd G$. Note also that if $x, y \in G$, then $x^{-1}y^{-1}xyG' = G'$, so $xyG' = yxG'$, and G/G' is abelian.

Theorem 5.2. If G is a group, then G has a universal pair (U, ε). In fact we may take U to be G/G' and $\varepsilon: G \to U$ to be the canonical quotient map.

Proof. Suppose $f: G \to A$ is a homomorphism, with A abelian. Since every commutator is in the kernel of f we have $G' \leq \ker f$. Thus if $xG' = yG'$ or $y^{-1}x \in G'$, then $f(y^{-1}x) = 1$, or $f(x) = f(y)$, and we may define

$g(xG') = f(x)$. Clearly $g: U \to A$ is a homomorphism (since f is a homomorphism) and $g\varepsilon = f$. If also $g_1: U \to A$ is a homomorphism such that $g_1\varepsilon = f$, then $g_1\varepsilon = g\varepsilon$ and so $g_1 = g$ since ε is an epimorphism. It follows that (U, ε) is a universal pair for G.

Exercise 5.3. (1) Find G' if $G = S_3, S_4$, or A_4.

(2) If $G' \leq H \leq G$ show that $H \lhd G$.

(3) Use Exercise 5.2 to show that if $K \lhd G$, then $K' \lhd G$.

(4) Suppose $f: G \to H$ is an epimorphism, with $\ker f = K$. Show that H is abelian if and only if $G' \leq K$. In particular, a quotient group G/K is abelian if and only if $G' \leq K$. Conclude that G/G' is a "maximal" abelian epimorphic image of G in the sense that G' is a minimal normal subgroup L for which G/L is abelian.

Since $G' \lhd G$ we see that $G'' = (G')' \lhd G$ by Exercise 5.3.3. Set $G^{(0)} = G$, $G^{(1)} = G'$, $G^{(2)} = G''$, ..., and in general $G^{(k+1)} = (G^{(k)})'$. Then $G^{(k)} \lhd G$ for all k, and the sequence

$$G = G^{(0)} \geq G^{(1)} \geq G^{(2)} \geq \cdots \geq G^{(k)} \geq \cdots$$

is called the *derived series* of G. A group G is called *solvable* if $G^{(k)} = 1$ for some $G^{(k)}$ in the derived series of G. Note that any subgroup of a solvable group is solvable.

For example, if G is abelian, then G is obviously solvable since then $G' = 1$. If $G = S_3$, then $G' = A_3$ and $G'' = A_3' = 1$, so S_3 is solvable. If $n \geq 5$, then A_n is nonabelian and simple, and $A_n' \lhd A_n$, so $A_n' = A_n$. Thus A_n is *not* solvable if $n \geq 5$.

Exercise 5.4. (1) Find the derived series for S_4 and conclude that S_4 is solvable.

(2) Show that $S_n' = A_n$ if $n \neq 2$. Conclude that S_n is not solvable if $n \geq 5$. [*Hint*: $(ijk) = (jrk)^{-1}(ijs)^{-1}(jrk)(ijs)$; see Proposition 3.5.]

A *subnormal series* for a group G is a sequence (finite or infinite)

$$G = G_0 \geq G_1 \geq G_2 \geq \cdots,$$

where $G_{i+1} \lhd G_i$ for all i. The subgroups G_i are called *subnormal* subgroups of G. A subnormal subgroup is not necessarily a normal subgroup; look at the group of symmetries of the square for an example. The successive quotient groups G_i/G_{i+1} are called the *factors* of the subnormal series.

A subnormal series $G = G_0 \geq G_1 \geq G_2 \geq \cdots$ is called a *normal series* if $G_i \lhd G$ for all i. Thus the derived series of a group G is a normal series, for example.

The *length* (finite or infinite) of a subnormal series for G is the number of nontrivial factors G_i/G_{i+1}, or equivalently the number of strict inclusions in the series.

Theorem 5.3. A group G is solvable if and only if it has a subnormal series $G = G_0 \geq G_1 \geq G_2 \geq \cdots \geq G_m = 1$ with abelian factors.

Proof. $\Rightarrow$: A segment of the derived series has the desired properties.

$\Leftarrow$: Suppose $G = G_0 \geq G_1 \geq \cdots \geq G_m = 1$ is a subnormal series with abelian factors. Since $G_0/G_1 = G/G_1$ is abelian we have $G' \leq G_1$ by Exercise 5.3.4. Similarly $G_2 \geq G_1' \geq (G')' = G^{(2)}$ since G_1/G_2 is abelian. Inductively $G^{(k)} \leq G_k$ for all k, so $G^{(m)} = 1$ and G is solvable.

Theorem 5.4. Suppose $K \lhd G$. Then G is solvable if and only if both K and G/K are solvable.

Proof. $\Rightarrow$: We observed earlier that any subgroup of a solvable group is solvable. Since the canonical quotient map η: $G \to G/K$ is an epimorphism every commutator $[xK, yK]$ in G/K is the image $\eta([x, y]) = [\eta x, \eta y]$ of a commutator in G. Thus $(G/K)' = \eta(G')$, and likewise $(G/K)^{(k)} = \eta(G^{(k)})$, all k, so G/K is solvable.

$\Leftarrow$: Choose subnormal series with abelian factors for K and for G/K, say $K = K_0 \geq K_1 \geq \cdots \geq K_m = 1$ and $G/K = G_0/K \geq G_1/K \geq \cdots \geq G_k/K = K$. Since $G_i/G_{i+1} \cong (G_i/K)/(G_{i+1}/K)$ by the Freshman Theorem (1.13) we see that

$$G = G_0 \geq \cdots \geq G_k = K = K_0 \geq K_1 \geq \cdots K_m = 1$$

is a subnormal series for G with abelian factors, so G is solvable by Theorem 5.3.

Exercise 5.5. Show that any finite p-group is solvable (use induction; see Proposition 2.4 and Theorem 5.4).

A subnormal series $G = G_0 \geq G_1 \geq G_2 \geq \cdots \geq G_m = 1$ is called a *composition series* for G if each G_{i+1} is a maximal proper normal subgroup of G_i. Equivalently, by the corollary to Proposition 1.12, each factor G_i/G_{i+1} is a nontrivial simple group. For example, $S_3 \geq A_3 \geq 1$ is a composition series, as is $S_n \geq A_n \geq 1$ for all $n \geq 5$. If G is a finite group it is clear (again using the corollary to Proposition 1.12) that any subnormal series with nontrivial factors can be "refined" (by inserting subgroups) to a composition series. For example, $S_4 \geq A_4 \geq 1$ can be refined to $S_4 \geq A_4 \geq K_1 \geq K_2 \geq 1$, where K_1 is the subgroup of order 4 in A_4 (Exercise 3.3), and $|K_2| = 2$.

Theorem 5.5 (The Jordan–Hölder Theorem). If G is a finite group and

$$G = G_0 \geq G_1 \geq \cdots \geq G_m = 1$$

and

$$G = H_0 \geq H_1 \geq \cdots \geq H_k = 1$$

are composition series, then $m = k$ and there is a 1–1 correspondence between the sets of factors so that corresponding factors are isomorphic.

Proof. *Induction on m*. The theorem holds when $m = 1$, for then G is a simple group. Assume that the theorem holds for groups having a composition series of length $m - 1$. If $G_1 = H_1$ the theorem holds for G by the induction hypothesis. If not, set $K_2 = G_1 \cap H_1$. Since G_1 and H_1 are maximal normal subgroups of G and $G_1 \neq H_1$ we have $G_1H_1 = G$, and, by the Isomorphism Theorem (2.6)

$$G/G_1 = G_1H_1/G_1 \cong H_1/H_1 \cap G_1 = H_1/K_2,$$
$$G/H_1 = G_1H_1/H_1 \cong G_1/G_1 \cap H_1 = G_1/K_2.$$

In particular K_2 is a maximal proper normal subgroup in both G_1 and H_1. Choose a composition series $K_2 \geq K_3 \geq \cdots \geq K_s = 1$ for K_2. Then

$$G = G_0 \geq G_1 \geq G_2 \geq G_3 \geq \cdots \geq G_m = 1,$$
$$G = G_0 \geq G_1 \geq K_2 \geq K_3 \geq \cdots \geq K_s = 1,$$
$$G = H_0 \geq H_1 \geq K_2 \geq K_3 \geq \cdots \geq K_s = 1,$$

and

$$G = H_0 \geq H_1 \geq H_2 \geq H_3 \geq \cdots \geq H_k = 1$$

are all composition series. By the induction hypothesis we may conclude that the first and second series have isomorphic factors (in some order), as do the third and fourth, and that $m - 1 = s - 1 = k - 1$. The second and third series have isomorphic factors by the remarks above, and the theorem follows.

For an example let G be the additive group $\mathbb{Z}_{12}$. It has composition series

$$G = \mathbb{Z}_{12} \geq G_1 = 2\mathbb{Z}_{12} \geq G_2 = 4\mathbb{Z}_{12} \geq G_3 = 0,$$
$$G = \mathbb{Z}_{12} \geq H_1 = 2\mathbb{Z}_{12} \geq H_2 = 6\mathbb{Z}_{12} \geq H_3 = 0,$$

and

$$G = \mathbb{Z}_{12} \geq K_1 = 3\mathbb{Z}_{12} \geq K_2 = 6\mathbb{Z}_{12} \geq K_3 = 0.$$

The resulting factors are

$$G/G_1 \cong \mathbb{Z}_2, \qquad G_1/G_2 \cong \mathbb{Z}_2, \qquad G_2/G_3 \cong \mathbb{Z}_3,$$
$$G/H_1 \cong \mathbb{Z}_2, \qquad H_1/H_2 \cong \mathbb{Z}_3, \qquad H_2/H_3 \cong \mathbb{Z}_2,$$

and

$$G/K_1 \cong \mathbb{Z}_3, \qquad K_1/K_2 \cong \mathbb{Z}_2, \qquad K_2/K_3 \cong \mathbb{Z}_2.$$

By the Jordan–Hölder Theorem each finite group G is associated with a finite collection of simple groups (viz., its composition factors), unique up to isomorphism. We say in general that a group A is an *extension* of a group B by another group C if $B \lhd A$ and $A/B \cong C$. Thus each finite group is obtained by a sequence of extensions from its uniquely determined set of simple composition factors. It should be clear then that knowledge of finite simple groups is of extreme importance for the study of finite groups in general. There are several infinite families of finite simple groups and a handful of "sporadic" simple groups. The finite simple groups have only very recently been completely classified; the classification will likely stand as the major triumph of algebra in this century.

6. PRODUCTS

If G_1 and G_2 are groups it is a routine matter to verify that the Cartesian product $G_1 \times G_2$ is also a group if we define a binary operation by setting $(x_1, x_2)(y_1, y_2) = (x_1 y_1, x_2 y_2)$. We find that $G_1 \times G_2$ has identity $(1, 1)$ and that $(x_1, x_2)^{-1} = (x_1^{-1}, x_2^{-1})$.

Our discussion of a "product" for an arbitrary nonempty family of groups will be in terms of the notion of universality encountered in the previous section. If $\{G_\alpha : \alpha \in A\}$ is any nonempty family of groups, then a *product* of the G_α is a group P together with a family $p_\alpha : P \to G_\alpha$, all $\alpha \in A$, of homomorphisms with the following universal property: given any group H and homomorphisms $f_\alpha : H \to G_\alpha$, all $\alpha \in A$, then there exists a unique homomorphism $f : H \to P$ such that $p_\alpha f = f_\alpha$ for all $\alpha \in A$, i.e., the diagrams

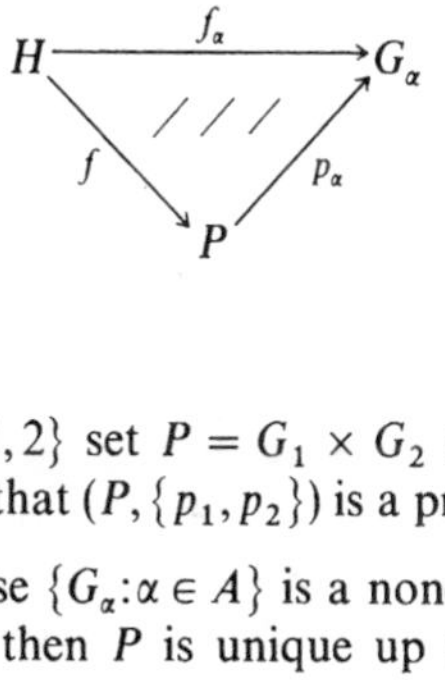

are all commutative.

Exercise 6.1. If $A = \{1, 2\}$ set $P = G_1 \times G_2$ and define $p_1(x_1, x_2) = x_1$ and $p_2(x_1, x_2) = x_2$. Show that $(P, \{p_1, p_2\})$ is a product of G_1 and G_2.

Proposition 6.1. Suppose $\{G_\alpha : \alpha \in A\}$ is a nonempty family of groups. If a product $(P, \{p_\alpha\})$ exists, then P is unique up to isomorphism and each $p_\alpha : P \to G_\alpha$ is an epimorphism.

Proof. Let $(P_1, \{p'_\alpha\})$ be another product. We have

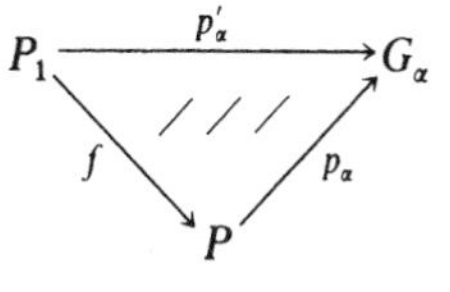

and

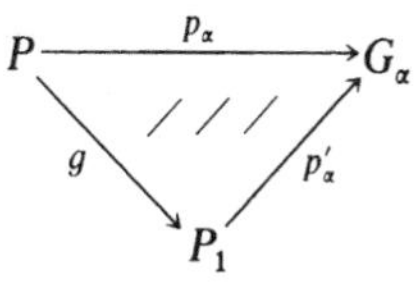

for each $\alpha \in A$, i.e., $p'_\alpha = p_\alpha f$ and $p_\alpha = p'_\alpha g = p_\alpha f g$ for all $\alpha \in A$. Thus we have

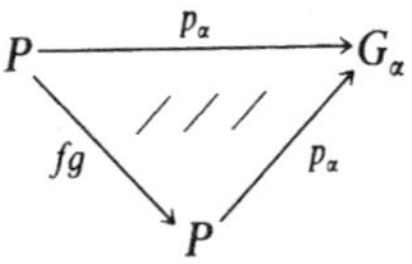

and

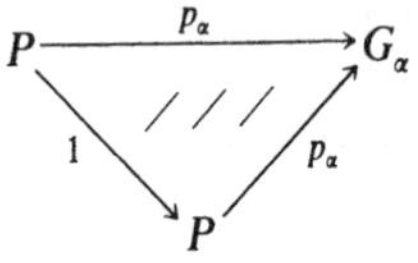

for all $\alpha \in A$, and so $fg = 1$ by the uniqueness in the definition of a product. Similarly $gf = 1$, so $f: P_1 \to P$ and $g: P \to P_1$ are a pair of inverse isomorphisms.

Fix G_α and define $f_\beta: G_\alpha \to G_\beta$ by setting $f_\beta(x) = 1 \in G_\beta$ for all $x \in G_\alpha$ if $\beta \neq \alpha$, but $f_\alpha = i_\alpha$, the identity map on G_α. Then we have

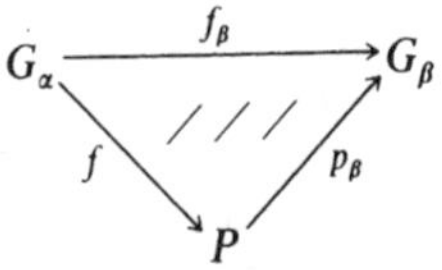

for a unique homomorphism f and for all β. In particular $p_\alpha f = f_\alpha = i_\alpha$, so p_α must be an epimorphism since i_α is an epimorphism.

Theorem 6.2. If $\{G_\alpha : \alpha \in A\}$ is any nonempty family of groups, then a product of the G_α exists.

Proof. Take P to be the Cartesian product, $P = \prod\{G_\alpha : \alpha \in A\}$. Write the elements of P as $(x_\alpha)_{\alpha \in A}$, or simply as (x_α). Define a binary operation on P by setting $(x_\alpha)(y_\alpha) = (x_\alpha y_\alpha)$. Then P is easily seen to be a group, with $(1)_{\alpha \in A}$ as identity and $(x_\alpha)^{-1} = (x_\alpha^{-1})$. For each $\alpha \in A$ define $p_\alpha: P \to G_\alpha$ by setting $p_\alpha\big((x_\beta)_{\beta \in A}\big) = x_\alpha \in G_\alpha$. Let H be any group and suppose that $f_\alpha: H \to G_\alpha$ is a homomorphism for all $\alpha \in A$. Define $f: H \to P$ by setting $f(h) = \big(f_\alpha(h)\big)_{\alpha \in A} \in P$, so that $p_\alpha f = f_\alpha$, all $\alpha \in A$. If also $g: H \to P$ is a homomorphism with $p_\alpha g = f_\alpha$, all $\alpha \in A$, then $p_\alpha g = p_\alpha f$, all α. Thus if $x \in H$ we have $p_\alpha g(x) = g(x)_\alpha = p_\alpha f(x) = f(x)_\alpha$, all $\alpha \in A$, and hence $f(x) = g(x)$ for all $x \in H$. Thus f is unique and P is a product.

If $A = \{1, 2, 3, \ldots, n\}$, or if $A = \{1, 2, 3, \ldots\}$, we often write $G_1 \times G_2 \times \cdots \times G_n$, or $G_1 \times G_2 \times G_3 \times \cdots$ for the product constructed above. Also, the product, as constructed, is often called the *direct* product. In general the homomorphism p_α is called the *projection* of $\prod G_\alpha$ on the *direct factor* G_α.

Theorem 6.3. Suppose $G_1, G_2 \lhd G$, $G_1 \cap G_2 = 1$, and $G_1 G_2 = G$. Then $G \cong G_1 \times G_2$. More generally, if $G_1, G_2, \ldots, G_n \lhd G$, if $G_i \cap \langle \bigcup\{G_j : j \neq i\}\rangle = 1$ for $1 \leq i \leq n$, and if $G_1 G_2 \cdots G_n = G$, then $G \cong G_1 \times G_2 \times \cdots \times G_n$.

Proof. We sketch the proof when $n = 2$ and leave the proof of the more general result as an exercise. Each $x \in G$ can be expressed as $x = x_1x_2$, with $x_i \in G_i$, uniquely since $G_1 \cap G_2 = 1$. Note that $[x_1,x_2] = x_1^{-1}x_2^{-1}x_1x_2 \in G_1 \cap G_2 = 1$, so $x_1x_2 = x_2x_1$. If we define $p_ix = x_i$, $i = 1,2$, then p_i is a homomorphism from G to G_i. If H is any group and $f_i: H \to G_i$ a homomorphism, $i = 1,2$, define $f: H \to G$ by setting $f(h) = f_1(h)f_2(h)$. Then $p_i f = f_i$, $i = 1,2$. If also $g: H \to G$ is a homomorphism satisfying $p_ig = f_i$, $i = 1,2$, then $p_if = p_ig$, $i = 1,2$. But then for any $x \in H$ we have

$$f(x) = p_1f(x)p_2f(x) = p_1g(x)p_2g(x) = g(x),$$

and so $f = g$ is unique. It follows from Proposition 6.1 and Theorem 6.2 that $G \cong G_1 \times G_2$ since G acts as a product for G_1 and G_2.

Exercise 6.2. Complete the proof of Theorem 6.3.

Under the circumstances of Theorem 6.3 we say that G is the *internal direct product* of its subgroups $G_1, G_2, \ldots, G_n$.

Exercise 6.3. Let G be the additive group $\mathbb{Q}$ of rational numbers. Show that G cannot be the internal direct product of two of its subgroups since any two subgroups have nontrivial intersection.

Exercise 6.4. If G is the internal direct product of subgroups G_1 and G_2 show that $G/G_1 \cong G_2$ and $G/G_2 \cong G_1$ (use the FHT).

Exercise 6.5. (1) Show that $Z(\prod_\alpha G_\alpha) = \prod_\alpha Z(G_\alpha)$.
(2) Show that $(G_1 \times G_2 \times \cdots \times G_n)' = G_1' \times G_2' \times \cdots \times G_n'$.
(3) Under what circumstances is $G_1 \times G_2 \times \cdots \times G_n$ solvable?

7. NILPOTENT GROUPS

The *ascending central series* of a group G is the sequence

$$1 = Z_0 \leq Z_1 = Z(G) \leq Z_2 \leq Z_3 \leq \cdots,$$

where $Z_{i+1}/Z_i = Z(G/Z_i)$ for all i. At times it may be necessary to stress the dependence on G and write $Z_i = Z_i(G)$. We see inductively, using the corollary to Proposition 1.12, that each Z_i is normal in G, so the definition makes sense. If $Z_n = G$ for some n, then G is called a *nilpotent* group.

Clearly every abelian group is nilpotent, since $Z_1 = Z(G) = G$. The symmetric groups S_n are *not* nilpotent if $n \geq 3$ since $Z(S_n) = 1$ and hence all $Z_i = 1$.

Proposition 7.1. If G is a finite p-group, then G is nilpotent.

Proof. We see by Proposition 2.4 that $Z_1 \neq 1$. If $Z_1 \neq G$, then G/Z_1 is also a p-group, so $Z(G/Z_1) \neq 1$, again by Proposition 2.4, so $Z_2 \geq Z_1$ and

$Z_2 \neq Z_1$. Likewise if $Z_2 \neq G$, then $Z_3 \geq Z_2$ and $Z_3 \neq Z_2$. Since G is finite we conclude that $Z_n = G$ for some n.

Proposition 7.2. If G is nilpotent, then G is solvable.

Proof. Since $Z_{i+1}/Z_i = Z(G/Z_i)$ it is abelian. Thus $G = Z_n \geq Z_{n-1} \geq \cdots \geq Z_0 = 1$ is a subnormal series with abelian factors so G is solvable by Theorem 5.3.

Corollary. Finite p-groups are solvable.

Exercise 7.1. Show by example that a solvable group need not be nilpotent.

Proposition 7.3. If G is nilpotent and $H \leq G$ but $H \neq G$, then $N_G(H) \neq H$.

Proof. Choose Z_i in the ascending central series such that $Z_i \leq H$ but $Z_{i+1} \nleq H$. We show that $Z_{i+1} \leq N_G(H)$. If $x \in Z_{i+1}$, then $xZ_i \in Z_{i+1}/Z_i = Z(G/Z_i)$, so for any $h \in H$ we have $xZ_ih^{-1}Z_i = h^{-1}Z_ixZ_i$, and hence $x^{-1}hxh^{-1} \in Z_i \leq H$. Consequently $x^{-1}hx \in H$, and so $x \in N_G(H)$.

Corollary. If G is nilpotent and $H \leq G$ is a maximal proper subgroup, then $H \lhd G$.

Proof. $N_G(H) = G$ since H is maximal.

Note that $H = \langle(12)\rangle$ is a maximal proper subgroup of S_3 that is not normal.

Proposition 7.4. If G is a finite group and $P \leq G$ is a p-Sylow subgroup, then $N_G(N_G(P)) = N_G(P)$.

Proof. This was Exercise 4.2.2. If $x \in N_G(N_G(P))$ then both P and P^x are p-Sylow subgroups of $N_G(P)$. By the second Sylow Theorem we have $P = (P^x)^y = P^{xy}$ for some $y \in N_G(P)$. But then $xy \in N_G(P)$ and hence $x \in N_G(P)$.

If H and K are subgroups of a group G denote by $[H, K]$ the subgroup generated by all commutators $[x, y]$, $x \in H$, $y \in K$. Thus, for example, in the derived series for G we have $G^{(k+1)} = [G^{(k)}, G^{(k)}]$. Note that $H \lhd G$ if and only if $[G, H] \leq H$.

Exercise 7.2. Show that $[H, K] = [K, H]$.

Proposition 7.5. Suppose $K \lhd G$ and $K \leq H \leq G$. Then $H/K \leq Z(G/K)$ if and only if $[H, G] \leq K$.

Proof. This is clear since $[y, x] = y^{-1}x^{-1}yx \in K$ if and only if $yxK = xyK$ for all $y \in H$, $x \in G$.

Define the *descending central series* of a group G by setting $L_0 = G$, $L_1 = [G, G]$, and in general $L_{k+1} = [G, L_k]$ for $k \geq 1$. We see inductively, by

Exercise 5.2, that $L_k \lhd G$ for each k, and hence that $L_{k+1} \leq L_k$ for all k. It follows from Proposition 7.5 that $L_k/L_{k+1} \leq Z(G/L_{k+1})$, which is the reason that $L_0 \geq L_1 \geq L_2 \geq \cdots$ is called a *central* series.

Theorem 7.6. A group G is nilpotent if and only if $L_n(G) = 1$ for some n.

Proof. $\Rightarrow$: Take n minimal so that $Z_n = G = L_0$. Then, since $Z_n/Z_{n-1} = Z(G/Z_{n-1})$, we see by Proposition 7.5 that $Z_{n-1} \geq [G, Z_n] = [G, L_0] = L_1$. The same reasoning shows inductively that $Z_{n-k} \geq L_k$, $0 \leq k \leq n$, and in particular $L_n \leq Z_0 = 1$.

$\Leftarrow$: Take n minimal so that $L_n = 1 = Z_0$. Then $[L_{n-1}, G] = L_n = 1$, so $L_{n-1} \leq Z(G) = Z_1$ by Proposition 7.5. Next, $[L_{n-2}, G] = L_{n-1} \leq Z_1$, and we may apply Proposition 7.5 with $K = Z_1$, $H = L_{n-2}Z_1$. We conclude that $L_{n-2}Z_1/Z_1 \leq Z(G/Z_1) = Z_2/Z_1$. Thus $L_{n-2}Z_1 \leq Z_2$, and therefore $L_{n-2} \leq Z_2$. Inductively $L_{n-k} \leq Z_k$ for $0 \leq k \leq n$, and $Z_n \geq L_0 = G$.

Exercise 7.3. Show that subgroups and homomorphic images of nilpotent groups are nilpotent.

Proposition 7.7. If H and K are nilpotent groups, then $G = H \times K$ is nilpotent.

Proof. Clearly $L_0(G) = L_0(H) \times L_0(K)$. If

$$L_k(G) = L_k(H) \times L_k(K)$$

for some k, then

$$\begin{aligned} L_{k+1}(G) &= [H \times K, L_k(H) \times L_k(K)] \\ &= [H, L_k(H)] \times [K, L_k(K)] \\ &= L_{k+1}(H) \times L_{k+1}(K). \end{aligned}$$

The proposition follows by Theorem 7.6.

Theorem 7.8. A finite group G is nilpotent if and only if it is the (internal) direct product of its Sylow subgroups. In particular if G is nilpotent, then each Sylow subgroup is normal, and hence is unique.

Proof. $\Leftarrow$: Use Propositions 7.1 and 7.7.

$\Rightarrow$: If P is p-Sylow in the nilpotent group G, then $N_G(N_G(P)) = N_G(P)$ by Proposition 7.4, and so $N_G(P) = G$ by Proposition 7.3. Thus $P \lhd G$, and P is unique by the second Sylow Theorem. Let $P_1, P_2, \ldots, P_n$ be the distinct (nontrivial) Sylow subgroups of G.

If $i \neq j$ and $x \in P_i$, $y \in P_j$, then $[x, y] \in P_i \cap P_j = 1$, so $xy = yx$. It follows that $P_i \cap \langle \bigcup P_j : j \neq i \rangle = 1$ (compare orders of elements), and also that $|P_1 P_2 \cdots P_n| = |G|$, so $G = P_1 P_2 \cdots P_n$. Thus $P_1, P_2, \ldots, P_n$ satisfy the hypotheses of Theorem 6.3, so G is their internal direct product.

8. FINITE ABELIAN GROUPS

If $G_1, G_2, \ldots, G_n$ are additive abelian groups it is customary to write $G_1 \oplus G_2 \oplus \cdots \oplus G_n$ for their direct product, and to call it the *direct sum* rather than product. In like manner we speak of an abelian group as being an *internal direct sum* of subgroups as in Theorem 6.3, in which case we also write $G = G_1 \oplus G_2 \oplus \cdots \oplus G_n$.

Theorem 8.1 (Frobenius and Stickelberger). If G is a finite abelian group, then G is a direct sum of cyclic subgroups, each of prime power order.

Proof. Since G is abelian, it is nilpotent, hence it is the direct sum of its Sylow subgroups, by Theorem 7.8. Thus we may as well assume that G is a p-group for some prime p. Use induction on $|G|$. Choose $a \in G$ of maximal order, say $|a| = p^k$, and choose $H \leq G$ maximal with respect to $H \cap \langle a \rangle = 0$. Set $G_1 = H \oplus \langle a \rangle \leq G$. If $G_1 \neq G$ we may choose an element $x \in G \backslash G_1$ such that $px \in G_1$ (this is Cauchy's Theorem, 2.7, applied to G/G_1). Say $px = h + ma$, with $h \in H$ and $m \in \mathbb{Z}$. Since p^k is the maximal order for elements of G we have

$$0 = p^k x = p^{k-1}(px) = p^{k-1}h + p^{k-1}ma,$$

and so $p^{k-1}ma = 0$ since $H \cap \langle a \rangle = 0$. But then $p \mid m$, say $m = pr$, since $|a| = p^k$. Consequently $h = px - ma = px - pra = p(x - ra) \in H$, but $x - ra \notin H$ (since $x \notin G_1$). By the maximality of H we must have $(H + \langle x - ra \rangle) \cap \langle a \rangle \neq 0$, so there exist $h_1 \in H$ and $s, t \in \mathbb{Z}$ such that $sa = h_1 + t(x - ra) \neq 0$. Thus $tx = -h_1 + (s + tr)a \in G_1$. Note that $p \nmid t$, for if $t = up$, $u \in \mathbb{Z}$, then

$$t(x - ra) = up(x - ra) = uh \in H,$$

and then $sa = h_1 + uh \in H$, contradicting the fact that $H \cap \langle a \rangle = 0$. Thus p and t are relatively prime, so there are integers m and n such that $mt + np = 1$. But then

$$x = (mt + np)x = m(tx) + n(px) \in G_1,$$

a final contradiction. We conclude that $G = G_1 = H \oplus \langle a \rangle$, and an application of the induction hypothesis to H completes the proof.

Theorem 8.2. Suppose G is an abelian group, $|G| = p^m$ for some prime p, and

$$G = G_1 \oplus G_2 \oplus \cdots \oplus G_r = H_1 \oplus H_2 \oplus \cdots \oplus H_s,$$

with each G_i and H_j cyclic and $1 < |G_i| \leq |G_{i+1}|$, $1 < |H_j| \leq |H_{j+1}|$ for all i and j. Then $r = s$ and $G_i \cong H_i$ for $1 \leq i \leq r$.

Proof. Let $H = \{x \in G : px = 0\}$. Each G_i, being cyclic, has a unique subgroup K_i of order p, so clearly $H = K_1 \oplus K_2 \oplus \cdots \oplus K_r$, and $|H| = p^r$. But similarly $|H| = p^s$, so $p^r = p^s$ and $r = s$, i.e., any two decompositions of the sort indicated must have the same number of cyclic summands.

Suppose now that $|G_i| = |H_i|$ for $1 \le i \le k - 1$, but that (say) $|G_k| < |H_k|$. If $|G_k| = q = p^t$, then $qG_k = 0$ but $qH_k \neq 0$, and we have

$$qG = qG_{k+1} \oplus \cdots \oplus qG_r = qH_k \oplus \cdots \oplus qH_r,$$

which contradicts the first part of the proof.

Theorem 8.3 (The Fundamental Theorem of Finite Abelian Groups). Suppose that G is a finite abelian group and that $p_1, p_2, \ldots, p_k$ are the primes that divide $|G|$. Then $G = G_1 \oplus G_2 \oplus \cdots \oplus G_k$, where G_i is a p_i-group, and

$$G_i = H_{i1} \oplus H_{i2} \oplus \cdots \oplus H_{im(i)}$$

for each i, with H_{ij} cyclic and $1 < |H_{ij}| \le |H_{i(j+1)}|$ for each j. The group G is determined to within isomorphism by the orders of the cyclic subgroups H_{ij}, $1 \le j \le m(i)$, $1 \le i \le k$.

Proof. The decomposition $G = G_1 \oplus G_2 \oplus \cdots \oplus G_k$ is the representation of G as the direct sum of its unique p_i-Sylow subgroups as in Theorem 7.8. The rest of the theorem is a consequence of Theorems 8.1 and 8.2.

Theorem 8.3 will be reestablished in considerably greater generality in Chapter IV.

EXAMPLES

1. Suppose G is abelian and $|G| = 100 = 2^2 \cdot 5^2$. Then $G = G_1 \oplus G_2$, with $|G_1| = 2^2$ and $|G_2| = 5^2$. The possibilities for G_1 are $\mathbb{Z}_4$ and $\mathbb{Z}_2 \oplus \mathbb{Z}_2$; the possibilities for G_2 are $\mathbb{Z}_{25}$ and $\mathbb{Z}_5 \oplus \mathbb{Z}_5$. Thus there are (up to isomorphism) four different abelian groups of order 100, viz., $\mathbb{Z}_4 \oplus \mathbb{Z}_{25}$, $\mathbb{Z}_2 \oplus \mathbb{Z}_2 \oplus \mathbb{Z}_{25}$, $\mathbb{Z}_4 \oplus \mathbb{Z}_5 \oplus \mathbb{Z}_5$, and $\mathbb{Z}_2 \oplus \mathbb{Z}_2 \oplus \mathbb{Z}_5 \oplus \mathbb{Z}_5$.
2. Suppose $|G| = p^2$, p a prime. Then G is abelian by Exercise 2.1 and either $G \cong \mathbb{Z}_{p^2}$ or $G \cong \mathbb{Z}_p \oplus \mathbb{Z}_p$.

Exercise 8.1. Describe all abelian groups of order n for $n = 24, 200, 1000$, p^3, and p^4, where p is a prime.

9. FREE GROUPS

Recall that a *semigroup* is a nonempty set with an associative binary operation. It will be convenient to discuss free semigroups before proceeding to the discussion of free groups. If X and Y are semigroups, then a *homomorphism* from X to Y is a function $f: X \to Y$ such that $f(x_1x_2) = f(x_1)f(x_2)$ for all $x_1, x_2 \in X$. We say that X and Y are *isomorphic* if there is a 1–1 homomorphism from X onto Y.

A semigroup X is *free* on a set S if there is a function $j: S \to X$ such that if we are given any semigroup Y and any function $k: S \to Y$, then there is a unique homomorphism $f: X \to Y$ for which the diagram

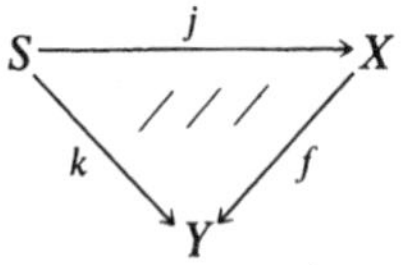

is commutative, i.e., $fj = k$.

Proposition 9.1. If a free semigroup exists on a set S, then it is unique up to isomorphism.

Proof. See the proofs of Propositions 5.1 and 6.1.

Exercise 9.1. If $S = \{1\}$ and $\mathbb{N} = \{1, 2, 3, \ldots\}$ is the semigroup of natural numbers under addition, verify that $\mathbb{N}$ is a free semigroup on S, with $j(1) = 1$.

Theorem 9.2. If S is any nonempty set then a free semigroup X on S exists.

Proof. Set $X = S \cup (S \times S) \cup (S \times S \times S) \cup \cdots$, the union of all the Cartesian powers of finitely many copies of S. If we define a binary operation on X by setting

$$(a_1, a_2, \ldots, a_m)(b_1, b_2, \ldots, b_k) = (a_1, \ldots, a_m, b_1, \ldots, b_k),$$

then it is immediate that the operation is associative, so X is a semigroup. Take $j: S \to X$ to be the inclusion map, $j(x) = x$. If Y is any semigroup and $k: S \to Y$ a function define $f: X \to Y$ by setting $f(a_1, \ldots, a_m) = k(a_1) \cdots k(a_m)$. Then f is a homomorphism and $fj = k$. If also $g: X \to Y$ is a homomorphism and $gj = k$, then

$$\begin{aligned} g(a_1, \ldots, a_m) &= g(ja_1\, ja_2 \cdots ja_m) = (gja_1) \cdots (gja_m) \\ &= (fja_1) \cdots (fja_m) = f(ja_1\, ja_2 \cdots ja_m) \\ &= f(a_1, \ldots, a_m), \end{aligned}$$

so $g = f$.

If A is a set and R is an equivalence relation on A let us write $\mathrm{cl}_R(x)$ for the R-equivalence class containing an element x of A. We write A/R to denote the set of all equivalence classes determined by R and call that set the *quotient* of A mod R. The function $q: A \to A/R$ defined by setting $q(x) = \mathrm{cl}_R(x)$ is called the *quotient map*. An example is provided by the canonical quotient map η from a group G to its quotient G/K by a normal subgroup K.

Proposition 9.3. Suppose Y is a semigroup and R is an equivalence relation on Y such that if xRy and zRw, then $xzRyw$. Then Y/R is a semigroup if we define $\mathrm{cl}_R(x) \cdot \mathrm{cl}_R(y) = \mathrm{cl}_R(xy)$. If Z is another semigroup and $h: Y \to Z$ is a homomorphism, and if $h(x) = h(y)$ whenever xRy, then there is a unique homomorphism $f: Y/R \to Z$ for which the diagram

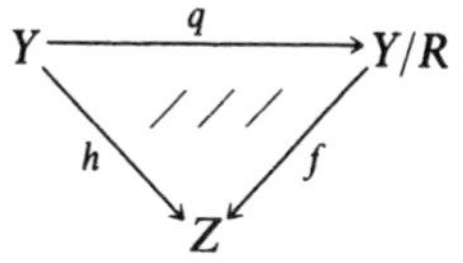

is commutative.

Proof. The condition imposed on R is precisely what is needed to insure that Y/R is a semigroup and q is a homomorphism. If $\text{cl}_R(x) = \text{cl}_R(y)$, then xRy, so $h(x) = h(y)$, and we may define $f(\text{cl}_R(x)) = f(q(x)) = h(x)$. It is clear then that f is a homomorphism; it is unique since q is an epimorphism.

Definition: A group F is *free* on a nonempty set S if there is a function $\phi: S \to F$ such that if G is any group and $\theta: S \to G$ is any function then there is a unique homomorphism $f: F \to G$ such that the diagram

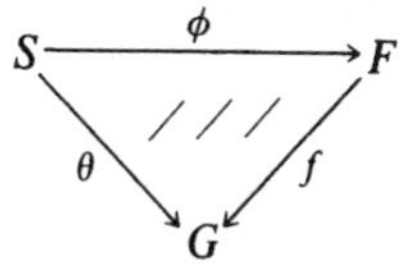

is commutative.

Exercise 9.2. If $S = \{a\}$ let F be a set of distinct "powers" a^n, $n \in \mathbb{Z}$, and define $a^m \cdot a^n = a^{m+n}$. Agree that $a^0 = 1$ and define $\phi(a) = a = a^1$. Show that F is a free group on S. Thus a free group on a single "generator" a can be taken to be an infinite cyclic group $\langle a \rangle$.

Proposition 9.4. If a free group F exists on a nonempty set S, then F is unique up to isomorphism, and $\phi: S \to F$ is a 1–1 function.

Proof. Uniqueness is proved as usual. If ϕ is not 1–1 suppose that $\phi(a) = \phi(b)$ but $a \neq b$ in S. Let $\langle a \rangle$ and $\langle b \rangle$ be free groups on $\{a\}$ and $\{b\}$, respectively, as in Exercise 9.2. Set $G = \langle a \rangle \times \langle b \rangle$, the direct product, and define $\theta: S \to G$ by setting $\theta(a) = (a, 1)$, $\theta(b) = (1, b)$, and $\theta(c) = (1, 1)$ for all c other than a or b. Then there is a homomorphism $f: F \to G$, with $\theta = f\phi$. Thus

$$(a, 1) = \theta(a) = f(\phi(a)) = f(\phi(b)) = \theta(b) = (1, b),$$

and so $a = 1$, a contradiction since $\langle a \rangle$ is infinite cyclic.

Theorem 9.5. If S is a nonempty set, then there is a free group on S.

Proof (Meyer [27]). Choose a set S' with $|S'| = |S|$ and $S \cap S' = \varnothing$, and let $s \mapsto s'$ denote a 1–1 correspondence between S and S'. If we also write

$(s')' = s'' = s$ for each $s \in S$, then $t \mapsto t'$ is a 1–1 map of $T = S \cup S'$ onto itself. Let X be the free semigroup on T, as constructed in Theorem 9.2. If G is a group and $g: X \to G$ is a homomorphism we say that g is *proper* if $g(s') = g(s)^{-1}$ for all $s \in S$ [it follows that $g(t') = g(t)^{-1}$ for all $t \in T$]. Define a relation R on X by agreeing that xRy if and only if $g(x) = g(y)$ whenever G is a group and $g: X \to G$ is a proper homomorphism. It is immediate that R is an equivalence relation, and that if xRy and zRw then $xzRyw$. Thus $F = X/R$ is a semigroup and the quotient map $q: X \to X/R$ is a homomorphism by Proposition 9.3. Write $\bar{x} = q(x)$ for each $x \in X$, so $F = \{\bar{x}: x \in X\}$.

We show first that F is in fact a group. Choose $a \in S$ and any $x \in X$. If $g: X \to G$ is any proper homomorphism, then $g(aa') = 1$, so $g(aa'x) = g(x) = g(xaa')$, and $\overline{aa'} \cdot \bar{x} = \bar{x} = \bar{x} \cdot \overline{aa'}$. Thus $\overline{aa'}$ is an identity for F (unique by Proposition 1.1), and we write $1 = 1_F = \overline{aa'}$. Write $x \in X$ as $x = a_1 a_2 \cdots a_k$, with $a_i \in T$, and set $y = a_k' \cdots a_2' a_1' \in X$. For any proper homomorphism $g: X \to G$ we have

$$g(xy) = g(a_1) \cdots g(a_k) g(a_k)^{-1} \cdots g(a_1)^{-1} = 1 = g(aa'),$$

so $xyRaa'$, or $\overline{xy} = \bar{x} \cdot \bar{y} = \overline{aa'} = 1_F$, and similarly $\bar{y} \cdot \bar{x} = 1$. Thus $\bar{y}$ is an inverse for $\bar{x}$ and F is a group.

Next, F is free on S. Let $i: S \to T$ and $j: T \to X$ be the inclusion maps and define $\phi: S \to F$ by setting $\phi = qji$. If G is any group and $\theta: S \to G$ any function we extend θ to $\theta': T \to G$ by setting $\theta'(s') = \theta(s)^{-1}$ for all $s \in S$. Since X is free on T there is a unique homomorphism $\gamma: X \to G$ such that $\gamma j = \theta'$, and by Proposition 9.3 there is a unique homomorphism $f: F \to G$ such that $fq = \gamma$. But then $f\phi = fqji = \gamma ji = \theta' i = \theta$. If there were another homomorphism $h: F \to G$ such that $h\phi = \theta$, or $hqji = \theta$, then $hqj = \theta'$. (Why?) Thus $hq = \gamma = fq$ since X is a free semigroup on T, and $h = f$ by Proposition 9.3 (see the diagram below). The proof is complete.

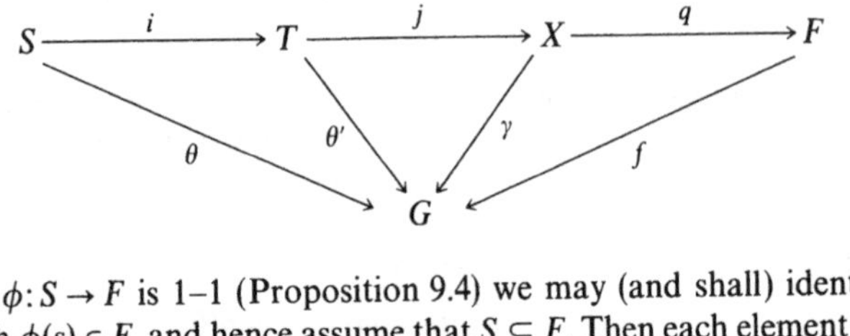

Since $\phi: S \to F$ is 1–1 (Proposition 9.4) we may (and shall) identify each $s \in S$ with $\phi(s) \in F$, and hence assume that $S \subseteq F$. Then each element of F is a product of elements of S and of elements $\phi(s')$, $s \in S$, which are the inverses in F of elements of S, so we have $F = \langle S \rangle$. We usually write $F = F_S$ and call it the free group with S as its set of *free generators*. The definition of freeness thus says that F_S has a set of generators (viz., S) such that any function $\phi: S \to G$, for any group G, can be extended uniquely to a homomorphism $f: F_S \to G$.

Theorem 9.6. Suppose S and U are nonempty sets. Then $F_S \cong F_U$ if and only if $|S| = |U|$.

Proof. $\Rightarrow$: Suppose first that one of the sets, say S, is finite. Since F_S and F_U are isomorphic they have the same number of subgroups of index 2, and each such subgroup is the kernel of a homomorphism onto $\mathbb{Z}_2$. Any such homomorphism is obtained by choosing a nonempty subset of S (resp. U) which is mapped to $\bar{1} \in \mathbb{Z}_2$, and mapping the complementary set to $\bar{0} \in \mathbb{Z}_2$. Thus $2^{|S|} - 1 = 2^{|U|} - 1$, and hence $|S| = |U|$.

If S is an infinite set of free generators set $T = S \cup S^{-1}$. Then $|T| = 2|S| = |S|$. Since any $x \in F_S$ is a product of finitely many elements from T we have

$$|F_S| \leq 1 + |T| + |T \times T| + |T \times T \times T| + \cdots = \aleph_0 |T| = |S|.$$

Consequently $|F_S| = |S|$. If S and U are infinite sets and $F_S \cong F_U$, then $|S| = |F_S| = |F_U| = |U|$.

$\Leftarrow$: Let $\theta: S \to U$ be a 1–1 correspondence. We may view θ as a function from S into F_U, so it extends to a homomorphism $f: F_S \to F_U$. Likewise θ^{-1}: $U \to S$ extends to a homomorphism $g: F_U \to F_S$. But then the restriction of gf to S is just $\theta^{-1}\theta = 1$, the identity map on S, and so $gf = 1$, the identity map on F_S, since $\langle S \rangle = F_S$. Similarly $fg = 1$, the identity map on F_U, so f and g are inverse isomorphisms.

As a consequence of Theorem 9.6 we may define the *rank* of a free group F to be the cardinality of any set of free generators. Thus two free groups are isomorphic if and only if they have the same rank.

A subgroup H of a free group F is also free (it is customary to agree that $H = 1$ is free on the empty set $\varnothing$; this meets the requirements of the definition with ϕ the "empty function"). However, even when the rank of F is finite a subgroup H may have infinite rank. For example, if F is free of rank 2, then its derived group F' has countably infinite rank. One source for proofs of these statements is Kurosh [22].

10. GENERATORS AND RELATIONS

Theorem 10.1. If G is any group and S is a subset of G that generates G, then G is a homomorphic image of the free group F_S.

Proof. Let $\phi: S \to F_S$ be the inclusion map and let $\theta: S \to G$ be the inclusion map. By the definition of F_S we may extend ϕ to a homomorphism $f: F_S \to G$ with $f(S) = S$. Since $\langle S \rangle = G$ it is clear that f is an epimorphism.

Suppose G is a group, $S \subseteq G$, and $\langle S \rangle = G$. As a consequence of Theorem 10.1 and the FHT we see that $G \cong F_S/K$ for some $K \lhd F_S$. If $T \subseteq K$, then each $t \in T$ is a "word" (i.e., product) in F_S involving the generators $s \in S$

and their inverses. The effect of the canonical quotient map is to map each $t \in T$ to $tK = K$, the identity element in F_S/K, or, speaking loosely, to "set $t = 1$." We say then that G has S as a set of *generators* which are subject to the *relations* $\{t = 1 : t \in T\}$.

The remarks above suggest a means of describing groups abstractly. Let S be a set, let $T \subseteq F_S$, and let K be the *normal* subgroup of F_S generated by T, i.e., the intersection of all normal subgroups of F_S that contain T. Then the *presentation* $\langle S \mid t = 1, \text{all } t \in T\rangle$ is defined to be the quotient group F_S/K. If G is a group and G has S as a set of generators subject to the relations $\{t = 1 : t \in T\}$, so that $G \cong F_S/K$, then we also say that G has the presentation $\langle S \mid t = 1, \text{all } t \in T\rangle$.

For an easy example a cyclic group $\langle a \rangle$ of order n has the presentation $\langle a \mid a^n = 1\rangle$.

Proposition 10.2. Suppose G_1 is a group with presentation $\langle S_1 \mid t = 1, \text{all } t \in T\rangle$, and that G_2 is a group. Suppose $S_2 \subseteq G_2$, $\langle S_2 \rangle = G_2$, and $s \mapsto s'$ is a function from S_1 onto S_2. Suppose further that the generators $s' \in S_2$ satisfy all the relations $t = 1, t \in T$, in the sense that if each $s \in S_1$ is replaced by the corresponding $s' \in S_2$ in each word $t \in T$, then the result is an element $t' \in G_2$ with $t' = 1$. Then there is a homomorphism from G_1 onto G_2.

Proof. We may assume that $G_1 = F_{S_1}/K_1$, where K_1 is the normal subgroup of F_{S_1} generated by T. The map $s \mapsto s'$ extends to a homomorphism f from F_{S_1} onto G_2 (since $\langle S_2 \rangle = G_2$), and $K_2 = \ker f$ contains K_1 since the elements of S_2 satisfy the relations for G_1. Since $K_1 \leq K_2$ the mapping $g: F_{S_1}/K_1 \to G_2$ defined by $g(xK_1) = f(x)$ is a well-defined homomorphism from $G_1 = F_{S_1}/K_1$ onto G_2.

For example, let $G_1 = \langle a \mid a^n = 1\rangle$, and let G_2 be cyclic of order m, generated by b, where $m \mid n$. Then $b^n = 1$, so the generator of G_2 satisfies the relation defining G_1 and $a \mapsto b$ extends to a homomorphism from G_1 onto G_2.

For another example consider the presentation

$$G = \langle a, b \mid a^3 = 1, b^2 = 1, abab = 1\rangle.$$

The third relation can be written in the form $ab = ba^{-1}$, or $ab = ba^2$. It follows that each element of G can be written in the form $b^i a^j$, with $0 \leq i \leq 1$ and $0 \leq j \leq 2$, and hence that $|G| \leq 6$. If we consider the elements $\sigma = (123)$ and $\tau = (12)$, which generate the symmetric group S_3, then $\sigma^3 = \tau^2 = 1$ and $\sigma\tau = (13) = \tau\sigma^2$, so they satisfy the relations that define G. By Proposition 10.2 there is a homomorphism from G onto S_3, and so $|G| \geq 6 = |S_3|$. Consequently $|G| = |S_3| = 6$, and $G \cong S_3$. Thus S_3 has the presentation $\langle a, b \mid a^3 = b^2 = 1, ab = ba^2\rangle$.

Exercise 10.1. (1) Show that Klein's 4-group has the presentation $\langle a, b \mid a^2 = b^2 = (ab)^2 = 1\rangle$.

(2) Show that the quaternion group $Q_2 = \{\pm 1, \pm i, \pm j, \pm k\}$ has the presentation $\langle a, b \mid a^4 = 1, a^2 = b^2, ab = ba^3 \rangle$ (try $a \mapsto i$, $b \mapsto j$).

Given a presentation for a group G it can often be difficult to determine whether G is finite or infinite, and even to determine whether or not $G = 1$. In particular it can be difficult to find $|G|$.

We present two more examples in some detail.

EXAMPLES

1. Let $G = \langle x, y \mid xy = y^2x, yx = x^2y \rangle$. The relations do not suggest immediately that x *and* y even have finite order, so it is conceivable that G is an infinite group. However,

$$y^{-1}xy = y^{-1}y^2x = yx = x^2y = xxy,$$

and by canceling xy we see that $y^{-1} = x$. Thus

$$xy = 1 = y^2x = y(yx) = y,$$

so also $x = 1$, and hence $G = 1$.

2. Let $G = \langle a, b \mid a^m = b^2 = 1, ab = ba^{-1} \rangle$. In this case all elements of G can be written in the form $b^i a^j$, with $0 \le i \le 1$ and $0 \le j \le m - 1$, so $|G| \le 2m$. If we can exhibit a group of order $2m$ with two generators that satisfy the relations defining G it will follow from Proposition 10.2 that $|G| = 2m$. We suggest two proofs that such a group exists.

Consider first the group D_m of symmetries of a regular m-gon in the plane (assume here that $m \ge 3$). It has a generating rotation σ through angle $2\pi/m$ and it has m reflections. If τ is one of the reflections then σ and τ generate D_m and satisfy the relations for G, and $|D_m| = 2m$, so $G \cong D_m$.

For another approach let H, as a set (but *not* as a group), be the Cartesian product $\mathbb{Z}_2 \times \mathbb{Z}_m$. Define an operation on H by setting

$$(\bar{i}, \bar{j}) \cdot (\bar{k}, \bar{n}) = (\bar{i} + \bar{k}, (-1)^k \bar{j} + \bar{n}).$$

It can be verified (Do so!) that H is a group, and it is clear that $|H| = 2m$. If we set $a = (\bar{0}, \bar{1})$ and $b = (\bar{1}, \bar{0})$, then a and b satisfy the relations for G. In fact, the operation defined on H was inspired by the fact that $(b^i a^j)(b^k a^n) = b^{i+k} a^{(-1)^k j + n}$ in G, as follows from the relations. Thus also $G \cong H$.

The first approach is perhaps more natural and is geometrically satisfying. The second is more tedious but may have the advantage that it could be applied in some cases where a geometric interpretation is not available. In any event the group $D_m = \langle \sigma, \tau \mid \sigma^m = \tau^2 = 1, \sigma\tau = \tau\sigma^{-1} \rangle$ is called the *dihedral* group of order $2m$ for any $m \ge 2$. When $m = 2$, then $D_m = D_2$ is isomorphic with Klein's 4-group.

Exercise 10.2. (1) Verify that the rotation σ and the reflection τ in D_m satisfy the relations defining G in the second example above.

(2) Verify all the statements made about the group H in the second example.

11. SOME FINITE GROUPS CLASSIFIED

In this section we classify several finite groups (up to isomorphism) according to arithmetic properties of their orders.

1. If p is a prime and $|G| = p^2$, then G is abelian, by Exercise 2.1. By Theorem 8.3 we conclude that $G \cong \mathbb{Z}_{p^2}$ or $G \cong \mathbb{Z}_p \oplus \mathbb{Z}_p$.

2. Suppose p and q are primes, with $q < p$, and suppose that $|G| = pq$. If $p \not\equiv 1 \pmod q$ we saw in Example 2, p. 21, that $G \cong \mathbb{Z}_{pq}$. Suppose then that $p \equiv 1 \pmod q$. Choose elements x and y in G of orders p and q, respectively, and set $p = \langle x \rangle$, $Q = \langle y \rangle$. Just as on p. 21 we see that $P \lhd G$, and that $[G:N_G(Q)] = 1$ or p. If $[G:N_G(Q)] = 1$, then $G = \langle xy \rangle$ is cyclic, just as before. Suppose then that $[G:N_G(Q)] = p$. Then $x^y = y^{-1}xy = x^n$ for some integer n, $2 \le n < p$. Thus $y^{-k}xy^k = x^{n^k}$ if $0 < k \in \mathbb{Z}$, and in particular $x = y^{-q}xy^q = x^{n^q}$, so $n^q \equiv 1 \pmod p$. Consequently G has generators x and y that satisfy the relations $x^p = y^q = 1$, $y^{-1}xy = x^n$, where $2 \le n < p$ and $n^q \equiv 1 \pmod p$. The group with that presentation, $\langle x, y \,|\, x^p = y^q = 1, y^{-1}xy = x^n \rangle$, with $2 \le n < p$ and $n^q \equiv 1 \pmod p$, does in fact have order pq, so G has that presentation.

With the aid of some elementary number theory it can be shown that if m is any other integer such that $m^q \equiv 1 \pmod p$ and $m \not\equiv 1 \pmod p$, then the presentation $\langle x, y \,|\, x^p = y^q = 1, y^{-1}xy = x^m \rangle$ defines a group isomorphic with G.

3. Suppose p is a prime and $|G| = p^3$.

(a) If G is abelian, then $G \cong \mathbb{Z}_{p^3}$, $\mathbb{Z}_p \oplus \mathbb{Z}_{p^2}$, or $\mathbb{Z}_p \oplus \mathbb{Z}_p \oplus \mathbb{Z}_p$ by the Fundamental Theorem of Finite Abelian Groups.

(b) Suppose p is odd and G is not abelian.

(i) Assume first that G has no element of order p^2. Set $Z(G) = Z$ and note that Z has order p by Exercise 2.1, for otherwise G/Z would be cyclic. Thus $|G/Z| = p^2$, so G/Z is abelian by Exercise 2.1, but is not cyclic (or else G would have an element of order p^2). Thus $G/Z \cong \mathbb{Z}_p \oplus \mathbb{Z}_p$, so it has generators xZ, yZ with $x^pZ = y^pZ = Z$ and $xyZ = yxZ$. Set $z = [x, y] = x^{-1}y^{-1}xy \in G$. If $z = 1$, then G is abelian, since $G = \langle x, y, Z \rangle$, and so $z \ne 1$. But $z \in Z$, so $Z = \langle z \rangle$ and G has the presentation

$$\langle x, y, z \,|\, x^p = y^p = z^p = 1, [x, y] = z, [x, z] = [y, z] = 1 \rangle.$$

(ii) Assume next that G has an element x of order p^2 and set $H = \langle x \rangle$. Then H is normal in G by the corollary to Proposition 7.3. Take $y \in G \backslash H$. Then $y^p \in H$ since $|G/H| = p$. Say $y^{-1}xy = x^r$, where $r \not\equiv 1 \pmod{p^2}$. Then $y^{-j}xy^j = x^{r^j}$ for all positive integers j, and in particular $x = y^{-p}xy^p = x^{r^p}$, so $r^p \equiv 1 \pmod{p^2}$. But also $r^p \equiv r \pmod p$ by Fermat's Little Theorem, so $r \equiv 1 \pmod p$, say $r - 1 = sp$, $s \in \mathbb{Z}$. Choose $j \in \mathbb{Z}$, $j > 0$, such that $js \equiv 1 \pmod p$, and set $z = y^j$. Then

$$z^{-1}xz = y^{-j}xy^j = x^{r^j} = x^{(1+sp)^j} = x^{1+jsp} = x^{1+(js-1)p+p} = x^{1+p}.$$

We still have $G/H = \langle zH \rangle$, so $z^p \in H$, say $z^p = x^t$. Then $p \mid t$, for otherwise $|z| = p^3$. Say $t = up$, so $z^p = x^{up}$. If $i \in \mathbb{Z}$ we have

$$z^{-1}x^iz = (z^{-1}xz)^i = x^{(1+p)i},$$

or

$$x^iz = zx^{(1+p)i}.$$

Thus

$$\begin{aligned}(zx^{-u})^p &= zx^{-u}zx^{-u}\cdots zx^{-u} = z^2x^{-u(1+(1+p))}zx^{-u}\cdots zx^{-u}\\ &= z^3x^{-u(1+(1+p)+(1+p)^2)}zx^{-u}\cdots zx^{-u}\\ &\ \vdots\\ &= z^px^{-u(1+(1+p)+\cdots+(1+p)^{p-1})}\\ &= z^px^{-u(1+(1+p)+(1+2p)+\cdots+(1+(p-1)p))}\\ &= z^px^{-u(p+p(1+2+3+\cdots+(p-1)))}\\ &= z^px^{-u(p+p^2(p-1)/2)} = z^px^{-up} = z^pz^{-p} = 1.\end{aligned}$$

Set $w = zx^{-u}$. Then $w^{-1}xw = x^uz^{-1}xzx^{-u} = x^{1+p}$, and G has the presentation $\langle x, w \mid x^{p^2} = w^p = 1, w^{-1}xw = x^{1+p}\rangle$.

(c) If $p = 2$ and G is not abelian, then G is isomorphic with the dihedral group D_4 or the quaternion group Q_2.

Exercise 11.1. Prove 2(c) above.

Exercise 11.2. Classify all groups of orders 12 and 20.

12. FURTHER EXERCISES

1. If G is a group and $f: G \to G$ is defined by $f(x) = x^{-1}$, all $x \in G$, show that f is a homomorphism if and only if G is abelian.
2. If a group G has a unique element x of order 2 show that $x \in Z(G)$.
3. Suppose G is finite, $K \lhd G$, $H \leq G$, and $|K|$ is relatively prime to $[G:H]$. Show that $K \leq H$.

4. If G is not abelian show that $Z(G)$ is properly contained in an abelian subgroup of G.

5. If G is a group and $|x| = 2$ for all $x \neq 1$ in G show that G is abelian. Can you say more?

6. Suppose G is a group, $H \leq G$, and $K \leq G$. Show that $H \cup K$ is not a group unless $H \leq K$ or $K \leq H$.

7. (Haber and Rosenfeld [12]). Show that a group G is the union of three proper subgroups if and only if there is an epimorphism from G to Klein's 4-group.

8. Suppose S is a subset of a finite group G, with $|S| > |G|/2$. If S^2 is defined to be $\{xy : x, y \in S\}$ show that $S^2 = G$.

9. If $A, B \leq G$ and both $[G:A]$ and $[G:B]$ are finite show that $[G:A \cap B] \leq [G:A][G:B]$, with equality if and only if $G = AB$.

10. If $[G:A]$ and $[G:B]$ are finite and relatively prime show that $G = AB$.

11. Suppose G acts on S, $x \in G$, and $s \in S$. Show that $\text{Stab}_G(xs) = x\,\text{Stab}_G(s)x^{-1}$.

12. If $A, B \leq G$ and $y \in G$ define the (A, B)-*double coset* $AyB = \{ayb : a \in A, b \in B\}$. Show that G is the disjoint union of its (A, B)-double cosets. Show that $|AyB| = [A^y : A^y \cap B]\,|B|$ if A and B are finite.

13. Suppose G is a permutation group on a set S, with $|S| > 1$. Say that G is *doubly transitive* on S if given any $(a, b), (c, d) \in S \times S$, with $a = b$ if and only if $c = d$, then $xa = c$ and $xb = d$ for some $x \in G$.

(1) If G is transitive on S show that G is doubly transitive if and only if $H = \text{Stab}_G(s)$ is transitive on $S \setminus \{s\}$ for each $s \in S$.

(2) If G is doubly transitive on S and $|S| = n$ show that $n(n-1) \mid |G|$.

14. Permutation groups G_1 and G_2 acting on sets S_1 and S_2 are called *permutation isomorphic* if there exist an isomorphism $\theta: G_1 \to G_2$ and a bijection $\phi: S_1 \to S_2$ such that $(\theta x)(\phi s) = \phi(xs)$ for all $x \in G_1$, all $s \in S_1$.

$$\begin{array}{ccc} S_1 & \xrightarrow{\quad x \quad} & S_1 \\ \phi \downarrow & & \downarrow \phi \\ S_2 & \xrightarrow[\quad \theta x \quad]{} & S_2 \end{array}$$

Define two actions of a group G on itself as follows:

(i) the action of $x \in G$ is left multiplication by x;
(ii) the action of $x \in G$ is right multiplication by x^{-1}.

Show that the two actions are permutation isomorphic.

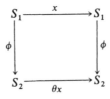

15. Suppose G is a transitive permutation group on a set S, and $H = \mathrm{Stab}_G(s)$ for some $s \in S$. Show that the usual permutation action of G on the set of left cosets of H is permutation isomorphic (see Exercise 14, above) with the original permutation action of G on S.

16. Suppose G is finite, p is the smallest prime dividing $|G|$, $H \le G$, and $[G:H] = p$. Show that $H \lhd G$.

17. Suppose $[G:H]$ is finite. Show that there is a normal subgroup K of G, with $K \le H$, such that $[G:K]$ is finite.

18. Suppose G is finite, $H \le G$, and $G = \bigcup\{H^x : x \in G\}$. Show that $H = G$.

19. Let G be the group $GL(2, \mathbb{C})$ of all 2×2 invertible complex matrices, and let H be the subgroup of all lower triangular matrices $\left[\begin{smallmatrix} a & 0 \\ b & c \end{smallmatrix}\right]$, $ac \ne 0$. Show that $G = \bigcup\{H^x : x \in G\}$ (compare with Exercise 18 above).

20. Let T be the set of $n - 1$ successive transpositions (12), (23), $(34), \ldots$, $(n-1\ n)$ in S_n. Show that $\langle T \rangle = S_n$.

21. If p is a prime, $H \le S_p$, and H contains a transposition σ and a p-cycle τ, show that $H = S_p$. (*Hint*: Show that there is no loss of generality in assuming $\sigma = (12)$ and $\tau = (123 \cdots p)$. Then conjugate σ by powers of τ.) Show by example that the result may fail if p is not prime.

22. Suppose $H \le S_n$ but $H \not\le A_n$. Show that $[H : A_n \cap H] = 2$. (*Hint*: Observe that $HA_n = S_n$.)

23. If $S = \{1, 2, 3, 4, \ldots\}$ let A_∞ denote the (infinite) group of all $\sigma \in \mathrm{Perm}(S)$ such that there is a finite subset $T \subseteq S$ for which σ restricts to an even permutation of T and $\sigma(s) = s$ for all $s \in S \backslash T$. Equivalently $A_\infty = \bigcup\{A_n : n = 1, 2, 3, \ldots\}$. Show that A_∞ is simple.

24. Let $\sigma = (12)$ and $\tau = (123 \cdots n)$ in S_n. Determine the centralizers of σ and τ in S_n. (*Hint*: What must their orders be?)

25. If $\sigma \in S_n$ has cycle type $(k_1, \ldots, k_n)$ show that the conjugacy class of σ has $n!/[(\prod i^{k_i})(\prod k_i!)]$ elements.

26. Suppose K is a conjugacy class in S_n of cycle type $(k_1, \ldots, k_n)$, and that $K \subseteq A_n$. If $\sigma \in K$ write L for the conjugacy class of σ in A_n.

(1) If either $k_{2m} > 0$ or $k_{2m+1} > 1$ for some m show that $L = K$.

(2) If $k_{2m} = 0$ and $k_{2m+1} \le 1$ for all m show that $K = L \cup L'$, where L' is also a conjugacy class in A_n and $|L'| = |L| = |K|/2$.

27. Show that the group of rotational symmetries of the tetrahedron is isomorphic with A_4.

28. Show that the group of rotational symmetries of the cube is isomorphic with S_4. (*Hint*: Each rotation permutes the four diagonals of the cube.)

29. If $0 < n \in \mathbb{Z}$, then a *partition* of n is a sequence $(k_1, k_2, \ldots, k_m)$ of positive integers such that $k_1 \leq k_2 \leq \cdots \leq k_m$ and $k_1 + k_2 + \cdots + k_m = n$. For example, $(2, 3)$ and $(1, 1, 1, 2)$ are partitions of $n = 5$. Write $P(n)$ for the number of distinct partitions of n [e.g., $P(3) = 3$, $P(5) = 7$, etc.]. Show that S_n has $P(n)$ conjugacy classes.

30. For any real number r write $[r]$ for the greatest integer not exceeding r. If p is a prime show that the p-Sylow subgroups of S_n have order p^m, where $m = [n/p] + [n/p^2] + [n/p^3] + \cdots$.

31. Suppose G is a finite group, $H \lhd G$, and P is a p-Sylow subgroup of H. Set $N = N_G(P)$. Show that $G = NH$. (*Hint*: If $x \in G$, then P^x is p-Sylow in H.)

32. If G is a finite p-group and $1 \neq H \lhd G$ show that $H \cap Z(G) \neq 1$. (*Hint*: H is a union of G-conjugacy classes.)

33. Suppose G is a finite p-group having a unique subgroup of index p. Show that G is cyclic. [Use induction and look at $G/Z(G)$.]

34. Suppose G is a finite p-group. Show that $Z(G)$ is cyclic if and only if G has exactly one normal subgroup of order p. (See Exercise 32, above.)

35. Show that there are no simple groups of orders 104, 176, 182, or 312.

36. There is a simple group G of order 168. Show that G has 48 elements of order 7.

37. If p and q are primes show that any group of order p^2q is solvable.

38. If G is a finite p-group show that all composition factors of G are isomorphic with $\mathbb{Z}_p$.

39. If A and B are subnormal subgroups of G show that $A \cap B$ is subnormal.

40. Show that an infinite abelian group G can not have a composition series.

41. Show that A_∞ (see Exercise 23, above) has the composition series $A_\infty \geq 1$, but the subgroup $H = \langle (12)(34), (56)(78), \ldots \rangle$ has no composition series.

42. If p is a prime, $|G| = p^3$, and G is not abelian show that $G' = Z(G)$.

43. A pair of homomorphisms $K \xrightarrow{f} G \xrightarrow{g} H$ is said to be *exact* at G if $\text{Im}(f) = \ker g$. A sequence $1 \to K \xrightarrow{f} G \xrightarrow{g} H \to 1$ is called a *short exact sequence* if it is exact at each of K, G, and H.

(1) Show that if $K \lhd G$, $f: K \to G$ is the inclusion map and $g: G \to G/K$ is the canonical quotient map, then $1 \to K \xrightarrow{f} G \xrightarrow{g} G/K \to 1$ is a short exact sequence.

(2) Show that $1 \to K \xrightarrow{f} G \xrightarrow{g} H \to 1$ is short exact if and only if f is 1–1, g is onto, and $\text{Im}(f) = \ker g$. Conclude that then K is isomorphic with a normal subgroup of G and that $G/f(K) \cong H$.

(3) Suppose $1 \to K \to G \to H \to 1$ is a short exact sequence. Show that G is solvable if and only if both K and H are solvable (see Theorem 5.4).

(4) Give an example of a short exact sequence $1 \to K \to G \to H \to 1$ for which K and H are nilpotent but G is not.

44. If G is a group and $x \in G$ define the *inner automorphism* f_x by setting $f_x(y) = xyx^{-1}$, all $y \in G$. Write $I(G)$ for the set of all inner automorphisms of G.

(1) Show that $I(G) \leq \text{Aut}(G)$.
(2) Show that $I(G) \cong G/Z(G)$.
(3) If $I(G)$ is abelian show that $G' \leq Z(G)$. Conclude that G is nilpotent.

45. Show that $\text{Aut}(S_3) \cong S_3$. [*Hint*: Any automorphism permutes the elements of order 2 in S_3, so $|\text{Aut}(S_3)| \leq 6$. Look at $I(S_3)$ (Exercise 44 above).]

46. If G is cyclic of order n show that $\text{Aut}(G)$ is abelian and has order $\phi(n)$. If G is infinite cyclic show that $|\text{Aut}(G)| = 2$.

47. If K is Klein's 4-group show that $\text{Aut}(K) \cong S_3$.

48. If $A \lhd G$ and $B \lhd G$ show that $G/(A \cap B)$ is isomorphic with a subgroup of $G/A \times G/B$.

49. If G is a finite p-group that is not cyclic show that there is a homomorphism from G onto $\mathbb{Z}_p \times \mathbb{Z}_p$. (*Hint*: Let A and B be distinct maximal subgroups of G and apply Exercise 48.)

50. If $A, B \leq G$ show that $[A, B] \lhd \langle A \cup B \rangle$.

51. If G is a finite group in which every maximal subgroup is normal show that G is nilpotent. (*Hint*: Suppose to the contrary that P is a nonnormal Sylow subgroup and choose $M \leq G$ maximal with $N_G(P) \leq M$. If $x \in G \backslash M$ consider P^x.)

52. Define the *generalized quaternion group* Q_m by the presentation

$$Q_m = \langle a, b \,|\, a^{2m} = 1, b^2 = a^m, ab = ba^{-1} \rangle \qquad \text{for} \quad m \geq 1.$$

Show that $|Q_m| = 4m$. Show that Q_1 is cyclic and Q_2 is the quaternion group introduced on p. 5. Show that Q_3 is not isomorphic with either the dihedral group D_6 or the alternating group A_4.

53. Find all positive integers n for which there is only one abelian group (up to isomorphism) of order n (e.g., $n = 5$, 6, 10, and 105 have that property).

54. Suppose there is only one group (up to isomorphism) of order n. Show that $(n, \phi(n)) = 1$. (The converse is also true; the proof is more difficult.)

55. Use generators $a = (12)$, $b = (23)$, and $c = (34)$ for S_4 and show that S_4 has the presentation

$$\langle a, b, c \,|\, a^2 = b^2 = c^2 = (ab)^3 = (ac)^2 = (bc)^3 = 1 \rangle.$$

56. If $G = \langle a, b \mid a^4 = b^3 = 1, ab = ba^3 \rangle$ show that G is cyclic of order 6.

57. A group G is called nilpotent of *class* n if $L_{n-1}(G) \neq 1$ but $L_n(G) = 1$.

(1) Show that G is nilpotent of class n if and only if $Z_{n-1} \neq G$ but $Z_n = G$.

(2) Show that the generalized quaternion group Q_{2^n} (Exercise 52) is nilpotent of class $n + 1$.

58. Compute the centers and derived groups of the dihedral groups D_m and the generalized quaternion groups Q_m (Exercise 52, above). For which values of m are the groups nilpotent?

59. Show that the dihedral group D_m has $(m + 3)/2$ conjugacy classes if m is odd and $(m + 6)/2$ conjugacy classes if m is even. Describe the classes explicitly.

60. Begin, at least, to classify all groups of order p^4, where p is a prime. When $p = 2$ there are 14 isomorphism types, when $p > 2$ there are more.

61. If H and K are subgroups of G, with $K \lhd G$, $K \cap H = 1$, and $KH = G$, then G is called a *semidirect product* (or *split extension*) of K by H.

(1) If $\sigma = (12) \in S_n$, $n \geq 2$, show that S_n is a semidirect product of A_n by $\langle \sigma \rangle$.

(2) Show that the dihedral group $D_n = \langle a, b \mid a^n = b^2 = 1, b^{-1}ab = a^{-1} \rangle$ is a semidirect product of $A = \langle a \rangle$ by $B = \langle b \rangle$.

(3) Show that the quaternion group Q_2 can not be expressed as a semidirect product (except with one of the subgroups trivial).

62. Suppose K and H are groups and $\phi\colon H \to \mathrm{Aut}(K)$ is a homomorphism. Let G be the Cartesian product $K \times H$ as a set, but with binary operation $(x, y)(u, v) = (x \cdot \phi(y)u, yv)$. Show that G is a group; denote it by $G = K \rtimes_\phi H$, and call it the *external semidirect product* of K by H relative to ϕ. Show that $K_1 = \{(x, 1) : x \in K\} \lhd G$, $H_1 = \{(1, y) : y \in H\} \leq G$, and that G is the semidirect product of K_1 by H_1 as in Exercise 61.

Chapter II | Rings

1. PRELIMINARIES: IDEALS AND HOMOMORPHISMS

A *ring* is an additive abelian group R with a second associative binary operation, multiplication, the two operations being related by the *distributive* laws

$$x(y + z) = xy + xz \qquad \text{and} \qquad (y + z)x = yx + zx$$

for all $x, y, z \in R$. In particular R is a semigroup with respect to multiplication. If $xy = yx$ for all $x, y \in R$ we call R a *commutative* ring. If R has a multiplicative identity $1 = 1_R \neq 0$, then we say that R is a *ring with* 1.

It is an easy consequence of the distributive laws that $x \cdot 0 = 0 \cdot x = 0$ for all x in a ring R.

A *subring* of a ring R is a subset S of R such that the binary operations on R restrict to binary operations on S so that S is also a ring.

EXAMPLES

1. Let $R = \mathbb{Z}$, the integers, with the usual addition and multiplication. Then R is a commutative ring with 1.

2. Likewise if $R = \mathbb{Q}$, $\mathbb{R}$, or $\mathbb{C}$, with the usual operations, then R is a commutative ring with 1.

3. If R is the additive group $\mathbb{Z}_n$ and if we define multiplication via $\bar{a} \cdot \bar{b} = \overline{ab}$, then R is a commutative ring with $1 = \bar{1}$.

4. Let $R = \mathbb{Z}_n$, $\mathbb{Z}$, $\mathbb{Q}$, $\mathbb{R}$, $\mathbb{C}$, or *any* ring with 1, and suppose $0 < n \in \mathbb{Z}$. Let $M_n(R)$ be the set of all $n \times n$ matrices having entries from R, with the usual

operations of addition and multiplication for matrices. Then $M_n(R)$ is a ring with $1 = I$, the $n \times n$ identity matrix. It is not commutative if $n \geq 2$ or if R is not commutative.

5. Let A be an additive abelian group and let $R = \text{End}(A)$, the set of all endomorphisms of A. We add endomorphisms in the obvious fashion, viz., if $\phi, \theta \in R$ and $a \in A$, then $(\phi + \theta)(a) = \phi(a) + \theta(a)$, and we take multiplication to be composition of functions. Then R is a ring with 1 (the identity function on A); it does not tend to be commutative.

Exercise 1.1. Verify that End(A) is a ring if A is an additive abelian group.

Proposition 1.1. If R is a ring and S is a nonempty subset of R, then S is a subring of R if and only if $x - y \in S$ and $xy \in S$ whenever $x, y \in S$.

Proof. $\Rightarrow$: Obvious.

$\Leftarrow$: S is an additive subgroup of R by Proposition I.1.4, and multiplication in R restricts to a binary operation on S. The associative law for multiplication and the two distributive laws hold for S since they hold for all elements of R, and so S is a ring.

For example, the set $S = 2\mathbb{Z}$ of all even integers is a subring of $R = \mathbb{Z}$; S is *not* a ring with 1. The ring $R = M_2(\mathbb{R})$ has a subring S consisting of all matrices $\left[\begin{smallmatrix} a & 0 \\ 0 & 0 \end{smallmatrix}\right]$, $a \in \mathbb{R}$. In this case both R and S are rings with 1, but $1_R \neq 1_S$.

For a more substantial example let $\mathbb{H}$ be the subset of $M_4(\mathbb{R})$ consisting of all matrices

$$\begin{bmatrix} a & -b & -c & -d \\ b & a & -d & c \\ c & d & a & -b \\ d & -c & b & a \end{bmatrix},$$

$a, b, c, d \in \mathbb{R}$. An application of Proposition 1.1 shows that $\mathbb{H}$ is a subring of $M_4(\mathbb{R})$. If we write $1 = I$, the identity matrix in $M_4(\mathbb{R})$, and set

$$i = \begin{bmatrix} 0 & -1 & 0 & 0 \\ 1 & 0 & 0 & 0 \\ 0 & 0 & 0 & -1 \\ 0 & 0 & 1 & 0 \end{bmatrix}, \quad j = \begin{bmatrix} 0 & 0 & -1 & 0 \\ 0 & 0 & 0 & 1 \\ 1 & 0 & 0 & 0 \\ 0 & -1 & 0 & 0 \end{bmatrix}, \quad k = \begin{bmatrix} 0 & 0 & 0 & -1 \\ 0 & 0 & -1 & 0 \\ 0 & 1 & 0 & 0 \\ 1 & 0 & 0 & 0 \end{bmatrix},$$

then each $x \in \mathbb{H}$ can be written uniquely in the form $x = a1 + bi + cj + dk$, where $a, b, c, d \in \mathbb{R}$. Note that $ij = k = -ji$, $jk = i = -kj$, $ki = j = -ik$, and $i^2 = j^2 = k^2 = -1$. The ring $\mathbb{H}$, called Hamilton's ring of *quaternions*, is of considerable importance in the history of algebra. It will be discussed further as we proceed. Observe that $Q_2 = \{\pm 1, \pm i, \pm j, \pm k\} \subseteq \mathbb{H}$ is the quaternion group introduced in Chapter I.

If R is a ring with 1, then $x \in R$ is called a *unit* if it has a multiplicative inverse, i.e., if $xy = yx = 1$ for some $y \in R$. The set of all units in R is denoted by $U(R)$ and is called the *group of units.*

Exercise 1.2. (1) Show that $U(R)$ is a group.
(2) Find $U(R)$ when $R = \mathbb{Z}$ and when $R = \mathbb{Z}_n$.
(3) If $R = M_2(\mathbb{Z})$ show that $U(R)$ is the group of all matrices $\left[\begin{smallmatrix} a & b \\ c & d \end{smallmatrix}\right]$ with integer entries such that $ad - bc = \pm 1$. Generalize.

If R is any ring we write R^* for $R \setminus \{0\}$. It should be noted that this notation is not universally used. Some authors, for example, write R^* for the group of units in R, which we have chosen to denote by $U(R)$.

If $x, y \in R^*$ and $xy = 0$ we say that x is a *left zero divisor* and y is a *right zero divisor*. The distinction between left and right disappears, of course, if R is commutative. Note that $\mathbb{Z}$ has no zero divisors, and that $\mathbb{Z}_n$ has zero divisors if and only if n is not a prime. If

$$x = \begin{bmatrix} 1 & -2 \\ -2 & 4 \end{bmatrix} \quad \text{and} \quad y = \begin{bmatrix} 6 & 2 \\ 3 & 1 \end{bmatrix}$$

in $R = M_2(\mathbb{Z})$, then $xy = 0$, so x is a left and y a right zero divisor.

A commutative ring $R \neq 0$ is called an *integral domain* if it has no zero divisors, or equivalently if R^* is a (multiplicative) semigroup. If R is a ring with 1 in which every nonzero element has a multiplicative inverse [i.e., $U(R) = R^*$], then R is called a *division ring*, or a *skew field*. A commutative division ring is called a *field*. Clearly a field is an integral domain.

Note that $\mathbb{Z}$ is an integral domain but not a field; $\mathbb{Z}_n$ is a field if and only if $n = p$, a prime; $\mathbb{Q}$, $\mathbb{R}$, and $\mathbb{C}$ are all fields. The next exercise will show that the ring $\mathbb{H}$ of quaternions is a division ring.

Exercise 1.3. (1) If $x = a1 + bi + cj + dk \in \mathbb{H}$, define $\bar{x} = a1 - bi - cj - dk$. Verify (by matrix multiplication) that $x\bar{x} = \bar{x}x = N(x)1$, where $N(x) = a^2 + b^2 + c^2 + d^2 \in \mathbb{R}$. Conclude that $\mathbb{H}$ is a division ring.
(2) Show that $\overline{xy} = \bar{y}\bar{x}$, and use that fact to show that $N(xy) = N(x)N(y)$ if $x, y \in \mathbb{H}$.

If R and S are rings then a function $f: R \to S$ is called a (ring) *homomorphism* if $f(x + y) = f(x) + f(y)$ and $f(xy) = f(x)f(y)$ for all $x, y \in R$. As usual we call f a *monomorphism* if it is 1–1 and an *epimorphism* if it is onto. If f is 1–1 and onto we call it an *isomorphism*. In that case f^{-1} is also an isomorphism from S to R; we say then that R and S are *isomorphic* and write $R \cong S$.

The *kernel* of a homomorphism $f: R \to S$ is $\ker f = \{x \in R : f(x) = 0\}$. It is easy to verify, with the aid of Proposition 1.1, that $\ker f$ and $\operatorname{Im}(f)$ are subrings of R and S, respectively. The kernel satisfies a further condition.

If $x \in \ker f$ and $y \in R$, then $f(yx) = f(y)f(x) = f(y) \cdot 0 = 0$, and similarly $f(xy) = 0$, so both yx and xy are in $\ker f$.

A subring I of a ring R is called a *left ideal* if given any $x \in I$ and any $y \in R$ we have $yx \in I$. Analogously I is a *right ideal* if $xy \in I$ whenever $x \in I$, $y \in R$. If I is both a left and a right ideal in R we say that I is an *ideal*, or a *two-sided ideal*, in R. Thus the kernel of a homomorphism is an ideal.

Proposition 1.2. A nonempty subset I of a ring R is an ideal in R if and only if $x - y \in I$, $zx \in I$, and $xz \in I$ for all $x, y \in I$ and all $z \in R$.

Exercise 1.4. (1) If $m \in \mathbb{Z}$ show that $I = m\mathbb{Z} = \{mk : k \in \mathbb{Z}\}$ is an ideal.

(2) If

$$R = M_2(\mathbb{Z}) \qquad \text{and} \qquad I = \left\{ \begin{bmatrix} a & 0 \\ c & 0 \end{bmatrix} : a, c \in \mathbb{Z} \right\}$$

show that I is a left ideal but not a right ideal.

(3) If R is a ring with 1 and I is an ideal (left, right, or two-sided) in R such that $I \cap U(R) \neq \varnothing$ show that $I = R$.

Proposition 1.3. If $\{I_\alpha\}_{\alpha \in A}$ is any collection of ideals in a ring R, then $I = \bigcap\{I_\alpha : \alpha \in A\}$ is an ideal in R.

Proof. Apply Proposition 1.2.

Clearly statements corresponding to Proposition 1.3 hold as well for collections of left ideals or of right ideals.

If X is a subset of a ring R, then, by Proposition 1.3, $\bigcap\{I : I$ is an ideal of R and $X \subseteq I\}$ is an ideal, the smallest ideal of R that contains X as a subset. We call it the ideal *generated by* X and denote it by (X). If an ideal J of R is generated by a single element $a \in R$ we write $J = (a)$ and say that J is a *principal* ideal. If R is a commutative ring with 1, then the principal ideal (a) consists of all multiples ra, $r \in R$, and we may write $(a) = Ra$. For example, in the ring $\mathbb{Z}$ of integers (5) is the principal ideal consisting of all multiples of 5.

Since an ideal I in a ring R is an additive subgroup the quotient $R/I = \{x + I : x \in R\}$ is also an additive group. Suppose $x + I = r + I$ and $y + I = s + I$ in R/I, i.e., that $x - r$ and $y - s$ are in I. Then

$$xy - rs = xy - ry + ry - rs = (x - r)y + r(y - s) \in I$$

since I is an ideal. Thus $xy + I = rs + I$ and we may define a multiplication on R/I by setting $(x + I)(y + I) = xy + I$. It is immediate that R/I is a ring; we call it the *quotient ring* or the *residue class ring* of R modulo I. The quotient map $\eta: R \to R/I$ is a ring homomorphism from R onto R/I, with $\ker \eta = I$.

Theorem 1.4 (The Fundamental Homomorphism Theorem for Rings, or the FHTR). If R and S are rings and $f: R \to S$ is a homomorphism, with

$\ker f = I$, then there is an isomorphism $g: R/I \to \mathrm{Im}(f)$ such that $g\eta = f$, i.e., the diagram

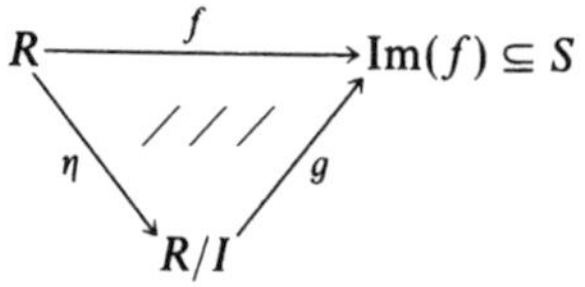

is commutative.

Proof. As far as the additive structure of R is concerned this is just the FHT for groups. All that remains is to verify that g is in fact a (ring) homomorphism. But

$$\begin{aligned} g((x + I)(y + I)) &= g(xy + I) = g(\eta(xy)) = f(xy) \\ &= f(x)f(y) = g(\eta(x))g(\eta(y)) \\ &= g(x + I)g(y + I), \end{aligned}$$

and the proof is complete.

Similar remarks apply to the analogs for rings of Proposition I.1.12, the Freshman Theorem (I.1.13), and the Isomorphism Theorem (I.2.6). We state the results and leave the completions of the proofs as an exercise.

Proposition 1.5. Suppose R and S are rings and $f: R \to S$ is an epimorphism, with $\ker f = I$. Then there is a 1–1 correspondence between the set of all ideals J in S and the set of all those ideals K in R such that $I \subseteq K$, given by $J \leftrightarrow f^{-1}(J) = K$. In particular each ideal in a quotient ring R/I has the form K/I for some ideal K, $I \subseteq K \subseteq R$.

Theorem 1.6 (The Freshman Theorem for Rings). Suppose R is a ring, I and J are ideals, and $J \subseteq I$. Then I/J is an ideal in R/J, and $(R/J)/(I/J) \cong R/\mathrm{I}$.

Theorem 1.7 (The Isomorphism Theorem for Rings). Suppose R is a ring and I and J are ideals. Then $I + J$ and $I \cap J$ are ideals and $(I + J)/I \cong J/(I \cap J)$.

A ring R is called *simple* if its only (two-sided) ideals are 0 and R. An ideal I of a ring R is called *maximal* if $I \neq R$ and the only ideals J of R for which $I \subseteq J \subseteq R$ are $J = I$ and $J = R$. Thus, by Proposition 1.5, I is maximal in R if and only if R/I is a simple ring.

Exercise 1.5. (1) If $I = (n)$, a principal ideal in the ring $\mathbb{Z}$ of integers, show that I is maximal if and only if $n = p$, a prime.

(2) Show that $M_2(\mathbb{R})$ is a simple ring. (*Hint:* Try working with the matrices $\begin{bmatrix} 1 & 0 \\ 0 & 0 \end{bmatrix}$, $\begin{bmatrix} 0 & 1 \\ 0 & 0 \end{bmatrix}$, $\begin{bmatrix} 0 & 0 \\ 1 & 0 \end{bmatrix}$, and $\begin{bmatrix} 0 & 0 \\ 0 & 1 \end{bmatrix}$.) Generalize.

Proposition 1.8. If R is a ring with 1 and $I \neq R$ is an ideal in R, then R has a maximal ideal M with $I \subseteq M$.

Proof. We apply Zorn's Lemma (see the Appendix). Let $\mathscr{S}$ be the collection of all proper ideals J in R with $I \subseteq J$, partially ordered by inclusion. Then $I \in \mathscr{S}$, so $\mathscr{S} \neq \varnothing$. If $\mathscr{C}$ is a chain in $\mathscr{S}$, it is easy to check that $L = \bigcup\{J : J \in \mathscr{C}\}$ is an ideal with $I \subseteq L$, and $L \neq R$ since $1 \notin L$. Thus $L \in \mathscr{S}$ and L is an upper bound for $\mathscr{C}$. By Zorn's Lemma there is a maximal element M in $\mathscr{S}$, and M is by definition a maximal ideal in R.

Proposition 1.9. Suppose R is a commutative ring with 1. Then R is simple if and only if R is a field.

Proof. $\Rightarrow$: If $0 \neq a \in R$, then $(a) = R$. In particular $1 \in (a)$, so $1 = ba$ for some $b \in R$. Thus a is a unit and $U(R) = R^*$, so R is a field.

$\Leftarrow$: Let $I \subseteq R$ be an ideal, $I \neq 0$. Choose $a \in I$, $a \neq 0$; then $a \in R^* = U(R)$, so $a^{-1}a = 1 \in I$. But then if $r \in R$ we have $r \cdot 1 = r \in I$, so $I = R$ and R is simple.

Corollary. Suppose R is a commutative ring with 1. Then an ideal I in R is maximal if and only if R/I is a field.

Proof. I is maximal if and only if R/I is simple if and only if R/I is a field.

Exercise 1.6. The corollary to Proposition 1.9 does not generalize to noncommutative rings in the way one might imagine. Give an example of a ring R with 1 and a maximal ideal I such that R/I is *not* a division ring.

Exercise 1.7. Show that a finite integral domain is a field.

If R is a commutative ring and $P \neq R$ is an ideal in R, then P is called a *prime ideal* if given any $a, b \in R$ with $ab \in P$, then either $a \in P$ or $b \in P$ (or both). For example, if $R = \mathbb{Z}$, then both 0 and (5) are prime ideals; (5) is also maximal but 0, of course, is not.

Exercise 1.8. Let R be a commutative ring.

(1) Show that an ideal P in R is prime if and only if R/P is an integral domain.

(2) If R is a ring with 1 show that every maximal ideal is prime.

2. THE FIELD OF FRACTIONS OF AN INTEGRAL DOMAIN

The ring $\mathbb{Z}$ of integers is an integral domain, the ring $\mathbb{Q}$ of rational numbers is a field, and $\mathbb{Z}$ is a subring of $\mathbb{Q}$. Each $r \in \mathbb{Q}$ can be written as a fraction $r = a/b$, with $a, b \in \mathbb{Z}$, and no proper subfield of $\mathbb{Q}$ has that property (the simplest reason being that $\mathbb{Q}$ has no proper subfields). The exact same remarks apply if we consider the ring $2\mathbb{Z}$ of even integers as a subring of $\mathbb{Q}$.

We discuss in this section the imbedding of an arbitrary integral domain R as a subring in a minimal field. It is clearly necessary that R be an integral domain if it is to be a subring of a field.

A *field of fractions* for an integral domain R is a field F_R with a monomorphism $\phi: R \to F_R$ such that if K is any field and $\theta: R \to K$ a monomorphism then there is a unique homomorphism (necessarily a monomorphism) $f: F_R \to K$ for which the diagram

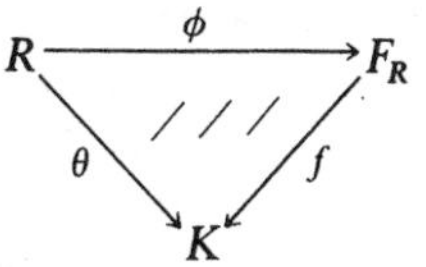

commutes, i.e., $\theta = f\phi$.

Proposition 2.1. If an integral domain $R \neq 0$ has a field of fractions F_R, then F_R is unique up to isomorphism.

Exercise 2.1. (1) If R is a field show that R itself is a field of fractions for R.

(2) Show that $\mathbb{Q}$ is a field of fractions for $\mathbb{Z}$ and for $2\mathbb{Z}$.

Theorem 2.2. If $R \neq 0$ is an integral domain, then R has a field of fractions.

Proof. Let X be the Cartesian product

$$R \times R^* = \{(a, b): a, b \in R, b \neq 0\}.$$

Define a relation $\sim$ on X by agreeing that $(a, b) \sim (c, d)$ if $ad = bc$. It is easily checked that $\sim$ is a reflexive and symmetric relation. If $(a, b) \sim (c, d) \sim (e, f)$, then $ad = bc$ and $cf = de$. Thus $adf = bcf = bde$, or $afd = bed$, and hence $(af - be)d = 0$. But $d \neq 0$ and R is an integral domain, so $af = be$, or $(a, b) \sim (e, f)$. Thus $\sim$ is transitive and hence is an equivalence relation. Denote the equivalence class containing (a, b) by a/b, and let F_R be the quotient set $X/\sim$, i.e., $F_R = \{a/b: a, b \in R, b \neq 0\}$.

If $a/b, c/d \in F_R$ we define $a/b + c/d = (ad + bc)/bd$ and $(a/b)(c/d) = ac/bd$. Routine but tedious calculations show that the resulting operations of addition and multiplication for F_R are well defined and associative. F_R is in fact a commutative ring with 1, with $0 = 0/b$ for any $b \in R^*$, $-(a/b) = -a/b$, and $1 = a/a$ for any $a \in R^*$. If $a/b \neq 0$ in F_R then necessarily $a \neq 0$ (Why?) and so $b/a \in F_R$. But then $(a/b) \cdot (b/a) = 1$ in F_R, so F_R is a field.

Define $\phi: R \to F_R$ by setting $\phi(r) = ra/a$ for any $a \in R^*$. Further routine but tedious calculations show that ϕ is a ring homomorphism and that ϕ is 1–1. Suppose then that K is a field and that $\theta: R \to K$ is a monomorphism. Define $f: F_R \to K$ by setting $f(a/b) = \theta(a)\theta(b)^{-1}$ (it is well defined). Then f is a monomorphism since θ is a monomorphism, and $f\phi = \theta$.

Suppose $g\colon F_R \to K$ is another monomorphism such that $g\phi = f\phi = \theta$. Then if $r/s \in F_R$ we have

$$\begin{aligned} g(r/s) &= g((ra/a) \cdot (sa/a)^{-1}) = g(ra/a)g(sa/a)^{-1} \\ &= g(\phi(r))g(\phi(s))^{-1} = \theta(r)\theta(s)^{-1} = f(r/s), \end{aligned}$$

so $g = f$ and F_R is a field of fractions for R.

Since $r \mapsto ra/a$ is a monomorphism from R into F_R we may identify r with ra/a and take the point of view that R is a subring of F_R. Then each element of F_R is a fraction r/s, where $r, s \in R$ and $s \neq 0$. It is clear from these remarks that F_R is (to within isomorphism) a minimal field F in which R is a subring.

The construction of F_R can be generalized considerably. Let R be any commutative ring and let S be a subset of R^* that is a multiplicative semigroup containing no zero divisors. Let X be the Cartesian product $R \times S$ and define a relation $\sim$ on X by agreeing that $(a, b) \sim (c, d)$ if $ad = bc$.

Exercise 2.2. (1) Show that the relation $\sim$ just defined is an equivalence relation on X.

(2) Denote the equivalence class of (a, b) by a/b and the set of all equivalence classes by R_S. Show that R_S is a commutative ring with 1.

(3) If $a \in S$ show that $\{ra/a : r \in R\}$ is a subring of R_S and that $r \mapsto ra/a$ is a monomorphism, so that R can be identified with a subring of R_S.

(4) Show that every $s \in S$ is a unit in R_S.

(5) Give a "universal" definition for the ring R_S and show that R_S is unique up to isomorphism.

The ring R_S is called the *localization* of R at S. The concept is important in algebraic number theory and algebraic geometry.

Exercise 2.3. Suppose R is an integral domain and $P \subseteq R$ is a prime ideal.

(1) Show that both P and $R \backslash P$ are multiplicative semigroups.

(2) If $S = R \backslash P$ show that $U(R_S) = R_S \backslash R_S P$. Conclude that $R_S P$ is the unique maximal ideal in R_S.

3. Polynomials

If R is a ring, then intuitively a polynomial in one variable x with coefficients in R is an expression of the form

$$f(x) = a_0 + a_1 x + a_2 x^2 + \cdots + a_n x^n,$$

where $a_i \in R$ and x is a "variable" that may be assigned values from R or from some ring that contains R as a subring. We present next a construction that makes these notions precise.

If R is a ring denote by $P(R)$ the set of all sequences $a = (a_i) = (a_0, a_1, a_2, \ldots)$, where each $a_i \in R$ and $a_i = 0$ for all but finitely many values of i, $0 \leq i \in \mathbb{Z}$. If $a, b \in P(R)$ we define

$$a + b = (a_i + b_i),$$

$$ab = \left(\sum_{j=0}^{i} a_j b_{i-j}\right) = (a_0 b_0, a_0 b_1 + a_1 b_0, a_0 b_2 + a_1 b_1 + a_2 b_0, \ldots).$$

Theorem 3.1. If R is a ring, then $P(R)$ is also a ring. It is commutative if and only if R is commutative and it is a ring with 1 if and only if R is a ring with 1, in which case $1_{P(R)} = (1_R, 0, 0, \ldots)$.

Exercise 3.1. Prove Theorem 3.1.

Suppose now that R is a ring with 1 and set $x = (0, 1, 0, 0, \ldots) \in P(R)$. Note that $x^2 = (0, 0, 1, 0, \ldots)$, $x^3 = (0, 0, 0, 1, 0, \ldots), \ldots$, and in general x^n has 1 as its $(n + 1)$st entry and 0s elsewhere. We agree that $x^0 = 1_{P(R)}$. The map $a \mapsto (a, 0, 0, \ldots)$ is a monomorphism from R into $P(R)$, and we may thus identify R with a subring of $P(R)$, with $1_R = 1_{P(R)}$. From that point of view we may write

$$a = (a_0, a_1, a_2, \ldots) = a_0 + a_1 x + a_2 x^2 + \cdots$$

for each $a \in P(R)$.

We call x an *indeterminate*, and write $R[x]$ for $P(R)$. We write $f(x)$ [or $g(x)$, etc.] for $a = a_0 + a_1 x + \cdots \in R[x]$, and call $f(x)$ a *polynomial* with *coefficients* $a_0, a_1, \ldots$ in R. If $a_n \neq 0$ but $a_m = 0$ for all $m > n$ we say that $f(x)$ has *degree n* and write $\deg f(x) = n$; in that case we call a_n the *leading coefficient* of $f(x)$. If $f(x)$ has leading coefficient 1 we say that $f(x)$ is a *monic* polynomial. The *zero polynomial* $(0, 0, 0, \ldots)$ is denoted by 0, and we agree that $\deg 0 = -\infty$. Polynomials of degree 0 or $-\infty$ are called *constants*–they are, of course, just the elements of R viewed as a subring of $R[x]$.

Proposition 3.2. Suppose R is a ring with 1 and $f(x), g(x) \in R[x]$. Then

(a) $\deg(f(x) + g(x)) \leq \max\{\deg f(x), \deg g(x)\}$, and
(b) $\deg(f(x)g(x)) \leq \deg f(x) + \deg g(x)$.

Equality always holds in (b) if R has no left or right zero divisors.

Proof. These statements are obvious consequences of the definitions of addition and multiplication in $R[x]$.

Corollary 1. If R has no left or right zero divisors, then $f(x) \in R[x]$ is a unit if and only if $f(x) = r$, a constant, with $r \in U(R)$.

Corollary 2. $R[x]$ is an integral domain if and only if R is an integral domain.

Exercise 3.2. Find $U(R)$ if $R = \mathbb{Z}_4[x]$.

Theorem 3.3 (The Division Algorithm). Suppose R is a commutative ring with 1 and $f(x), g(x) \in R[x]$. If $g(x)$ has leading coefficient b, then there exist a nonnegative integer k and $q(x), r(x) \in R[x]$ such that

$$b^k f(x) = q(x)g(x) + r(x),$$

with $\deg r(x) < \deg g(x)$. If b is not a zero divisor in R, then $q(x)$ and $r(x)$ are unique. If $b \in U(R)$ we may take $k = 0$.

Proof. If $\deg f(x) < \deg g(x)$ we may take $k = 0$, $q(x) = 0$, and $r(x) = f(x)$. Assume then that $\deg f(x) = m \geq \deg g(x) = n$, and say $f(x)$ has leading coefficient $a \in R$. Proceed by induction on m. Set $bf(x) - ax^{m-n}g(x) = f_1(x)$. Clearly $\deg f_1(x) < m$, so we may write $b^{k-1}f_1(x) = p(x)g(x) + r(x)$, where $k - 1$ is some nonnegative integer, $p(x), r(x) \in R[x]$, and $\deg r(x) < \deg g(x)$. Thus

$$\begin{aligned} b^k f(x) &= b^{k-1}ax^{m-n}g(x) + b^{k-1}f_1(x) \\ &= b^{k-1}ax^{m-n}g(x) + p(x)g(x) + r(x) \\ &= (b^{k-1}ax^{m-n} + p(x))g(x) + r(x). \end{aligned}$$

The main result follows if we set $q(x) = b^{k-1}ax^{m-n} + p(x)$.

Suppose b is not a zero divisor. If also $b^k f(x) = q_1(x)g(x) + r_1(x)$, with $\deg r_1(x) < \deg g(x)$, then

$$(q(x) - q_1(x))g(x) = r_1(x) - r(x).$$

If $q(x) \neq q_1(x)$, then the polynomial on the left-hand side has degree at least n since b is the leading coefficient of $g(x)$, but the polynomial on the right has degree less than n. Thus $q(x) = q_1(x)$, and consequently $r(x) = r_1(x)$.

Finally, if $b \in U(R)$ we may simply multiply through by b^{-k} and replace $q(x)$ and $r(x)$ by $b^{-k}q(x)$ and $b^{-k}r(x)$, respectively.

Note that the proof actually suggests the "long division" algorithm learned in high school algebra. The polynomials $q(x)$ and $r(x)$ are called the *quotient* and *remainder* upon division of $f(x)$ by $g(x)$. The division algorithm also holds when R is not commutative if we assume that b is a unit (Exercise 8.15).

From this point on we will discuss polynomial rings $R[x]$ only for R a commutative ring with 1. Noncommutative versions of the next results can be stated and proved but we shall only have occasion to use polynomials having coefficients from a commutative ring in later chapters.

Theorem 3.4 (Substitution). Suppose R and S are commutative rings with 1, that $\phi: R \to S$ is a homomorphism with $\phi(1_R) = 1_S$, and that $a \in S$. Then there is a unique homomorphism $E_a: R[x] \to S$ such that $E_a(r) = \phi(r)$ for all $r \in R$ and $E_a(x) = a$.

Proof. If $f(x) = r_0 + r_1x + \cdots + r_nx^n \in R[x]$ set

$$E_a(f(x)) = \phi(r_0) + \phi(r_1)a + \phi(r_2)a^2 + \cdots + \phi(r_n)a^n \in S.$$

It is an easy exercise to verify that E_a is a homomorphism with the right properties. If also $F: R[x] \to S$ is a homomorphism with $F(r) = \phi(r)$ for all $r \in R$ and $F(x) = a$, then

$$\begin{aligned} F(f(x)) &= F(r_0 + r_1x + \cdots + r_nx^n) \\ &= F(r_0) + F(r_1)F(x) + \cdots + F(r_n)F(x)^n \\ &= \phi(r_0) + \phi(r_1)a + \cdots + \phi(r_n)a^n = E_a(f(x)). \end{aligned}$$

Theorem 3.4 describes a universal mapping property of $R[x]$ that could have served as a definition prior to the construction. As usual we may conclude uniqueness up to isomorphism.

A particular case of Theorem 3.4 occurs when R is a subring of S, $1_R = 1_S$, and ϕ is the inclusion map. Then $E_a(f(x)) = r_0 + r_1a + \cdots + r_na^n$. This is simply the result of replacing, or "substituting," each occurrence of the indeterminate x by the element a of S. We write $f(a)$ for the result. The image of E_a is a subring of S which we shall denote by $R[a]$. Thus $R[a] = \{f(a): f(x) \in R[x]\}$.

Proposition 3.5 (The Remainder Theorem). Suppose R is a commutative ring with 1, $f(x) \in R[x]$, and $a \in R$. Then the remainder upon division of $f(x)$ by $g(x) = x - a$ is $r = f(a)$.

Proof. Write $f(x) = q(x)(x - a) + r$ by the Division Algorithm [note that r is a constant since $\deg g(x) = 1$]. Substitute a for x to see that

$$f(a) = q(a)(a - a) + r = q(a) \cdot 0 + r = r.$$

Corollary (The Factor Theorem). Suppose R is a commutative ring with 1, $f(x) \in R[x]$, $a \in R$, and $f(a) = 0$. Then $x - a$ is a factor of $f(x)$, in the sense that $f(x) = q(x) \cdot (x - a)$ for some $q(x) \in R[x]$.

If R and S are commutative rings with $R \subseteq S$ and $1_R = 1_S$, then an element $a \in S$ is called *algebraic* over R if $f(a) = 0$ for some nonzero polynomial $f(x) \in R[x]$. If $a \in S$ is not algebraic over R we call it *transcendental* over R (equivalently $a \in S$ is transcendental over R if E_a is a monomorphism).

For example, $\sqrt{2} \in \mathbb{R}$ is algebraic over $\mathbb{Z}$ since $f(\sqrt{2}) = 0$ for $f(x) = x^2 - 2 \in \mathbb{Z}[x]$. However, $\pi \in \mathbb{R}$ is transcendental over $\mathbb{Z}$—the proof will be given later in Chapter III.

Exercise 3.3. Show that $\sqrt{2} + \sqrt{3} \in \mathbb{R}$ is algebraic over $\mathbb{Z}$.

If R is an integral domain with 1, then the field of fractions of $R[x]$ is called the field of *rational functions* over R. It is commonly denoted by $R(x)$. Thus

$$R(x) = \{f(x)/g(x): f(x), g(x) \in R[x], g(x) \neq 0\}.$$

Exercise 3.4. If R is a commutative ring with 1, $f(x) \in R[x]$, and $a \in R$, then we may substitute a for x and obtain $f(a) \in R$. Thus $f(x)$ determines a *polynomial function* $f:R \to R$.

(1) If R is finite show that there must exist polynomials $f(x)$ and $g(x)$ in $R[x]$, with $f(x) \neq g(x)$, such that the associated polynomial functions f and g are identical, i.e., $f(a) = g(a)$, all $a \in R$.

(2) Find explicit examples of the phenomenon in (1) when $R = \mathbb{Z}_n$.

(3) If R is an infinite integral domain show that the mapping $f(x) \mapsto f$ assigning to each polynomial in $R[x]$ its corresponding polynomial function is 1–1.

4. POLYNOMIALS IN SEVERAL INDETERMINATES

Let $I = \{0, 1, 2, 3, \ldots\}$ and write I^n for the Cartesian product of n copies of I, $1 \leq n \in \mathbb{Z}$. If R is a ring denote by $P_n(R)$ the set of all functions $a: I^n \to R$ having finite support, i.e., having nonzero value for only finitely many elements of I^n. If $n = 1$ such a function is just a sequence indexed by I and having only finitely many nonzero entries, so it is a polynomial as discussed in the previous section. Thus $P_1(R) = P(R)$.

Let us write 0 for the element $(0, 0, \ldots, 0) \in I^n$, and if $i = (i_1, i_2, \ldots, i_n) \in I^n$ and $j = (j_1, j_2, \ldots, j_n) \in I^n$ define

$$i + j = (i_1 + j_1, i_2 + j_2, \ldots, i_n + j_n) \in I^n.$$

We define operations of addition and multiplication on $P_n(R)$ by setting

$$(a + b)(i) = a(i) + b(i),$$
$$(ab)(i) = \sum\{a(j)b(k): j, k \in I^n, j + k = i\}$$

for all $a, b \in P_n(R)$, $i \in I^n$.

Theorem 4.1. If R is a ring, then $P_n(R)$ is a ring; it is commutative if and only if R is commutative and it is a ring with 1 if and only if R is a ring with 1.

Proof. The proof is straightforward. We show that multiplication is associative as a sample and leave the rest as an exercise. If $a, b, c \in P_n(R)$ and $i \in I^n$, then

$$\begin{aligned}((ab)c)(i) &= \sum\{(ab)(j)c(k): j, k \in I^n, j + k = i\}\\ &= \sum\{(\sum\{a(u)b(v): u, v \in I^n, u + v = j\})c(k): j, k \in I^n, j + k = i\}\\ &= \sum\{(a(u)b(v))c(k): u, v, k \in I^n, u + v + k = i\},\end{aligned}$$

and

$$\begin{aligned}(a(bc))(i) &= \sum\{(a(u)(bc)(j): u, j \in I^n, u + j = i\} \\ &= \sum\{a(u)\sum\{b(v)c(k): v, k \in I^n, v + k = j\}: u, j \in I^n, u + j = i\} \\ &= \sum\{a(u)(b(v)c(k)): u, v, k \in I^n, u + v + k = i\} = ((ab)c)(i).\end{aligned}$$

We remark that if R is a ring with 1, then the function $1: I^n \to R$ defined by $1(0) = 1 \in R$, $1(i) = 0 \in R$ if $0 \neq i \in I^n$, is the multiplicative identity of $P_n(R)$.

Exercise 4.1. Complete the proof of Theorem 4.1.

For each $r \in R$ define a function $a_r \in P_n(R)$ by setting $a_r(0) = r$ and $a_r(i) = 0$ if $0 \neq i \in I^n$. Then $a_r + a_s = a_{r+s}$ and $a_r a_s = a_{rs}$ for all $r, s \in R$. Clearly the map $r \mapsto a_r$ is 1–1 from R into $P_n(R)$, so it is a monomorphism. As usual we identify $r \in R$ with $a_r \in P_n(R)$ and view R as subring of $P_n(R)$.

Suppose now that R is a ring with 1. Let e_k be the n-tuple in I^n having 1 as its kth entry and 0's elsewhere. Define $x_k \in P_n(R)$ by setting $x_k(e_k) = 1$ and $x_k(i) = 0$ if $e_k \neq i \in I^n$. We see by multiplying that $x_k^2(2e_k) = 1$ (where $2e_k = e_k + e_k$), and $x_k^2(i) = 0$ otherwise, and in general that $x_k^m(me_k) = 1$, $x_k^m(i) = 0$ otherwise, $1 \leq m \in \mathbb{Z}$. We agree as usual that $x_k^0 = 1 \in P_n(R)$. The functions x_k all commute with one another. If we take $i = (i_1, \ldots, i_n)$ in I^n and $r \in R$ and consider the element $rx_1^{i_1} \cdots x_n^{i_n} \in P_n(R)$ we find that its value at i is r and its value is 0 elsewhere. Since any nonzero $a \in P_n(R)$ can be written uniquely as a sum of functions having distinct one-point supports it follows that each nonzero $a \in P_n(R)$ can be written uniquely as a sum of elements each of the form

$$rx_1^{i_1}x_2^{i_2}\cdots x_n^{i_n}, \qquad 0 \neq r \in R, \quad i = (i_1, i_2, \ldots, i_n) \in I^n.$$

An element of that form is called a *monomial* in $P_n(R)$. The *degree* of $a = rx_1^{i_1} \cdots x_n^{i_n}$ is $\deg a = i_1 + i_2 + \cdots + i_n$. More generally if $a \in P_n(R)$, $a \neq 0$, and a is written as $a = a_1 + \cdots + a_m$, its unique expression as a sum of monomials, then we define the *degree* of a as $\deg a = \max\{\deg a_i : 1 \leq i \leq m\}$. As usual we agree that $\deg 0 = -\infty$. If all the monomial summands of $a \in P_n(R)$ have the same degree we say that a is *homogeneous.*

The elements of $P_n(R)$ are called *polynomials* in the n commuting *indeterminates* $x_1, x_2, \ldots, x_n$. In accordance with tradition we write $R[x_1, x_2, \ldots, x_n]$ for $P_n(R)$, and denote its elements by $f(x_1, x_2, \ldots, x_n)$, etc. It is often typographically convenient to write X for the n-tuple $(x_1, x_2, \ldots, x_n)$ of indeterminates, and write $f(x_1, \ldots, x_n)$ as $f(X)$ when the number of indeterminates is clear from the context. When $n = 2$ or 3 it is customary to write $x_1 = x$, $x_2 = y$, and $x_3 = z$.

For example, $f(x, y) = x^2y^3 - 3x^4y \in \mathbb{Z}[x, y]$ is homogeneous of degree 5.

Proposition 4.2. Suppose R is a ring with 1 and $f(X)$, $g(X) \in R[x_1, \cdots, x_n]$. Then

(a) $\deg(f(X) + g(X)) \leq \max\{\deg f(X), \deg g(X)\}$, and
(b) $\deg(f(X)g(X)) \leq \deg f(X) + \deg g(X)$.

Equality always holds in (b) if R has no zero divisors.

Theorem 4.3 (Substitution). Suppose R and S are commutative rings with 1 and $\phi: R \to S$ is a homomorphism with $\phi(1_R) = 1_S$. If $a_1, a_2, \ldots, a_n \in S$, then there is a unique homomorphism $E = E_{(a_1,\ldots,a_n)}$ from $R[x_1, x_2, \ldots, x_n]$ to S such that $E(r) = \phi(r)$, all $r \in R$, and $E(x_i) = a_i$, $1 \leq i \leq n$.

Proof. We define

$$E(rx_1^{i_1}x_2^{i_2}\cdots x_n^{i_n}) = \phi(r)a_1^{i_1}a_2^{i_2}\cdots a_n^{i_n}$$

for each monomial, extend to arbitrary polynomials in the obvious fashion, and proceed as for polynomials in one indeterminate.

Thus $R[x_1, \ldots, x_n]$ could also have been defined abstractly in terms of the universal mapping property in Theorem 4.3, and is unique in that respect up to isomorphism. Other constructions are possible. Perhaps the most popular is first to construct $R[x]$ by some means, then to construct $(R[x])[y]$, and then to say "et cetera" (i.e., apply induction). Because of the uniqueness one construction is as good as another, but the latter seems to have an aesthetic deficiency in that the indeterminates do not arise in a symmetric fashion.

The finiteness of the cardinal number n played no essential role in the construction of the polynomial ring $P_n(R)$. Thus the discussion above would have served, with minor changes in notation, to define $R[x_\alpha : \alpha \in A]$ for any nonempty index set A and commuting indeterminates x_α.

When $R \subseteq S$, with $1_R = 1_S$, and $\phi: R \to S$ is the inclusion map, then the homomorphism E of Theorem 4.3 allows us to substitute elements $a_1, \ldots, a_n \in S$ in any polynomial $f(x_1, \ldots, x_n) \in R[x_1, \ldots, x_n]$ and obtain $f(a_1, \ldots, a_n) \in S$. The image of E is a subring $R[a_1, \ldots, a_n]$ of S.

Elements $a_1, \ldots, a_n \in S$ are called *algebraically dependent* over R if $f(a_1, \ldots, a_n) = 0$ for some nonzero polynomial $f(X) \in R[x_1, \ldots, x_n]$; otherwise they are *algebraically independent*.

Exercise 4.2. Show that $a_1 = \sqrt{3}$ and $a_2 = \sqrt{5}$ are algebraically dependent over $\mathbb{Z}$.

When 0 exponents occur in monomials it is customary to suppress the corresponding indeterminate, in accordance with the usual convention that $x_i^0 = 1$. Thus, for example, we simply write $5x_1^0x_2^1x_3^0x_4^5$ as $5x_2x_4^5$. With that notational convention we have

$$R[x_1] \subseteq R[x_1, x_2] \subseteq R[x_1, x_2, x_3] \subseteq \cdots.$$

For an infinite sequence $x_1, x_2, \ldots$ of distinct commuting indeterminates we may write $R[x_1, x_2, x_3, \ldots] = \bigcup\{R[x_1, \ldots, x_k] : 1 \leq k \in \mathbb{Z}\}$, with the obvious addition and multiplication, and obtain the polynomial ring in $x_1, x_2, x_3, \ldots$ over R. It will be useful for examples.

The field of fractions of $R[x_1, \ldots, x_n]$ consists of all *rational functions* $f(x_1, \ldots, x_n)/g(x_1, \ldots, x_n)$, $g(x_1, \ldots, x_n) \neq 0$. It is denoted by $R(x_1, \ldots, x_n)$, and is defined, of course, only if R is an integral domain.

5. DIVISIBILITY AND FACTORIZATION

We assume throughout this section that R is an integral domain with 1. If $a, b \in R$ we say that *a divides b*, or b is a *multiple* of a, if $b = ac$ for some $c \in R$, and then we write $a \mid b$. If $a \mid b$ and $b \mid a$ we say that a and b are *associates* and write $a \sim b$. Note that the only associate of 0 is 0, and the associates of 1 are just the units in R.

Proposition 5.1. Suppose R is an integral domain with 1 and $a, b \in R$. Then a and b are associates ($a \sim b$) if and only if $a = bu$ for some $u \in U(R)$.

Proof. We may assume $a \neq 0$. $\Rightarrow$: Since $a \sim b$ we have $a = bc$ and $b = ad$ for some $c, d \in R$. Thus $a = (ad)c = a(dc)$, so $dc = 1$ and c is a unit.
$\Leftarrow$: If $a = bu$, with $u \in U(R)$, then $b \mid a$, but also $b = au^{-1}$, so $a \mid b$.

It is clear that the relation of being associates is an equivalence relation. The equivalence class containing $a \in R$ is just $aU(R)$.

Note that if $b \in R$ and $u \in U(R)$, then $u \mid b$ since $b = u(u^{-1}b)$. If $b \in R$ is not a unit and the only divisors of b are units and associates of b, then b is called *irreducible.*

Suppose that $0 \neq p \in R$, that p is not a unit, and that if $p \mid ab$ for some $a, b \in R$, then necessarily $p \mid a$ or $p \mid b$. Then p is called a *prime* in R.

Proposition 5.2. If $p \in R$ is prime, then p is irreducible.

Proof. If not we may write $p = ab$, with $a, b \notin U(R)$. Since p is prime $p \mid a$ or $p \mid b$, say $p \mid a$, with $a = pc$, $c \in R$. But then $p = ab = (pc)b = p(cb)$, so $cb = 1$ since $p \neq 0$. But then $b \in U(R)$, a contradiction.

It is important to note that the converse to Proposition 5.2 may fail to hold, as the following exercise shows.

Exercise 5.1. Let $R = \{a + b\sqrt{-5} : a, b \in \mathbb{Z}\} \subseteq \mathbb{C}$.

(1) Show that R is an integral domain with 1 (it is a subring of $\mathbb{C}$).
(2) Show that $U(R) = \{\pm 1\}$.
(3) Show that 3 is irreducible in R.
(4) Show that $a = 2 + \sqrt{-5}$ and $b = 2 - \sqrt{-5}$ are both irreducible in R.
(5) Conclude that $3 \nmid 2 + \sqrt{-5}$ and $3 \nmid 2 - \sqrt{-5}$ in R.
(6) Conclude that 3 is irreducible but *not* prime in R.

Recall that a principal ideal in a commutative ring R with 1 is an ideal (a) generated by a single element $a \in R$, consisting of all multiples ra, $r \in R$, so $(a) = Ra$. If R is an integral domain with 1 in which every ideal is principal, then R is called a *principal ideal domain*, abbreviated PID.

The most prominent example of a PID is the ring $\mathbb{Z}$ of integers. If I is a nonzero ideal in $\mathbb{Z}$ let a be the smallest positive element in I. If $b \in I$ we may write $b = aq + r$, with $q, r \in \mathbb{Z}$ and $0 \leq r < a$. But then $r = b - aq \in I$ and $r = 0$ by the minimality of a, so $b = qa \in \mathbb{Z}a = (a)$, and $I = (a)$.

A *common divisor* of two elements a and b in an integral domain R is an element $c \in R$ such that $c \mid a$ and $c \mid b$. A common divisor d of a and b is a *greatest common divisor* (or GCD) if $c \mid d$ for every common divisor c of a and b. For example $a = 143$ and $b = 154$ have $d = -11$ as a GCD in the ring $\mathbb{Z}$ of integers.

Proposition 5.3. Suppose R is a PID and $a, b \in R$ are not both zero. Then a and b have a GCD, denoted by $d = (a, b)$. It is unique up to associates in R, and it can be expressed as $d = (a, b) = xa + yb$ for some $x, y \in R$.

Proof. Let $I = (a, b)$, the ideal generated by a and b. Then $I = \{ua + vb : u, v \in R\}$. Since R is a PID we may write $I = (d)$ for some $d \in I$, say $d = xa + yb$. Since $a \in I$ and $b \in I$ we have $d \mid a$ and $d \mid b$. If c is any common divisor of a and b, then $c \mid xa + yb$, or $c \mid d$, so d is a GCD for a and b. If d_1 is another, then $d \mid d_1$ and $d_1 \mid d$, so d and d_1 are associates.

Corollary. If R is a PID, then every irreducible element of R is prime.

Proof. Let $p \in R$ be irreducible and suppose $p \mid ab$ for some $a, b \in R$. If $p \nmid a$, then $(p, a) = 1$, so we may write $1 = xa + yp$ for some $x, y \in R$. Thus

$$b = (xa + yp)b = x(ab) + (yb)p,$$

so $p \mid b$.

Suppose $m \in \mathbb{Z}, m \neq 0$ or 1. For an important class of examples consider the subring $\mathbb{Q}[\sqrt{m}]$ of $\mathbb{C}$ obtained by substituting $\sqrt{m}$ for x *in* $\mathbb{Q}[x]$. If $m = k^2 n$, with $k, n \in \mathbb{Z}$ and $k > 1$, then clearly $\mathbb{Q}[\sqrt{m}] = \mathbb{Q}[\sqrt{n}]$, so we will only consider the case where m is square-free, i.e., not divisible by any square $k^2 > 1$ in $\mathbb{Z}$.

Exercise 5.2. Show that $\mathbb{Q}[\sqrt{m}] = \{r + s\sqrt{m} : r, s \in \mathbb{Q}\}$, and that $\mathbb{Q}[\sqrt{m}]$ is a field. It is thus its own field of fractions, and we will write $\mathbb{Q}(\sqrt{m})$ rather than $\mathbb{Q}[\sqrt{m}]$.

For any square-free integer $m(\neq 0, 1)$ we define the ring R_m of *algebraic integers* in $\mathbb{Q}(\sqrt{m})$ as follows. If $m \equiv 2$ or $3 \pmod 4$, then $R_m = \{a + b\sqrt{m} : a, b \in \mathbb{Z}\}$, and if $m \equiv 1 \pmod 4$, then

$$R_m = \{(a + b\sqrt{m})/2 : a, b \in \mathbb{Z} \text{ and } a, b \text{ have the same parity}\}$$

(having the same parity means that a and b are either both even or both odd).

Exercise 5.3. (1) Show that R_m is an integral domain with 1.

(2) Show that $\mathbb{Q}(\sqrt{m})$ is the field of fractions for R_m.

(3) Show that R_m is the set of all those $r + s\sqrt{m} \in \mathbb{Q}(\sqrt{m})$ that are roots of a monic quadratic polynomial $x^2 + cx + d \in \mathbb{Z}[x]$. [This is the reason for the variation in the definition of R_m when $m \equiv 1 \pmod 4$.]

For any $x = r + s\sqrt{m} \in \mathbb{Q}(\sqrt{m})$ we define the *norm* of x to be $N(x) = r^2 - ms^2$. Note that $N(x) = (r + s\sqrt{m})(r - s\sqrt{m})$. Thus if $x, y \in \mathbb{Q}(\sqrt{m})$ we have $N(xy) = N(x)N(y)$.

Exercise 5.4. (1) Show that $N(x) \in \mathbb{Z}$ if $x \in R_m$.

(2) Show that $u \in U(R_m)$ if and only if $N(u) = \pm 1$.

(3) Use (2) to show that $U(R_{-1}) = \{\pm 1, \pm i\}$, $U(R_{-3}) = \{\pm 1, \pm(1 \pm \sqrt{-3})/2\}$, and $U(R_m) = \{\pm 1\}$ for all other negative square-free m in $\mathbb{Z}$.

We remark that if $m \in \mathbb{Z}$ is square-free and $m \geq 2$, then $U(R_m)$ is infinite (see Exercises 8.32 and 8.33).

Proposition 5.4 (Dedekind and Hasse, see [31]). Let m be square-free in $\mathbb{Z}$, $m \neq 0, 1$, and let R_m be the ring of algebraic integers in $\mathbb{Q}(\sqrt{m})$. Suppose that for all nonzero $x, y \in R_m$ such that $y \nmid x$ and $|N(x)| \geq |N(y)|$ there exist u, $v \in R_m$ such that $xu \neq yv$ and $|N(xu - yv)| < |N(y)|$. Then R_m is a PID.

Proof. Let $I \neq 0$ be an ideal in R_m. Choose $y \in I$, $y \neq 0$, so that $|N(y)|$ is minimal. Given any $x \in I$ we wish to show that $y \mid x$. If not then $y \nmid x$ and $|N(x)| \geq |N(y)|$ so we may choose $u, v \in R_m$ such that $xu - yv \neq 0$ but $|N(xu - yv)| < |N(y)|$. But that is a contradiction since $xu - yv \in I$ and $|N(y)|$ was minimal. Thus $y \mid x$, so $I = (y)$.

Corollary. If $m = -19$, then R_m is a PID.

Proof (Wilson [39]). Since $-19 \equiv 1 \pmod 4$ we know that R_{-19} consists of all $(a + b\sqrt{-19})/2$ for $a, b \in \mathbb{Z}$, a and b having the same parity. Note that $N(x) \geq 0$ for all $x \in \mathbb{Q}(\sqrt{-19})$. Suppose x, $y \in R_{-19}$, $y \nmid x$, and $N(x) \geq N(y)$. We need $u, v \in R_{-19}$ such that $xu \neq yv$ and $N(xu - yv) < N(y)$, or equivalently $N((x/y)u - v) < 1$. Since $x/y \in \mathbb{Q}(\sqrt{-19}) \backslash R_{-19}$ we may write $x/y = (a + b\sqrt{-19})/c$, where $a, b, c \in \mathbb{Z}$ are relatively prime and $c > 1$. There are four cases to consider.

(i) Suppose $c = 2$. Then a and b are of opposite parity since $x/y \notin R_{-19}$. But then $v = ((a-1) + b\sqrt{-19})/2$ is in R_{-19}, so we may take $u = 1$ and obtain $(x/y)u - v = \frac{1}{2}$, with $0 < N(\frac{1}{2}) = \frac{1}{4} < 1$.

(ii) Suppose $c = 3$. Then either $3 \nmid a$ or $3 \nmid b$, so $a^2 + b^2 \not\equiv 0 \pmod 3$. Since $a^2 + 19b^2 \equiv a^2 + b^2 \pmod 3$ we may divide by 3 and obtain

$$a^2 + 19b^2 = 3q + r, r = 1 \text{ or } 2.$$

Set $u = a - b\sqrt{-19}$ and $v = q$. Then

$$\begin{aligned}(x/y)u - v &= (a + b\sqrt{-19})(a - b\sqrt{-19})/3 - q \\ &= (a^2 + 19b^2)/3 - q = r/3,\end{aligned}$$

and $N(r/3) = \frac{1}{9}$ or $\frac{4}{9}$.

(iii) Suppose $c = 4$. Then a and b cannot both be even since $(a, b, c) = 1$. If they are of opposite parity, then $a^2 + 19b^2 \equiv a^2 - b^2 \equiv \pm 1 \pmod 4$, so we may divide by 4 and write $a^2 + 19b^2 = 4q + r$, $r = 1$ or 3. In that case set $u = a - b\sqrt{-19}$ and $v = q$, and obtain $(x/y)u - v = r/4$, of norm $\frac{1}{16}$ or $\frac{9}{16}$. If a and b are both odd, then $a^2 + 19b^2 \equiv a^2 + 3b^2 \equiv 4 \pmod 8$ and we may write $a^2 + 19b^2 = 8q_1 + 4$. In that case set $u = (a - b\sqrt{-19})/2$ and $v = q_1$, so that

$$(x/y)u - v = (a^2 + 19b^2)/8 - q_1 = \tfrac{1}{2},$$

of norm $\frac{1}{4}$.

(iv) Suppose $c \geq 5$. Since $(a, b, c) = 1$ we may choose $d, e, f \in \mathbb{Z}$ such that $ad + be + cf = 1$, then we may divide $ae - 19bd$ by c to obtain $q, r \in \mathbb{Z}$ such that $ae - 19bd = cq + r$, with $|r| \leq c/2$. (Why?) Set $u = e + d\sqrt{-19}$ and $v = q - f\sqrt{-19}$. Then

$$\begin{aligned}(x/y)u - v &= (a + b\sqrt{-19})(e + d\sqrt{-19})/c - (q - f\sqrt{-19}) \\ &= (ae - 19bd - qc)/c + (ad + be + cf)\sqrt{-19}/c \\ &= (r + \sqrt{-19})/c,\end{aligned}$$

and $N\big((r + \sqrt{-19})/c\big) = (r^2 + 19)/c^2$. If $c > 5$, then

$$\begin{aligned}(r^2 + 19)/c^2 &\leq (c^2/4 + 19)/c^2 = \tfrac{1}{4} + 19/c^2 \\ &\leq \tfrac{1}{4} + \tfrac{19}{36} = \tfrac{7}{9} < 1,\end{aligned}$$

and if $c = 5$, then $|r| \leq 2$, so

$$(r^2 + 19)/c^2 \leq (4 + 19)/25 < 1.$$

Exercise 5.5. Use Proposition 5.4 to show that R_{-1}, the ring of *Gaussian integers*, is a PID (this is easier than the corollary above).

An integral domain R with 1 is called *Euclidean* if there is a function $d: R^* \to \mathbb{Z}$, with $d(r) \geq 0$ for all $r \in R^*$, such that

(i) if $a, b \in R^*$ and $a \mid b$, then $d(a) \leq d(b)$, and
(ii) if $a, b \in R$, with $b \neq 0$, then there are $q, r \in R$ such that $a = bq + r$, with either $r = 0$ or $d(r) < d(b)$.

Clearly $R = \mathbb{Z}$ is Euclidean if we define $d(r) = |r|$ for all $r \in \mathbb{Z}^*$. For another example take $R = F[x]$, where F is a field, defining $d(f(x))$ to be $\deg f(x)$. Then R is Euclidean by Proposition 3.2 and Theorem 3.3.

Proposition 5.5. If R is Euclidean, then $U(R) = \{x \in R^*: d(x) = d(1)\}$.

Proof. Observe that if $a \sim b$ in R^*, then $d(a) \le d(b)$ and $d(b) \le d(a)$, so $d(a) = d(b)$. If $u \in U(R)$, then $u \sim 1$, so $d(u) = d(1)$. If $x \in R^*$ and $d(x) = d(1)$ write $1 = qx + r$, with $q \in R$ and $r = 0$ or $d(r) < d(x) = d(1)$. But if $r \neq 0$, then $d(1) \le d(r)$ since $1 \mid r$. Thus $r = 0$, $qx = 1$, and $x \in U(R)$.

Proposition 5.6. If R is Euclidean, then R is a PID.

Proof. Let I be any nonzero ideal in R. Choose $b \in I^*$ with $d(b)$ minimal. If $a \in I$ we may write $a = bq + r$, with $r = 0$ or $d(r) < d(b)$. But $r = a - bq \in I$, so $d(r) < d(b)$ is impossible, and so $r = 0$. Thus $a = qb \in (b)$, and $I = (b)$.

Exercise 5.6. (1) Show that the ideal $I = (3, 2 + \sqrt{-5})$ is not principal in R_{-5} (see Exercise 5.1). Conclude that R_{-5} is not Euclidean.

(2) If S is an integral domain with 1 set $R = S[x, y]$ and let $I = (x, y)$. Show that I is not a principal ideal, so R is not Euclidean.

Proposition 5.7. If $m = -2, -1, 2$, or 3, then R_m is Euclidean with $d(r) = |N(r)|$ for all $r \in R_m^*$.

Proof. Take $a, b \in R_m$, with $b \neq 0$. Then $a/b \in \mathbb{Q}(\sqrt{m})$, say $a/b = s + t\sqrt{m}$ with $s, t \in \mathbb{Q}$. Choose $c, d \in \mathbb{Z}$ nearest to s and t, respectively, so that $|s - c| \le \frac{1}{2}$ and $|t - d| \le \frac{1}{2}$. Set $q = c + d\sqrt{m}$ and $r = a - bq$, so that

$$r/b = a/b - q = (s - c) + (t - d)\sqrt{m}.$$

Thus $N(r/b) = (s - c)^2 - m(t - d)^2$. If $m = -2$ or -1, then

$$0 \le N(r/b) \le \tfrac{1}{4} - m/4 < 1,$$

so either $r = 0$ or $d(r) < d(b)$. If $m = 2$ or 3, then $-\frac{3}{4} \le N(r/b) \le \frac{1}{4}$, so $|N(r/b)| \le \frac{3}{4}$, and again either $r = 0$ or $d(r) < d(b)$.

Exercise 5.7. Extend Proposition 5.7 by showing that R_m is Euclidean [with $d(r) = N(r)$] if $m = -3, -7$, or -11. [*Hint*: Choose $d \in \mathbb{Z}$ nearest to $2t$ and then choose $c \in \mathbb{Z}$ so that c is as near to $2s$ as possible with c and d of the same parity. Then set $q = (c + d\sqrt{m})/2$.]

Proposition 5.8. Suppose $m \in \mathbb{Z}$ is negative and square-free, but $m \neq -1$, $-2, -3, -7$, or -11. Then R_m is not Euclidean.

Proof. Suppose, by way of contradiction, that R_m is Euclidean, with function d. We know by Exercise 5.6 that $m \neq -5$, so $m = -6$ or -10 or else $m \le -13$. Choose $b \in R_m^* \setminus U(R_m)$ so that $d(b)$ is minimal. Given any $a \in R_m$ there are $q, r \in R_m$ such that $a = bq + r$ and either $r = 0$ or $d(r) < d(b)$. By the minimality of $d(b)$ we may conclude that either $r = 0$ or $r = \pm 1$, since $U(R_m) = \{\pm 1\}$ by Exercise 5.4. Thus for any $a \in R_m$ we may write either $a = bq$ or $a = bq \pm 1$, and hence either $b \mid a$ or $b \mid a \pm 1$. Taking $a = 2$ we see in particular that $b \mid 2$ or $b \mid 3$.

Let us see that 2 and 3 are both irreducible in R_m. For example, if $m \equiv 1 \pmod 4$ and if 3 were reducible in R_m we could write

$$3 = ((u + v\sqrt{m})/2) \cdot ((x + y\sqrt{m})/2), \qquad \text{with} \quad u, v, x, y \in \mathbb{Z}$$

and neither factor a unit in R_m. Thus

$$N(3) = 3^2 = ((u^2 - mv^2)/4) \cdot ((x^2 - my^2)/4),$$

and so $(u^2 - mv^2)/4 = (x^2 - my^2)/4 = 3$ (since neither factor is a unit). But then $u^2 - mv^2 = 12$, which has no solutions in integers since $m = -6$ or -10 or $m \leq -13$ (try $v = 0, 1, 2$), and we have a contradiction. The proof is similar that 2 is irreducible, and the proofs are similar and easier when $m \equiv 2$ or $3 \pmod 4$. Since 2 and 3 are irreducible we may conclude that $b = \pm 2$ or $b = \pm 3$.

But now if $m \equiv 1 \pmod 4$ take $a = (1 + \sqrt{m})/2$, so $a + 1 = (3 + \sqrt{m})/2$ and $a - 1 = (-1 + \sqrt{m})/2$. Clearly none of $a, a + 1$, or $a - 1$ is divisible by 2 or 3 [since $(1 + \sqrt{m})/4 \notin R_m$, etc.], and we have a contradiction. If $m \equiv 2$ or $3 \pmod 4$ take $a = 1 + \sqrt{m}$. Again it is clear that none of $a, a + 1$, or $a - 1$ is divisible by 2 or 3, and the proof is complete.

Corollary. R_{-19} is a PID that is not Euclidean.

We remark that 2 is reducible in R_{-7} and 3 is reducible in R_{-11}.

An ascending chain $I_1 \subseteq I_2 \subseteq I_3 \subseteq \cdots$ of ideals in a ring R is said to *terminate* if it is either finite or there is some index k such that $I_j = I_k$ for all $j \geq k$. A commutative ring R is called *Noetherian* if it satisfies the *ascending chain condition* (ACC), i.e., if every ascending chain of ideals of R terminates.

Proposition 5.9. If R is a PID, then R is Noetherian.

Proof. If $I_1 \subseteq I_2 \subseteq I_3 \subseteq \cdots$ is any infinite ascending chain of ideals then $I = \bigcup\{I_k : 1 \leq k \in \mathbb{Z}\}$ is easily checked to be an ideal in R. Thus $I = (a)$ for some $a \in R$. But then $a \in I_k$ for some k, so $I = I_j = I_k$ for all $j \geq k$, and the chain terminates.

We define a *unique factorization domain* (UFD) to be an integral domain R with 1 in which every nonzero nonunit element is a product of irreducible elements and every irreducible element is prime.

Proposition 5.10. If R is a PID then R is a UFD.

Proof. Let X be the set of all elements in $R^* \backslash U(R)$ that can not be written as products of irreducible elements. We wish to show that X is empty. If not choose $a_1 \in X$. If possible choose $a_2 \in X$ such that $(a_1) \subseteq (a_2)$ but $(a_1) \neq (a_2)$. Then choose $a_3 \in X$ such that $(a_2) \subseteq (a_3)$ but $(a_2) \neq (a_3)$, etc. The process must terminate, by Proposition 5.9, with some $a_n \in X$ such that (a_n) is not properly contained in any (x) for $x \in X$. Since $a_n \in X$ it is not irreducible, and we may write $a_n = a_{n+1}b$, with a_{n+1} and b in $R^*\setminus U(R)$. One of a_{n+1} and b must be in X, or else a_n would not be in X, so assume $a_{n+1} \in X$. But then $(a_n) \subseteq (a_{n+1})$ and $(a_n) \neq (a_{n+1})$, a contradiction, so in fact $X = \varnothing$. An application of the corollary to Proposition 5.3 completes the proof.

Theorem 5.11 (Unique Factorization). If R is a UFD and $a \in R$ is not 0 and not a unit, then $a = p_1 p_2 \cdots p_k$, where each p_i is a prime. The representation of a as a product of primes is unique in the sense that if also $a = q_1 q_2 \cdots q_m$, with each q_i prime, then $m = k$, and by relabeling if necessary we have $p_i \sim q_i$, $1 \leq i \leq k$.

Proof. Only the uniqueness needs to be proved, and we use induction on k. Uniqueness is clear when $k = 1$ for then a is irreducible, so we assume the uniqueness of representation for a product of $k - 1$ primes. Since $p_1 p_2 \cdots p_k = q_1 q_2 \cdots q_m$ we have $p_1 \mid q_1 q_2 \cdots q_m$ and hence $p_1 \mid q_i$ for some i, since p_1 is prime. Relabel the qs if necessary so that $p_1 \mid q_1$. But then $p_1 \sim q_1$ since both are irreducible, so $q_1 = p_1 u$ for some $u \in U(R)$. Thus $p_1 p_2 \cdots p_k = p_1 u q_2 q_3 \cdots q_m$, and hence $p_2 \cdots p_k = q_2' q_3 \cdots q_m$, where $q_2' = u q_2 \sim q_2$. By the induction hypothesis $k - 1 = m - 1$, so $k = m$, and we may relabel if necessary so that $p_2 \sim q_2' \sim q_2, p_3 \sim q_3, \ldots, p_k \sim q_k$.

If R is a UFD it follows easily from Theorem 5.11 that each nonzero nonunit $a \in R$ can be written as

$$a = u p_1^{e_1} p_2^{e_2} \cdots p_n^{e_n},$$

where $n, e_1, e_2, \ldots, e_n \in \mathbb{Z}$ are unique and positive, and $p_1, p_2, \ldots, p_n$ are distinct primes in R, unique up to associates, and $u \in U(R)$.

As a consequence of Proposition 5.10 and Theorem 5.11 we see that unique factorization into primes holds in $\mathbb{Z}$ (that fact is often called the Fundamental Theorem of Arithmetic), in $F[x]$ for any field F, and in $R_{-19}, R_{-2}, R_{-1}, R_2$, and R_3. Unique factorization does *not* hold in R_{-5}.

If $m < 0$ it has been shown that R_m is a PID if and only if $m = -1, -2, -3, -7, -11, -19, -43, -67$, or -163. Final details were supplied by Stark [36]. If $m > 0$, then R_m is Euclidean [with $d(r) = |N(r)|$] if and only if $m = 2, 3, 5, 6, 7, 11, 13, 17, 19, 21, 29, 33, 37, 41, 57$, or 73. It is not known precisely which R_m, $m > 0$, are PIDs. Further historical information regarding these problems can be found in Hardy and Wright [14].

Proposition 5.12. If R is a UFD and if $a, b \in R$ are not both 0, then a and b have a GCD d that is unique up to associates.

Proof. If $a = 0$, then b is a GCD for a and b. If a is a unit, then 1 is a GCD for a and b. If neither a nor b is 0 or a unit we may write

$$a = p_1^{e_1} p_2^{e_2} \cdots p_k^{e_k} \quad \text{and} \quad b = u p_1^{f_1} p_2^{f_2} \cdots p_k^{f_k},$$

where $p_1, \ldots, p_k$ are distinct primes in R, $0 \leq e_i \in \mathbb{Z}$, $0 \leq f_i \in \mathbb{Z}$, and $u \in U(R)$. Set $g_i = \min\{e_i, f_i\}$, $1 \leq i \leq k$, and set

$$d = p_1^{g_1} p_2^{g_2} \ldots p_k^{g_k}.$$

Clearly $d \mid a$ and $d \mid b$. If $c \mid a$ and $c \mid b$, then only the primes $p_1, \ldots, p_k$ (and their associates) can be prime divisors of c, so we may write

$$c = v p_1^{h_1} p_2^{h_2} \cdots p_k^{h_k},$$

with $0 \leq h_i \in \mathbb{Z}$ and $v \in U(R)$. But then $h_i \leq g_i$, $1 \leq i \leq k$, since $c \mid a$ and $c \mid b$, and consequently $c \mid d$. Hence d is a GCD for a and b. If d_1 is another then $d \mid d_1$ and $d_1 \mid d$, so $d_1 \sim d$.

Recall that an ideal P in a commutative ring R is a *prime* ideal if $P \neq R$ and if $ab \notin P$ for all $a, b \in R \backslash P$, or equivalently if R/P is an integral domain. If R is an integral domain with 1, then a principal ideal $P = (p)$ is prime if and only if p is a prime element in R, as follows easily from the fact that $p \mid ab$ if and only if $p \mid a$ or $p \mid b$.

If R is a commutative ring with 1 and I as an ideal in R write $\eta: R \to R/I$ for the canonical quotient map. It is a consequence of Theorem 3.4 that there is a homomorphism from $R[x]$ onto $(R/I)[x]$ given by

$$\begin{aligned} f(x) = r_0 + r_1 x + \cdots + r_n x^n &\mapsto \hat{f}(x) \\ &= \eta(r_0) + \eta(r_1)x + \cdots + \eta(r_n)x^n. \end{aligned}$$

That homomorphism is commonly called *reduction of coefficients* modulo I.

Exercise 5.8. Describe the kernel of the homomorphism $f(x) \mapsto \hat{f}(x)$ given by reduction of coefficients modulo an ideal I.

Suppose R is a UFD. If

$$0 \neq f(x) = a_0 + a_1 x + \cdots + a_n x^n \in R[x]$$

we define the *content* of $f(x)$ to be $d = \mathrm{GCD}(a_0, a_1, \ldots, a_n)$. The content is of course only determined up to associates in R. If $f(x)$ has content 1 we say that $f(x)$ is *primitive*. Observe that for any $f(x) \in R[x]$ we may write $f(x) = d f_1(x)$, where d is the content of $f(x)$ and $f_1(x)$ is primitive. In particular if $f(x)$ is irreducible and nonconstant in $R[x]$, then $f(x)$ is primitive.

Theorem 5.13 (Gauss's Lemma). Suppose R is a UFD and $f(x), g(x) \in R[x]$ are both primitive. Then $f(x)g(x)$ is primitive.

Proof. If not then there is a prime $p \in R$ dividing all coefficients of $f(x)g(x)$. Set $S = R/(p)$. Since p is prime (p) is a prime ideal, and hence S is an integral domain. We may reduce coefficients mod(p) in each polynomial in $R[x]$ and obtain a homomorphism $h(x) \mapsto \hat{h}(x)$ from $R[x]$ onto $S[x]$. But $S[x]$ is an integral domain (since S is), and $(f(x)g(x))\hat{} = \hat{f}(x)\hat{g}(x) = 0$ in

$S[x]$, so $\hat{f}(x) = 0$ or $\hat{g}(x) = 0$. That means, however, that either p divides all coefficients of $f(x)$ or p divides all coefficients of $g(x)$, contradicting the fact that $f(x)$ and $g(x)$ are primitive.

Proposition 5.14. Suppose R is a UFD and $F = F_R$ is its field of fractions. Suppose $f(x)$ and $g(x)$ are primitive in $R[x]$, and that $f(x)$ and $g(x)$ are associates in $F[x]$. Then they are associates in $R[x]$.

Proof. Since $U(F[x]) = F^*$ we may write $f(x) = ag(x)$ for some $a \in F^*$. Write $a = b/c$, with $b, c \in R$, then $cf(x) = bg(x) \in R[x]$. The content of $cf(x)$ is c and the content of $bg(x)$ is b since $f(x)$ and $g(x)$ are primitive. Thus $b \sim c$ in R, and so $a = b/c$ is a unit in R. But that means that $f(x) \sim g(x)$ in $R[x]$.

Proposition 5.15. Suppose R is a UFD, F is its field of fractions, and $f(x) \in R[x]$ is irreducible. Then $f(x)$ remains irreducible when viewed as an element of $F[x]$.

Proof. Since $f(x)$ is irreducible in $R[x]$ it is primitive. Suppose we could write $f(x) = f_1(x)f_2(x)$, with $f_1(x)$, $f_2(x) \in F[x]$, neither being constant. By "clearing denominators" and factoring common terms from coefficients we may write $f_i(x) = a_i g_i(x)$, where $a_i \in F$, $g_i(x) \in R[x]$, and $g_i(x)$ is primitive, $i = 1, 2$. Thus $f(x) = a_1 a_2 g_1(x) g_2(x)$, so $f(x) \sim g_1(x)g_2(x)$ in $F[x]$, and $g_1(x)g_2(x)$ is primitive by Gauss's Lemma. Thus $f(x) \sim g_1(x)g_2(x)$ in $R[x]$ by Proposition 5.14, which means that $f(x) = ug_1(x)g_2(x)$ for some $u \in U(R)$, contradicting the irreducibility of $f(x)$ in $R[x]$.

Theorem 5.16. If R is a UFD, then $R[x_1, x_2, \ldots, x_n]$ is a UFD.

Proof. Since

$$R[x_1, x_2, \ldots, x_n] \cong (R[x_1, x_2, \ldots, x_{n-1}])[x_n]$$

we may assume that $n = 1$; the general result follows by induction. We show first that each nonzero nonunit $f(x) \in R[x]$ is a product of irreducibles, using induction on $m = \deg f(x)$. The case $m = 0$ is clear since R is a UFD. If $m > 0$ we assume the factorization possible for polynomials of degree less than m. Write $f(x) = af_1(x)$, where a is the content of $f(x)$ and $f_1(x)$ is primitive. Then $a \in R$ is either a unit or a product of irreducibles. If $f_1(x)$ is irreducible the factorization of $f(x)$ is accomplished. If not write $f_1(x) = f_2(x)f_3(x)$, with $\deg f_i(x) < \deg f(x) = m$, $i = 2, 3$. By the induction hypothesis each of $f_2(x)$ and $f_3(x)$ is a product of irreducibles, and hence $f(x)$ is a product of irreducibles.

It remains to be shown that every irreducible in $R[x]$ is prime. Suppose $f(x)$ is irreducible (and hence primitive), and suppose $f(x)|g(x)h(x)$ in $R[x]$. By Proposition 5.15 $f(x)$ is irreducible as an element of $F[x]$, where F is the field of fractions of R. But $F[x]$ is Euclidean, so it is a UFD, and hence $f(x)$ *is* prime in $F[x]$. We may assume then that $f(x) | g(x)$ in $F[x]$, say $g(x) = f(x)k(x)$, with $k(x) \in F[x]$. By clearing denominators in $k(x)$ and factoring out contents we may write

$$g(x) = ag_1(x) = (b/c)f(x)k_1(x),$$

with $g_1(x)$ and $k_1(x)$ primitive in $R[x]$, and so $g_1(x) \sim f(x)k_1(x)$ in $F[x]$. But $f(x)k_1(x)$ is primitive by Gauss's Lemma, and by Proposition 5.14 we have $g_1(x) \sim f(x)k_1(x)$ in $R[x]$, so $g_1(x) = uf(x)k_1(x)$ for some $u \in U(R)$. Thus $f(x) | g_1(x)$ in $R[x]$ and hence $f(x) | g(x)$ in $R[x]$ since $g_1(x) | g(x)$.

Corollary. If R is a UFD, then $R[x, y]$ is a UFD that is not a PID.

Proof. See Exercise 5.6.2.

Proposition 5.17. If R is a PID, then every nonzero prime ideal P in R is maximal.

Proof. Let I be an ideal, with $P \subseteq I \subseteq R$, and say $P = (p)$, $I = (a)$. Then $p \in (a)$, so $p = ab$ for some $b \in R$. Thus $p | a$ or $p | b$ since p is a prime. If $p | a$, then $a \in (p)$ and $I = P$. Otherwise say that $cp = b, c \in R$. Then $p = ab = acp$, so $ac = 1$. Thus a $\in U(R)$ and $I = R$.

Corollary. If R is a PID and $p \in R$ is a prime, then $R/(p)$ is a field.

Theorem 5.18 (The Eisenstein Criterion). Suppose R is a PID and

$$f(x) = a_0 + a_1x + a_2x^2 + \cdots + a_nx^n$$

is primitive in $R[x]$. Suppose there is a prime $p \in R$ such that $p | a_i$ for $0 \leq i \leq n - 1$, but $p \nmid a_n$ and $p^2 \nmid a_0$. Then $f(x)$ is irreducible.

Proof. Set $S = R/(p)$; it is a field by the corollary to Proposition 5.17. Then $S[x]$ is Euclidean, and hence is a UFD. Reduce coefficients mod(p) to obtain a homomorphism $k(x) \mapsto \hat{k}(x)$ from $R[x]$ onto $S[x]$. Suppose $f(x) = g(x)h(x)$, with

$$g(x) = b_0 + b_1x + \cdots + b_kx^k \qquad \text{and} \qquad h(x) = c_0 + c_1x + \cdots + c_mx^m,$$

neither being a unit in $R[x]$. In particular, both $g(x)$ and $h(x)$ must have positive degree since $f(x)$ is primitive. Then $\hat{f}(x) = \hat{g}(x)\hat{h}(x) = \eta(a_n)x^n$, since $p | a_i$, $0 \leq i \leq n - 1$. Consequently $x | \hat{g}(x)$ and $x | \hat{h}(x)$ in $S[x]$, which means that $\eta(b_0) = \eta(c_0) = 0$, or $p | b_0$ and $p | c_0$. But then $p^2 | b_0c_0 = a_0$, a contradiction.

We remark that Eisenstein's Criterion holds, more generally, when R is a UFD (see Exercise 8.30).

Note that under the hypotheses of Eisenstein's Criterion $f(x)$ is irreducible not only in $R[x]$ but also in $F[x]$, where F is the field of fractions of R, by

Proposition 5.15. For example $f(x) = 6 + 2x + 4x^3 + 7x^5$ is irreducible in $\mathbb{Q}[x]$ since $f(x) \in \mathbb{Z}[x]$ and $f(x)$ satisfies the hypotheses of Eisenstein's Criterion with $p = 2$.

6. THE CHINESE REMAINDER THEOREM

If R and S are rings (not necessarily commutative), then they are in particular additive groups, so their direct sum $R \oplus S$ is defined. If we define $(r_1, s_1)(r_2, s_2) = (r_1 r_2, s_1 s_2)$, then it is an easy matter to verify that $R \oplus S$ is a ring. Clearly these remarks extend to any finite number of rings $R_1, R_2, \ldots, R_n$. The direct sum $R_1 \oplus R_2 \oplus \cdots \oplus R_n$ is commutative if and only if each R_i is commutative, and has a 1 $[= (1, 1, \ldots, 1)]$ if and only if each R_i has a 1.

Exercise 6.1. Show that $\mathbb{Z}_6$ and $\mathbb{Z}_2 \oplus \mathbb{Z}_3$ are isomorphic rings.

Recall that if G is a group and $H \leq G$ we write $x \equiv y(\text{mod } H)$ for $x, y \in G$ if $y^{-1}x \in H$ (or $x - y \in H$ if G is additive and abelian). In particular if R is a ring and I an ideal in R we shall write $x \equiv y(\text{mod } I)$ if $x - y \in I$.

If I and J are ideals in a ring R, then their *product* IJ is defined to be the ideal generated by all ab, $a \in I$, $b \in J$. Clearly IJ consists of all finite sums $a_1 b_1 + a_2 b_2 + \cdots + a_k b_k$, where $a_i \in I$, $b_i \in J$. Note that $IJ \subseteq I \cap J$ since I and J are ideals. For an example note that $(6)(9) = (54)$ in the ring $\mathbb{Z}$ of integers, and that $(6) \cap (9) = (18)$. These remarks extend in an obvious fashion to any finite collection $I_1, I_2, \ldots, I_n$ of ideals in R. We write $\prod_{j=1}^{n} I_j$ for their product. This concept should not, of course, be confused with the notion of *direct* product.

Proposition 6.1. Suppose R is a ring with 1 and I, J are ideals in R such that $I + J = R$. Then for any $r_1, r_2 \in R$ we may find an element $r \in R$ such that $r - r_1 \in I$ and $r - r_2 \in J$, i.e., $r \equiv r_1(\text{mod } I)$ and $r \equiv r_2(\text{mod } J)$.

Proof. Write $1 = a + b$, with $a \in I$ and $b \in J$, and set $r = r_2 a + r_1 b$. Then

$$r - r_1 = r_2 a + r_1(b - 1) = r_2 a - r_1 a = (r_2 - r_1)a \in I$$

and

$$r - r_2 = r_2(a - 1) + r_1 b = r_1 b - r_2 b = (r_1 - r_2)b \in J.$$

Theorem 6.2 (The Chinese Remainder Theorem). Suppose R is a ring with 1 and $I_1, I_2, \ldots, I_n$ are ideals in R such that $I_j + I_k = R$ if $j \neq k$. Then given any $r_1, r_2, \ldots, r_n \in R$ we may find an element $r \in R$ such that $x = r$ is a simultaneous solution for all the congruences $x \equiv r_j(\text{mod } I_j)$, $1 \leq j \leq n$. Any two solutions are congruent modulo $I_1 \cap I_2 \cap \cdots \cap I_n$.

Proof. For $1 < j \leq n$ write $1 = a_j + b_j$; with $a_j \in I_1$ and $b_j \in I_j$. Then

$$1 = 1^{n-1} = (a_2 + b_2)(a_3 + b_3)\cdots(a_n + b_n) \in I_1 + \prod_{j=2}^{n} I_j,$$

and hence $I_1 + \prod_{j=2}^{n} I_j = R$. By Proposition 6.1 we may choose $s_1 \in R$ such that $s_1 \equiv 1(\text{mod } I_1)$ and $s_1 \equiv 0(\text{mod } \prod_{j=2}^{n} I_j)$, and hence also $s_1 \equiv 0(\text{mod } I_k)$ if $2 \leq k \leq n$, since $\prod_{j=2}^{n} I_j \subseteq I_k$. Similarly we may find $s_j \in R$ such that $s_j \equiv 1(\text{mod } I_j)$ and $s_j \equiv 0(\text{mod } I_k)$ if $k \neq j$ for all j, $1 \leq j \leq n$. Set $r = \sum_{k=1}^{n} r_k s_k$. Then

$$r - r_j = \sum\{r_i s_i : i \neq j\} + r_j(s_j - 1) \equiv 0(\text{mod } I_j), \qquad 1 \leq j \leq n.$$

If s is another solution, then $s \equiv r_j \equiv r(\text{mod } I_j)$ for all j, so

$$s \equiv r(\textit{mod} \bigcap_{j=1}^{n} I_j).$$

Note that the proof of the Chinese Remainder Theorem is constructive when applied to $\mathbb{Z}$ (or to any Euclidean ring). In fact if $I_j = (m_j)$, then since $(m_j, \prod\{m_k : k \neq j\}) = 1$ we can write $1 = c_j m_j + d_j \prod\{m_k : k \neq j\}$ by means of the Euclidean algorithm (see Exercise 8.26). We may then set $s_j = d_j \prod\{m_k : k \neq j\}$ and $r = r_1 s_1 + \cdots + r_n s_n$.

Exercise 6.2. Solve the congruences

$$x \equiv 1 \text{ (mod 8)}, \qquad x \equiv 3\text{(mod 7)}, \qquad x \equiv 9\text{(mod 11)}$$

simultaneously in the ring $\mathbb{Z}$ of integers.

Theorem 6.3. Suppose $I_1, I_2, \ldots, I_n$ are ideals in a ring R. Then there is a monomorphism

$$f: R/(I_1 \cap I_2 \cap \cdots \cap I_n) \to R/I_1 \oplus R/I_2 \oplus \cdots \oplus R/I_n.$$

If R is a ring with 1 and if $I_j + I_k = R$ when $j \neq k$, then f is an isomorphism.

Proof. Define a map $g: R \to R/I_1 \oplus \cdots \oplus R/I_n$ by setting $g(r) = (r + I_1, \ldots, r + I_n)$. Then g is clearly a homomorphism, and $\ker g = I_1 \cap I_2 \cap \cdots \cap I_n$. Thus the existence of f is a consequence of the FHTR. If R has a 1 and $I_j + I_k = R$ for $j \neq k$, then g is onto by the Chinese Remainder Theorem, so f is an isomorphism.

If $R = \mathbb{Z}$ in Theorem 6.3, with $I_j = (m_j)$, then

$$I_j + I_k = (m_j) + (m_k) = ((m_j, m_k)),$$

the principal ideal generated by the GCD of m_j and m_k. Thus $I_j + I_k = \mathbb{Z}$ if and only if $(m_j, \mathrm{m}_k) = 1$, and in that case $I_j \cap I_k = I_j I_k = (m_j m_k)$. More generally $I_1 \cap I_2 \cap \cdots \cap I_n = (m_1 m_2 \cdots m_n)$. As a consequence we see that if $(m_j, m_k) = 1$ for $j \neq k$, then $\mathbb{Z}_{m_1 m_2 \cdots m_n} \cong \mathbb{Z}_{m_1} \oplus \mathbb{Z}_{m_2} \oplus \cdots \oplus \mathbb{Z}_{m_n}$.

7. THE HILBERT BASIS THEOREM

Analytic geometry involves the description of geometric objects by means of equations. For example, the set of common solutions to the equations $x + y + z = 1$ and $2x - y - z = 3$ is a line in the three-dimensional space $\mathbb{R}^3$. In general if F is a field, then an (affine) *algebraic variety* V in the space F^n is the set of common solutions to a family of polynomial equations $f_a(x_1, x_2, \ldots, x_n) = 0$, where a ranges over some index set A and each $f_a(X) \in F[x_1, x_2, \ldots, x_n]$. If I is the ideal in $F[x_1, \ldots, x_n]$ generated by all $f_a(X)$, $a \in A$, observe that every $g \in I$ vanishes on the points in V, and that V could be described as the variety determined by the polynomials in I. This simple observation is the beginning of the rather deep subject called *algebraic geometry*, in which methods of abstract algebra are brought to bear on the study of geometry.

Suppose I is an ideal in $F[x_1, \ldots, x_n]$ and V is the corresponding algebraic variety in F^n. The main result in this section is the Hilbert Basis Theorem, which states that the ideal I is finitely generated and has the consequence that the variety V can be described as the set of common solutions to a finite set of polynomial equations.

A few preliminaries are necessary.

Recall that a commutative ring is *Noetherian* if it satisfies the ascending chain condition (ACC) for ideals, i.e., every ascending chain $I_0 \subseteq I_1 \subseteq I_2 \subseteq \cdots$ of ideals of R must terminate. For example, every PID is Noetherian (Proposition 5.9).

A ring R satisfies the *maximal condition* (for ideals) if every nonempty set $\mathcal{S}$ of ideals in R contains an element I_0 that is maximal with respect to set inclusion, i.e., if $I \in \mathcal{S}$ and $I_0 \subseteq I$, then $I_0 = I$.

Proposition 7.1. A commutative ring R is Noetherian if and only if it satisfies the maximal condition.

Exercise 7.1. Prove Proposition 7.1.

Proposition 7.2. A commutative ring R is Noetherian if and only if every ideal in R is finitely generated.

Proof. $\Rightarrow$: If I is any ideal in R let $\mathcal{S}$ be the set of all finitely generated ideals of R that are contained in I (e.g., $0 \in \mathcal{S}$). Let I_0 be a maximal element of $\mathcal{S}$, say with I_0 generated by $r_1, \ldots, r_k$. If $I_0 \neq I$ choose $r \in I \backslash I_0$ and let J be the ideal generated by $r_1, \ldots, r_k$ and r. Then $J \in \mathcal{S}$ but $I_0 \subseteq J$ and $I_0 \neq J$, contradicting maximality. Thus $I = I_0$ is finitely generated.

$\Leftarrow$: If $I_0 \subseteq I_1 \subseteq I_2 \subseteq \cdots$ is any ascending chain of ideals, then $I = \bigcup_j I_j$ is an ideal. Say I is generated by $r_1, \ldots, r_k$, and say that $r_i \in I_{j(i)}$, $1 \leq i \leq k$. We may assume the r_i labeled so that $j(i) \leq j(k)$ for all i, and hence all $r_i \in I_{j(k)}$. But then $I = I_{j(k)}$ and the chain terminates at $j(k)$.

Suppose R is a commutative ring with 1. If I is an ideal in $R[x]$ and m is a nonnegative integer denote by $I(m)$ the set of all leading coefficients of polynomials of degree m in I, together with 0.

Exercise 7.2. (1) Show that $I(m)$ is an ideal in R.
(2) Show that $I(m) \subseteq I(m+1)$ for all m.
(3) If J is an ideal with $I \subseteq J$ show that $I(m) \subseteq J(m)$ for all m.

Proposition 7.3. Suppose R is a commutative ring with 1, I and J are ideals in $R[x]$ with $I \subseteq J$, and $I(m) = J(m)$ for all nonnegative $m \in \mathbb{Z}$. Then $I = J$.

Proof. If not, choose $f(x) \in J \backslash I$ of minimal degree $m \geq 0$. Since $I(m) = J(m)$ there is some $g(x) \in I$ of degree m having the same leading coefficient as $f(x)$. But then $0 \neq f(x) - g(x) \in J \backslash I$, and $\deg(f(x) - g(x)) < m$, contradicting the minimality of $\deg f(x)$.

Theorem 7.4 (The Hilbert Basis Theorem). Suppose R is a commutative ring with 1. If R is Noetherian, then the polynomial ring $R[x_1, \ldots, x_n]$ is also Noetherian.

Proof. By induction it will suffice to assume that $n = 1$ and write $x_1 = x$. Let $I_0 \subseteq I_1 \subseteq I_2 \subseteq \cdots$ be an ascending chain of ideals in $R[x]$, and let $\mathscr{S}$ be the set of ideals

$$\{I_n(m) : 0 \leq n \in \mathbb{Z}, 0 \leq m \in \mathbb{Z}\}$$

in R. By Proposition 7.1 there is a maximal element $I_r(s) \in \mathscr{S}$. Consider the infinite rectangular array

$$\begin{array}{ccccccc}
I_0(0) & \subseteq & I_0(1) & \subseteq \cdots \subseteq & I_0(s-1) & \subseteq & \cdots \\
\cap\!| & & \cap\!| & & \cap\!| & & \\
I_1(0) & \subseteq & I_1(1) & \subseteq \cdots \subseteq & I_1(s-1) & \subseteq & \cdots \\
\cap\!| & & \cap\!| & & \cap\!| & & \\
I_2(0) & \subseteq & I_2(1) & \subseteq \cdots \subseteq & I_2(s-1) & \subseteq & \cdots \\
\cap\!| & & \cap\!| & & \cap\!| & & \\
\vdots & & \vdots & & \vdots & &
\end{array}$$

The jth column is an ascending chain of ideals in R so it terminates, say at $I_{f(j)}(j)$. Thus if $i \geq f(j)$, then $I_i(j) = I_{f(j)}(j)$. Set

$$u = \max\{f(0), f(1), \ldots, f(s-1), r\},$$

and suppose $i \geq u$. If $0 \leq j \leq s-1$, then $I_i(j) = I_u(j) = I_{f(j)}(j)$ since $i \geq u \geq f(j)$ and the jth column terminates at $f(j)$. If $j \geq s$, then $I_i(j) \supseteq I_i(s) \supseteq I_r(s)$, so $I_i(j) = I_r(s)$ by maximality; and $I_u(j) \supseteq I_r(j) \supseteq I_r(s)$, so $I_u(j) = I_r(s)$

by maximality, and $I_i(j) = I_u(j)$. We have thus shown that $I_i(j) = I_u(j)$ for all $j \geq 0$ and hence, by Proposition 7.3, that $I_i = I_u$ for all $i \geq u$. Thus the chain $I_0 \subseteq I_1 \subseteq I_2 \subseteq \cdots$ terminates at u and $R[x]$ is Noetherian.

Corollary. Suppose S is a ring with 1, R is a commutative Noetherian subring with $1_S \in R$, and $s_1, s_2, \ldots, s_n$ are in the center of S. If $R_1 = R[s_1, s_2, \ldots, s_n]$, the (commutative) subring of S generated by R and $\{s_1, \ldots, s_n\}$, then R_1 is Noetherian.

Proof. Clearly a homomorphic image of a Noetherian ring is Noetherian. We may map the Noetherian ring $R[x_1, \ldots, x_n]$ homomorphically onto R_1 by means of $f(x_1, \ldots, x_n) \mapsto f(s_1, \ldots, s_n)$ (see Theorem 4.3).

Exercise 7.3. If R is a commutative ring with 1 and $\{x_a : a \in A\}$ is an infinite set of distinct (commuting) indeterminates show that the polynomial ring $R[\{x_a : a \in A\}]$ is *not* Noetherian.

8. FURTHER EXERCISES

1. If R is a commutative ring and $a \in R$ show that the principal ideal (a) has the form $(a) = \{ra + na : r \in R, n \in \mathbb{Z}\}$. Describe the elements of (a) explicitly if R is not necessarily commutative.

2. Show that there is no ring R with 1 whose additive group is isomorphic with $\mathbb{Q}/\mathbb{Z}$.

3. If R is any ring denote by R_1 the additive group $R \oplus \mathbb{Z}$, with multiplication defined by setting

$$(r, n)(s, m) = (rs + mr + ns, nm).$$

Show that R_1 is a ring with 1. If $r \in R$ is identified with $(r, 0) \in R$, show that R is a subring of R_1. Conclude that every ring is a subring of a ring with 1.

4. (The Binomial Theorem). Suppose R is a commutative ring, $a, b \in R$, and $0 < n \in \mathbb{Z}$. Show that

$$(a + b)^n = \sum\left\{\binom{n}{k} a^{n-k} b^k : 0 \leq k \leq n\right\},$$

where $\binom{n}{k} = n!/k!(n-k)!$ is the usual binomial coefficient. (*Hint*: Use induction on n.)

5. Show that every ideal in $\mathbb{Z}_n$ is principal.

6. If F is any field show that the ring $M_n(F)$ of $n \times n$ matrices over F is a simple ring.

7. Suppose R is a commutative ring, I_1 and I_2 are ideals in R, P is a prime ideal in R, and $I_1 \cap I_2 \subseteq P$. Show that $I_1 \subseteq P$ or $I_2 \subseteq P$.

8. Suppose R is a commutative ring with 1 and $x \in \bigcap\{M : M$ is a maximal ideal in $R\}$. Show that $1 + x \in U(R)$.

9. An element a of a ring R is called *nilpotent* if $a^n = 0$ for some positive integer n. Show that the set of nilpotent elements in a commutative ring R is an ideal of R.

10. Find all nilpotent elements in $\mathbb{Z}_{p^k}$, then more generally in $\mathbb{Z}_n$. (See 9 above.)

11. Suppose R is a ring with 1, $u \in U(R)$, a is a nilpotent element of R, and $ua = au$. Show that $u + a \in U(R)$. In particular $1 + a \in U(R)$ for every nilpotent a. (*Hint*: Write $(u + a)^{-1}$ suggestively as $1/(u + a) = u^{-1}/(1 + u^{-1}a)$ and expand in a "power series." Then verify directly that the resulting element of R is an inverse for $u + a$.)

12. Determine the endomorphism rings of the additive groups $\mathbb{Z}$, $\mathbb{Z}_n$, and $\mathbb{Q}$.

13. Give an example of a ring R with a prime ideal $P \neq 0$ that is not maximal.

14. Show that the ideal $I = (2, x)$ is not principal in $\mathbb{Z}[x]$.

15. Suppose R is a ring with 1 and $f(x), g(x) \in R[x]$, $g(x) \neq 0$. If the leading coefficient b of $g(x)$ is a unit in R show that there are unique polynomials $q(x), r(x) \in R[x]$ such that $f(x) = g(x)q(x) + r(x)$, with $\deg r(x) < \deg g(x)$.

16. Suppose R is a commutative ring with 1 and that $f(x) \in R[x]$ is a zero divisor. Show that there is an element $a \in R$, $a \neq 0$, such that $af(x) = 0$.

17. Determine $U(\mathbb{Z}_n[x])$. It may help to warm up by assuming that $n = p^k$ for a prime p.

18. Suppose R and S are commutative rings with $R \subseteq S$ and $1_R = 1_S$, and that R is an integral domain. If $a \in S$ is transcendental over R and $g(x)$ is a nonconstant polynomial in $R[x]$ show that $g(a)$ is transcendental over R.

19. (Lagrange Interpolation). Suppose F is a field, $a_1, a_2, \ldots, a_n$ are n distinct elements of F, and $b_1, b_2, \ldots, b_n$ are n arbitrary elements of F. Set $p_i(x) = \prod\{x - a_j : j \neq i\}$ and set

$$f(x) = \sum\{b_i p_i(x)/p_i(a_i) : 1 \leq i \leq n\}.$$

Show that $f(x)$ is the unique polynomial of degree $n - 1$ or less over F for which $f(a_i) = b_i$, $1 \leq i \leq n$.

20. Find $f(x) \in \mathbb{Q}[x]$ of degree 3 or less such that $f(0) = f(1) = 1$, $f(2) = 3$, and $f(3) = 4$.

21. If R is a commutative ring with 1 and $S = R[x, y]$ show that $a_1 = x + y$ and $a_2 = xy$ are algebraically independent over R. If $a_3 = x - y$ show that a_1, a_2, and a_3 are algebraically dependent over R.

22. Suppose m and n are square-free integers, neither being 0 or 1, and that $m \neq n$. Show that R_m and R_n are not isomorphic rings.

23. If π is a prime element of R_m show that π is a divisor of exactly one prime $p \in \mathbb{Z}$. (*Hint*: Observe that $\pi \mid N(\pi) \in \mathbb{Z}$.)

24. If R is a PID show that every nonzero prime ideal P in R is a maximal ideal.

25. Suppose R is a Euclidean domain, $a, b \in R^*$, $a \mid b$, and $d(a) = d(b)$. Show that a and b are associates.

26. (The Euclidean Algorithm). Suppose R is a Euclidean domain, $a, b \in R$, and $ab \neq 0$. Write

$$\begin{aligned} a &= bq_1 + r_1, & d(r_1) &< d(b), \\ b &= r_1 q_2 + r_2, & d(r_2) &< d(r_1), \\ r_1 &= r_2 q_3 + r_3, & d(r_3) &< d(r_2), \\ &\vdots \\ r_{k-2} &= r_{k-1} q_k + r_k, & d(r_k) &< d(r_{k-1}), \\ r_{k-1} &= r_k q_{k+1}, \end{aligned}$$

with all $r_i, q_j \in R$. Show that $r_k = (a, b)$, and "solve" for r_k in terms of a and b, thereby expressing (a, b) in the form $ua + vb$, with $u, v \in R$.

27. Use the Euclidean Algorithm (Exercise 26) to find $d = (a, b)$ and to write $d = ua + vb$ in the following cases:

(1) $a = 29041$, $b = 23843$, $R = \mathbb{Z}$;
(2) $a = x^3 - 2x^2 - 2x - 3$, $b = x^4 + 3x^3 + 3x^2 + 2x$, $R = \mathbb{Q}[x]$;
(3) $a = 7 - 3i$, $b = 5 + 3i$, $R = R_{-1}$.

28. Suppose $f(x) = 1 + x + x^2 + \cdots + x^{p-1}$, where p is a prime in $\mathbb{Z}$.

(1) Show that $f(x)$ is irreducible in $\mathbb{Q}[x]$. (*Hint*: Write $f(x) = (x^p - 1)/(x - 1)$; substitute $x + 1$ for x.)

(2) Show that $\binom{p}{k} = \sum_{i=1}^{k+1} \binom{p-i}{p-k-1}$ for all $k < p$.

29. If $p \in \mathbb{Z}$ is prime and $1 < m \in \mathbb{Z}$ show that $f(x) = x^m - p$ is irreducible in $\mathbb{Q}[x]$. Conclude that $p^{1/m}$ is irrational.

30. Establish the Eisenstein Criterion for a polynomial $f(x)$ over a UFD. [*Hint*: Assume $f(x) = g(x)h(x)$, with p dividing the constant term of $g(x)$. Investigate the first coefficient of $g(x)$ *not* divisible by p.]

31. (Lipka [25]). Suppose p is a prime in $\mathbb{Z}$,

$$f(x) = \pm p + a_1 x + a_2 x^2 + \cdots + a_n x^n \in \mathbb{Z}[x],$$

and $\sum_{i=1}^{n} |a_i| < p$. Show that $f(x)$ is irreducible in $\mathbb{Q}[x]$. Conclude, for example, that $x^n + x + 3$ is irreducible over $\mathbb{Q}$ if $n > 2$. [*Hint*: Show that every root of $f(x)$ in $\mathbb{C}$ is larger than 1 in absolute value.]

32. (1) Show that $1 + \sqrt{2} \in U(R_2)$.

(2) If $u \in U(R_2)$ show that either $u \leq 1$ or $u \geq 1 + \sqrt{2}$ (use the fact that if $1 < u < 1 + \sqrt{2}$, then $-1 < u^{-1} < 1$).

(3) Use (2) to show that if $u \in U(R_2)$ and $u > 0$, then $u = (1 + \sqrt{2})^n$ for some $n \in \mathbb{Z}$.

(4) Conclude that $U(R_2) \cong \{\pm 1\} \times \langle 1 + \sqrt{2} \rangle$.

33. Determine $U(R_{10})$.

34. Let R be the set of rational numbers a/b with b odd. Then R is an integral domain.

(1) Find $U(R)$.
(2) Show that $P = R \backslash U(R)$ is a maximal ideal in R.
(3) Find all primes in R.
(4) Find all ideals in R and show that R is a PID. Is it Euclidean?

(*Remark*: R can be obtained from $\mathbb{Z}$ by localizing at the prime ideal P.)

35. Suppose $P \neq 0$ is a prime ideal in R_m.

(1) Show that $P \cap \mathbb{Z}$ is a prime ideal in $\mathbb{Z}$, so $P \cap \mathbb{Z} = (p)$ for some prime p in $\mathbb{Z}$.

(2) Set $I = pR_m \subseteq P$ and form the quotient ring R/I. Show that R/I, as an additive group, is generated by two elements of finite order; hence R/I is finite.

(3) Show that there is an epimorphism from R/I to R/P and conclude that R/P is finite.

(4) Conclude that every prime ideal in R_m is maximal.

36. If $R_1, R_2, \ldots, R_n$ are rings with 1 show that

$$U(R_1 \oplus R_2 \oplus \cdots \oplus R_n) = U(R_1) \times U(R_2) \times \cdots \times U(R_n).$$

37. Solve the congruences

$$x \equiv i (\mathrm{mod}\ 1 + i), \qquad x \equiv 1 (\mathrm{mod}\ 2 - i), \qquad x \equiv 1 + i (\mathrm{mod}\ 3 + 4i)$$

simultaneously for x in the ring R_{-1} of Gaussian integers.

38. Solve the congruences

$$f(x) \equiv 1 (\mathrm{mod}\ x - 1), \qquad f(x) \equiv x (\mathrm{mod}\ x^2 + 1), \qquad f(x) \equiv x^3 (\mathrm{mod}\ x + 1)$$

simultaneously for $f(x)$ in $F[x]$, where F is a field in which $1 + 1 \neq 0$.

39. If R is the subring of $\mathbb{Z}[x]$ consisting of all

$$f(x) = a_0 + a_1 x + \cdots + a_n x^n$$

such that a_k is even for $1 \leq k \leq n$, show that R is not Noetherian. [*Hint*: Consider the ideal generated by $\{2x^k : 1 \leq k \in \mathbb{Z}\}$.]

Chapter III | Fields and Galois Theory

1. FIELD EXTENSIONS

Suppose F and K are fields and that F is a subring of K. Then we say that F is a *subfield* of K or that K is an *extension field* of F. In that case K is a vector space over F. The dimension of K as an F-vector space is called the *degree* of K over F and is denoted by $[K:F]$. If $[K:F]$ is finite we say that K is a *finite extension* of F.

If K is an extension of F and S is any subset of K denote by $F(S)$ the intersection of all extension fields L of F such that $S \subseteq L \subseteq K$. Clearly $F(S)$ is the minimal field between F and K containing S as a subset. We call $F(S)$ the extension of F *generated by* S. If $a \in K$, then $F(a) = F(\{a\})$ is the *simple* extension of F generated by a, and a is called a *primitive element* for $F(a)$ over F. Note that $F(a)$ is the field of fractions of the ring $F[a]$ obtained by substitution of a for x in $F[x]$. Thus

$$F(a) = \{f(a)/g(a) : f(x), g(x) \in F[x], g(a) \neq 0\}.$$

Proposition 1.1. Suppose L is an extension field of F and K is an extension field of L. If A is a basis for L as an F-vector space and B is a basis for K as an L-vector space, then $AB = \{ab : a \in A, b \in B\}$ is a basis for K as an F-vector space. Consequently $[K:F] = [K:L][L:F]$, and K is finite over F if and only if K is finite over L and L is finite over F.

Proof. If $c \in K$ we may write $c = u_1 b_1 + u_2 b_2 + \cdots + u_k b_k$, with $u_j \in L$, $b_j \in B$, and we may write

$$u_j = v_{j1} a_1 + v_{j2} a_2 + \cdots + v_{jm(j)} a_{m(j)},$$

with $v_{ji} \in F, a_i \in A$. Thus $c = \sum_j u_j b_j = \sum_{j,i} v_{ji} a_i b_j$, and the F-span of AB is all of K.

Suppose $\sum_{i,j} v_{ij} a_i b_j = 0$ for some finite set $\{a_i b_j\} \subseteq AB$ and v_{ij} in F. For each j set $u_j = \sum_i v_{ij} a_i \in L$, so that $\sum_j u_j b_j = 0$. Then each $u_j = 0$ since B is linearly independent. But then all $v_{ij} = 0$ since A is linearly independent, so AB is linearly independent. Thus AB is an F-basis for K. In particular the elements ab in AB are all distinct and $|AB| = |A||B|$.

If F is a field let F_0 be the intersection of all the subfields of F. Thus F_0 is the unique minimal subfield of F; it is called the *prime field* of F. There is a (ring) homomorphism $f: \mathbb{Z} \to F_0$ defined by setting $f(n) = n \cdot 1, 1 \in F$, for all $n \in \mathbb{Z}$. If f is 1–1, then $\text{Im}(f)$ is an isomorphic copy of $\mathbb{Z}$ in F_0, so F_0, being minimal, must be isomorphic with $\mathbb{Q}$, the field of fractions of $\mathbb{Z}$. We say then that F has *characteristic* 0. In many cases we shall identify F_0 with $\mathbb{Q}$ so F is an extension of $\mathbb{Q}$.

If f is not 1–1, then $\ker f$ is an ideal in $\mathbb{Z}$, say $\ker f = (n)$. If $a, b \in \mathbb{Z}$ and $n \mid ab$, then $0 = (ab)1 = (a1)(b1)$, and so either $a1 = 0$ or $b1 = 0$. But then $n \mid a$ or $n \mid b$, so n is a prime. We take $n > 0$ and write $n = p$. Thus $\text{Im}(f) \cong \mathbb{Z}_p$, a field, and so $\text{Im}(f) = F_0 \cong \mathbb{Z}_p$. We say then that F has *characteristic p*, and often identify F_0 with $\mathbb{Z}_p$.

If K is an extension field of F recall that $a \in K$ is called *algebraic* over F if $f(a) = 0$ for some nonzero $f(x) \in F[x]$ (otherwise a is *transcendental* over F). If $a \in K$ is algebraic over F choose a monic polynomial $m_a(x) \in F[x]$ of minimal (positive) degree such that $m_a(a) = 0$.

Proposition 1.2. If $a \in K$ is algebraic over F, then the polynomial $m_a(x) \in F[x]$ is irreducible and unique. If $f(x) \in F[x]$ and $f(a) = 0$, then $m_a(x) \mid f(x)$ in $F[x]$.

Proof. If $m_a(x)$ is not irreducible write $m_a(x) = g(x)h(x)$, with neither factor a unit in $F[x]$. Then $0 = m_a(a) = g(a)h(a)$, so $g(a) = 0$ or $h(a) = 0$, contradicting the minimality of $\deg m_a(x)$.

Divide $f(x)$ by $m_a(x)$ and write $f(x) = m_a(x)q(x) + r(x)$, where $q(x), r(x) \in F[x]$ and $\deg r(x) < \deg m_a(x)$. Substitute a for x to see that $r(a) = 0$ and hence that $r(x) = 0$ since $\deg m_a(x)$ is minimal. Thus $m_a(x) \mid f(x)$. The uniqueness follows, for if $k(x)$ were another polynomial in $F[x]$ with the same properties, then $m_a(x) \mid k(x)$ and $k(x) \mid m_a(x)$, so $k(x) \sim m_a(x)$ in $F[x]$. But two *monic* polynomials are associates only if they are equal.

The polynomial $m_a(x)$ is called the *minimal polynomial* of a over F. We shall write $m_{a,F}(x)$ for $m_a(x)$ when it is important to emphasize the underlying field F.

Proposition 1.3. Suppose K is an extension field of F and $a \in K$ is algebraic over F, with minimal polynomial $m(x) = m_{a,F}(x)$. If $\deg m(x) = n$, then $[F(a):F] = n$ and $\{1, a, a^2, \ldots, a^{n-1}\}$ is an F-basis for $F(a)$.

Proof. If

$$b_0 1 + b_1 a + b_2 a^2 + \cdots + b_{n-1} a^{n-1} = 0, \qquad \text{with} \quad b_i \in F,$$

then $f(a) = 0$, where

$$f(x) = b_0 + b_1 x + \cdots + b_{n-1} x^{n-1} \in F[x].$$

That is in conflict with the definition of the minimal polynomial $m(x)$ unless all $b_i = 0$. Thus $\{1, a, \ldots, a^{n-1}\}$ is linearly independent. We know that

$$F(a) = \{f(a)/g(a) : f(x), g(x) \in F[x], g(a) \neq 0\}.$$

If $f(a)/g(a) \in F(a)$, then $(m(x), g(x)) = 1$ since $m(x)$ is irreducible and $g(a) \neq 0$, so we may write $1 = b(x)m(x) + c(x)g(x)$ for some $b(x), c(x) \in F[x]$. Substitute a for x to see that $1 = c(a)g(a)$ and hence $f(a)/g(a) = f(a)c(a)$. Thus $F(a) = \{h(a) : h(x) \in F[x]\}$. But if $h(x) \in F[x]$ we may write $h(x) = m(x)q(x) + r(x)$, with $q(x)$ and $r(x)$ in $F[x]$ and $\deg r(x) < \deg m(x) = n$. Substitute a for x to see that $h(a) = r(a)$. Thus $\{1, a, \ldots, a^{n-1}\}$ spans $F(a)$ and the proof is complete.

Note as a consequence that if a is algebraic over F, then $F(a) = F[a]$.

An extension K of F is called *algebraic* over F if every $a \in K$ is algebraic over F.

Proposition 1.4. If K is a finite extension of F, then K is algebraic over F.

Proof. Let $m = [K:F] \in \mathbb{Z}$. If $a \in K$, then the set $\{1, a, a^2, \ldots, a^m\}$ is linearly dependent, so there are $b_0, b_1, \ldots, b_m \in F$, not all 0, such that

$$b_0 1 + b_1 a + b_2 a^2 + \cdots + b_m a^m = 0.$$

But then if we set

$$f(x) = b_0 + b_1 x + \cdots + b_m x^m \in F[x]$$

we have $f(a) = 0$, so a is algebraic over F.

Proposition 1.5. If L is an algebraic extension of F and K is an algebraic extension of L, then K is an algebraic extension of F.

Proof. If $a \in K$ there is a nonzero polynomial $f(x) = b_0 + \cdots + b_k x^k$ in $L[x]$ such that $f(a) = 0$. By Propositions 1.1 and 1.3 each of the fields

$$F(b_0), F(b_0, b_1), \ldots, F(b_0, b_1, \ldots, b_k) = L'$$

is finite over F. But a is algebraic over L', so $L'(a)$ is finite over L', and hence over F. Thus $L'(a)$ is algebraic over F by Proposition 1.4, and in particular a is algebraic over F.

Proposition 1.6. If K is an extension field of F set

$$E = \{a \in K : a \text{ is algebraic over } F\}.$$

Then E is a field.

Proof. If $a, b \in E$, then $a \pm b$, ab, and a/b (if $b \neq 0$) are all in $F(a, b)$, which is finite over F. Thus $a \pm b$, ab, and a/b (if $b \neq 0$) are all in E and E is a field.

For example, if $F = \mathbb{Q}$ and $K = \mathbb{C}$, then the elements $a \in \mathbb{C}$ that are algebraic over $\mathbb{Q}$ are called *algebraic numbers*. By Proposition 1.6 the algebraic numbers constitute a field, which will be denoted by $\mathbb{A}$.

Exercise 1.1. Show that $[\mathbb{A}:\mathbb{Q}]$ is infinite.

If $f(x) \in F[x]$, K is an extension field of F, $a \in K$, and $f(a) = 0$, then we say that a is a *root* of $f(x)$ in K. Recall from the Factor Theorem (Proposition II.3.5, Corollary) that if a is a root of $f(x)$ in K, then $x - a \mid f(x)$ in $K[x]$. If $(x - a)^k \mid f(x)$ but $(x - a)^{k+1} \nmid f(x)$ we say that a is a root of $f(x)$ with *multiplicity* k.

Proposition 1.7. If K is an extension of F and $f(x) \in F[x]$, with $\deg f(x) = n$, then $f(x)$ has at most n roots in K.

Proof. Since $K[x]$ is a UFD we may factor $f(x)$ in $K[x]$ as a product of irreducible factors, which are unique up to associates, and $a \in K$ is a root of $f(x)$ if and only if $x - a$ (or an associate) is one of the irreducible factors of $f(x)$. Since the sum of the degrees of the irreducible factors is n the proposition follows.

If F is a field, $f(x) \in F[x]$ has degree $n > 0$, and K is an extension of F such that $f(x)$ has n roots (counting multiplicities) in K, we say that $f(x)$ *splits* over K. Note that then $f(x)$ is a product of n linear factors (i.e., factors of degree 1) in $K[x]$. It will be convenient to agree as a matter of convention that any constant polynomial in $F[x]$ splits over every extension K of F.

Proposition 1.8. If F is a field and $f(x) \in F[x]$ has degree $n \geq 1$, then there is an extension K of F such that $f(x)$ has a root a in K and $[K:F] \leq n$.

Proof. Let $g(x)$ be a nonconstant irreducible factor of $f(x)$ in $F[x]$, say $f(x) = g(x)h(x)$. Then $(g(x))$ is a prime ideal, and hence a maximal ideal, in $F[x]$ by Proposition II.5.17. Consequently $K = F[x]/(g(x))$ is a field. The map $b \mapsto b + (g(x))$ is a monomorphism from F into K so we may (and do) identify F with its image, a subfield of K. Set $a = x + (g(x)) \in K$. Then

$$f(a) = f(x) + (g(x)) = g(x)h(x) + (g(x)) = 0$$

in K. Note that K is in fact the simple extension $F(a)$, so

$$[K:F] = \deg g(x) \leq \deg f(x) = n$$

by Proposition 1.3.

Corollary. If $f(x) \in F[x]$ has degree n, then there is an extension K of F, with $[K:F] \leq n!$, such that $f(x)$ splits over K.

Suppose F is a field and $\mathcal{F}$ is a set of polynomials in $F[x]$. An extension K of F is called a *splitting field* for $\mathcal{F}$ over F if every $f(x) \in \mathcal{F}$ splits over K, and K is minimal in that respect. Equivalently K is a splitting field for $\mathcal{F}$ over F if every $f(x) \in \mathcal{F}$ splits over K, and if S is the set of all roots in K of all $f(x) \in \mathcal{F}$ then $K = F(S)$.

Note that if $\mathcal{F} = \{f(x)\}$ contains a single polynomial $f(x)$ of degree n, then by the corollary to Proposition 1.8 there is a splitting field K for $f(x)$ over F with $[K:F] \leq n!$.

Exercise 1.2. Find a splitting field $K \subseteq \mathbb{C}$ for $f(x) \in \mathbb{Q}[x]$ if $f(x) = x^3 - 1$.

A field F is called *algebraically closed* if every nonconstant $f(x) \in F[x]$ has a root in F, and consequently splits over F. The so-called Fundamental Theorem of Algebra, which will be proved later, asserts that the field $\mathbb{C}$ of complex numbers is algebraically closed.

An *algebraic closure* of a field F is an algebraic extension K of F that is algebraically closed. Thus for example $\mathbb{A}$ is an algebraic closure of $\mathbb{Q}$ and $\mathbb{C}$ is an algebraic closure of $\mathbb{R}$.

Exercise 1.3. Show that K is an algebraic closure of F if and only if K is a splitting field over F for $\mathcal{F} = F[x]$.

It is an easy consequence of Theorem II.3.4 that if F_1 and F_2 are fields and $\phi: F_1 \to F_2$ is an isomorphism, then ϕ extends to an isomorphism (also called ϕ) between $F_1[x]$ and $F_2[x]$, with $\phi(x) = x$.

The next proposition, which is rather technical, will be needed in discussing isomorphism of splitting fields.

Proposition 1.9. Suppose F_1 and F_2 are fields and $\phi: F_1 \to F_2$ is an isomorphism, which we assume extended to an isomorphism between $F_1[x]$ and $F_2[x]$ via $\phi(x) = x$. Suppose K_1 is an extension of F_1 and $a_1 \in K_1$ is algebraic over F_1, with minimal polynomial $m_1(x) \in F_1[x]$. Set $m_2(x) = \phi(m_1(x)) \in F_2[x]$ and suppose K_2 is an extension of F_2 in which $m_2(x)$ has a root a_2. Then $F_1(a_1) \cong F_2(a_2)$; in fact ϕ can be extended to an isomorphism by means of $\phi(a_1) = a_2$.

Proof. Note that $m_2(x)$ is the minimal polynomial of a_2 over F_2. Let us indicate two proofs, the first somewhat abstract, the second quite explicit.

I. Note that $(m_i(x))$ is the kernel of the substitution epimorphism E_i: $F_i[x] \to F_i(a_i)$ obtained by substitution of a_i for x, so that

$$F_i(a_i) \cong F_i[x]/(m_i(x)), \qquad i = 1, 2.$$

Let

$$\eta_i: F_i[x] \to F_i[x]/(m_i(x))$$

be the canonical quotient map, $i = 1, 2$. In the diagram

$$\begin{array}{ccc} F_1[x] & \xrightarrow{\ \phi\ } & F_2[x] \\ {\scriptstyle \eta_1}\downarrow & & \downarrow{\scriptstyle \eta_2} \\ F_1[x]/(m_1(x)) & \xrightarrow{\ \theta\ } & F_2[x]/(m_2(x)) \end{array}$$

note that $\ker(\eta_2\phi) = (m_1(x))$, so by the FHTR there is an *isomorphism* θ as indicated. It follows that $F_1(a_1) \cong F_2(a_2)$.

II. To be more explicit note that by Proposition 1.3 we have

$$F_1(a_1) = \{b_0 + b_1 a_1 + \cdots + b_{n-1} a_1^{n-1} : b_i \in F_1\}$$

and

$$F_2(a_2) = \{c_0 + c_1 a_2 + \cdots + c_{n-1} a_2^{n-1} : c_i \in F_2\},$$

where $n = \deg m_1(x) = \deg m_2(x)$. We may extend ϕ to an isomorphism between $F_1(a_1)$ and $F_2(a_2)$ by setting $\phi(a_1) = a_2$, and hence

$$\phi(b_0 + \cdots + b_{n-1} a_1^{n-1}) = \phi(b_0) + \cdots + \phi(b_{n-1}) a_2^{n-1}.$$

In the special case where $F_1 = F_2 = F$, $K_1 = K_2 = K$, and ϕ is the identity map on F we say that a_1 and a_2 are *conjugates* over F. Equivalently $a_1, a_2 \in K$ are conjugates over F if and only if they have the same minimal polynomial over F.

Corollary to Proposition 1.9. If K is an extension field of F and $a_1, a_2 \in K$ are conjugates over F, then there is an isomorphism $\phi: F(a_1) \to F(a_2)$, with $\phi(a_1) = a_2$ and $\phi(b) = b$ for all $b \in F$.

Theorem 1.10. Suppose F_1 and F_2 are fields and $\phi: F_1 \to F_2$ is an isomorphism. Suppose $f_1(x) \in F_1[x]$, $f_2(x) = \phi(f_1(x)) \in F_2[x]$, and K_1, K_2 are splitting fields for $f_1(x)$ and $f_2(x)$, respectively. Then ϕ can be extended to an isomorphism $\theta: K_1 \to K_2$.

Proof. Induction on $\deg f_1(x)$. The result is clear if $\deg f_1(x) \le 1$, so suppose $\deg f_1(x) > 1$ and assume the result for polynomials of lower degree. Let $a_1 \in K_1$ be a root of a monic irreducible divisor of $f_1(x)$ in $F_1[x]$ and let a_2 be a root of the corresponding irreducible factor of $f_2(x)$, $a_2 \in K_2$. By Proposition 1.9 we may extend ϕ to an isomorphism $\phi_1: F_1(a_1) \to F_2(a_2)$. Write $f_1(x) = (x - a_1)g_1(x)$ and $f_2(x) = (x - a_2)g_2(x)$, with $g_i(x) \in F_i(a_i)[x]$. Then $\phi_1(g_1(x)) = g_2(x)$, K_i is a splitting field for $g_i(x)$, and $\deg g_1(x) < \deg f_1(x)$. By the induction hypothesis ϕ_1 (and hence ϕ) can be extended to an isomorphism $\theta: K_1 \to K_2$.

If K_1 and K_2 are extensions of a field F, and if there is an isomorphism θ: $K_1 \to K_2$ such that $\theta(b) = b$ for all $b \in F$ we say that θ is an *F-isomorphism* from K_1 to K_2.

Corollary to Theorem 1.10. Suppose K_1 and K_2 are splitting fields over F for a polynomial $f(x) \in F[x]$. Then there is an F-isomorphism $\theta: K_1 \to K_2$.

Theorem 1.11. Suppose F_1 and F_2 are fields and $\phi: F_1 \to F_2$ is an isomorphism. Suppose $\mathscr{F}_1 \subseteq F_1[x]$, $\mathscr{F}_2 = \phi(\mathscr{F}_1) \subseteq F_2[x]$, and K_1, K_2 are splitting fields for $\mathscr{F}_1$, $\mathscr{F}_2$, respectively. Then ϕ can be extended to an isomorphism $\theta: K_1 \to K_2$.

Proof. Let $\mathscr{S}$ be the set of all ordered pairs (F_α, ϕ_α), where F_α is a field, $F_1 \subseteq F_\alpha \subseteq K_1$, $\phi_\alpha: F_\alpha \to K_2$ is a monomorphism, and $\phi_\alpha | F_1 = \phi$. Then $\mathscr{S}$ is partially ordered by means of $(F_\alpha, \phi_\alpha) \leq (F_\beta, \phi_\beta)$ if and only if $F_\alpha \subseteq F_\beta$ and $\phi_\beta | F_\alpha = \phi_\alpha$; $\mathscr{S} \neq \varnothing$ since $(F_1, \phi) \in \mathscr{S}$. By Zorn's Lemma there is a maximal element (F_0, θ) in $\mathscr{S}$. If $F_0 \neq K_1$, then there is a polynomial $f_1(x) \in \mathscr{F}_1$ that does *not* split over F_0. Since $f_1(x)$ does split over K_1 we may view $f_1(x)$ as an element of $F_0[x]$ to see that there is a splitting field L_1 for $f_1(x)$ over F_0, with $F_0 \subseteq L_1 \subseteq K_1$, and correspondingly there is a splitting field L_2 for $\theta(f_1(x)) = f_2(x)$ over $\theta(F_0)$, with $\theta(F_0) \subseteq L_2 \subseteq K_2$. By Theorem 1.10 we may extend θ to $\theta': L_1 \to L_2$. But then $(F_0, \theta) \leq (L_1, \theta')$ and $(F_0, \theta) \neq (L_1, \theta')$, contradicting the maximality of (F_0, θ). Thus $F_0 = K_1$. But then $\theta(K_1)$ is a splitting field for $\theta(\mathscr{F}_1) = \mathscr{F}_2$ within K_2, so $\theta(K_1) = K_2$ by the minimality of splitting fields, and the theorem is proved.

Corollary. If F is a field and K_1, K_2 are both algebraic closures of F, then there is an F-isomorphism $\theta: K_1 \to K_2$.

Proof. An algebraic closure is just a splitting field for $\mathscr{F} = F[x]$ (see Exercise 1.3).

Exercise 1.4. Suppose K is an algebraic extension of F.

(1) If F is a finite field (e.g., $F = \mathbb{Z}_p$) show that K is finite or countable.

(2) If F is infinite show that $|K| = |F|$. (This exercise may require some review of set theory. For a start see Theorem 2, p. 3 of Kamke [19].)

Theorem 1.12. If F is a field, then F has an algebraic closure.

Proof. Consider first a deceptively easy but invalid proof. Let $\mathscr{S}$ be the set of all algebraic extension fields K of F, partially ordered by set inclusion. Then $\mathscr{S} \neq \varnothing$ since $F \in \mathscr{S}$. By Zorn's Lemma there is a maximal element $\bar{F}$ in $\mathscr{S}$. If $f(x) \in \bar{F}[x]$ is not constant, then $f(x)$ has a root a in some extension of $\bar{F}$. If $a \notin \bar{F}$, then $\bar{F}(a)$ is an algebraic extension of F by Proposition 1.5, but $\bar{F}(a) \neq \bar{F}$, and the maximality of $\bar{F}$ is violated. Thus $\bar{F}$ is an algebraic closure for F.

Surprisingly (perhaps) the difficulty with the above "proof" is that $\mathscr{S}$ is not necessarily a set (see Exercise 8.9). In order to repair the proof it is necessary to obtain a class $\mathscr{S}$ of algebraic extensions of F that is large enough to apply Zorn's Lemma, but so that $\mathscr{S}$ *is* a set. To that end we choose an uncountable set X such that $F \subseteq X$ and $|F| < |X|$ (e.g., we could take $X = F \cup 2^F \cup \mathbb{R}$). We then impose field structures on subsets K of X in all possible ways such that K becomes an algebraic extension of F. Note that if L is *any* algebraic extension of F, then by Exercise 1.4 there is an algebraic extension K of F, with $F \subseteq K \subseteq X$, and $L \cong K$. We may now take $\mathscr{S} = \{K \subseteq X : K$ is an algebraic extension field of $F\}$, which really is a set, and the proof can proceed as above.

2. THE FUNDAMENTAL THEOREM OF GALOIS THEORY

If K is an extension of a field F, then an *F-automorphism* of K is an automorphism ϕ of K such that $\phi(b) = b$ for all $b \in F$. The set $G = G(K:F)$ of all F-automorphisms of K is a group, called the *Galois group* of K over F.

For example, if $F = \mathbb{R}$ and $K = \mathbb{C}$, then $|G| = 2$; the nonidentity element ϕ in G is the complex conjugation automorphism, i.e., $\phi(a + bi) = a - bi$. For another example let $F = \mathbb{Q}$ and $K = \mathbb{Q}(\sqrt[3]{2})$. Any $\phi \in G$ must carry $\sqrt[3]{2}$ to a conjugate, i.e., to another root of $m(x) = x^3 - 2$. But also $\phi(\sqrt[3]{2})$ must be real since $K \subseteq \mathbb{R}$, and the other two roots of $m(x)$ are not real. Thus $\phi(\sqrt[3]{2}) = \sqrt[3]{2}$, so $\phi = 1$, and $G = 1$.

If L is an intermediate field, $F \subseteq L \subseteq K$, we set

$$\mathscr{G}L = G(K:L) = \{\phi \in G : \phi(a) = a \text{ for all } a \in L\}.$$

Note that $\mathscr{G}L \leq G = G(K:F)$. For any $H \leq G$ we set

$$\mathscr{F}H = \{a \in K : \phi(a) = a \text{ for all } \phi \in H\},$$

the *fixed field* of H. Clearly $\mathscr{F}H$ is a field intermediate between F and K. Note that $\mathscr{G}F = G$, $\mathscr{G}K = 1$, and $\mathscr{F}1 = K$. Observe, however, that $\mathscr{F}G$ can be larger than F [see the example above, with $K = \mathbb{Q}(\sqrt[3]{2})$].

If K is an extension of F with Galois group G we say that K is a *Galois* extension of F if $\mathscr{F}G = F$. Equivalently K is Galois over F if and only if for each $a \in K \setminus F$ there is an element $\phi \in G(K:F)$ such that $\phi(a) \neq a$.

Exercise 2.1. If $m \in \mathbb{Z}$ is square-free and $m \neq 0, 1$ show that $K = \mathbb{Q}(\sqrt{m})$ is Galois over $F = \mathbb{Q}$.

If $F \subseteq L \subseteq K$, then the field $\mathscr{F}\mathscr{G}L$ is called the *closure* of L. We say that L is *closed* if $L = \mathscr{F}\mathscr{G}L$. Dually, if $1 \leq H \leq G = G(K:F)$, then the *closure* of H is $\mathscr{G}\mathscr{F}H$, and H is *closed* if $H = \mathscr{G}\mathscr{F}H$.

Proposition 2.1. Suppose $F \subseteq E \subseteq L \subseteq K$ and $1 \leq J \leq H \leq G(K:F)$. Then

(1) $\mathscr{G}L \leq \mathscr{G}E$ and $\mathscr{F}H \subseteq \mathscr{F}J$,
(2) $H \leq \mathscr{G}\mathscr{F}H$ and $L \subseteq \mathscr{F}\mathscr{G}L$, and
(3) $\mathscr{G}L = \mathscr{G}\mathscr{F}\mathscr{G}L$ and $\mathscr{F}H = \mathscr{F}\mathscr{G}\mathscr{F}H$.

Proof. We leave the proof of (1) and (2) as an (easy) exercise. Note that $\mathscr{G}L \leq \mathscr{G}\mathscr{F}(\mathscr{G}L)$ by (2). But also $L \subseteq \mathscr{F}\mathscr{G}L$ by (2), so $\mathscr{G}(\mathscr{F}\mathscr{G}L) \leq \mathscr{G}L$ by (1), and hence $\mathscr{G}L = \mathscr{G}\mathscr{F}\mathscr{G}L$. The proof that $\mathscr{F}H = \mathscr{F}\mathscr{G}\mathscr{F}H$ is entirely analogous.

Proposition 2.2. If K is an extension of F and $G = G(K:F)$, then all subgroups $\mathscr{G}L$ are closed and all intermediate fields $\mathscr{F}H$ are closed. Moreover $\mathscr{F}$ is a 1–1 inclusion–reversing correspondence between the set of all closed subgroups of G and the set of all closed intermediate fields between F and K.

Proof. See Proposition 2.1 above. Note that $\mathscr{F}$ is a 1–1 correspondence as indicated since it has $\mathscr{G}$ as an inverse.

Proposition 2.3. Suppose $F \subseteq E \subseteq L \subseteq K$, and suppose L is finite over E. Then $[\mathscr{G}E:\mathscr{G}L] \leq [L:E]$.

Proof. Induction on $n = [L:E]$. The result is obvious when $n = 1$. If there is a field M properly between E and L, then by Lagrange's Theorem, the induction hypothesis, and Proposition 1.1 we have

$$[\mathscr{G}E:\mathscr{G}L] = [\mathscr{G}E:\mathscr{G}M][\mathscr{G}M:\mathscr{G}L] \leq [M:E][L:M] = [L:E].$$

Assume then that there are no fields properly between E and L. Thus if $a \in L \backslash E$ we have $L = E(a)$. Let $m(x) = m_{a,E}(x)$, so $\deg m(x) = n$. If $\phi \in \mathscr{G}E$ form the left coset $\phi\mathscr{G}L$. If $\theta \in \mathscr{G}L$, then $\theta(a) = a$, so $\phi\theta(a) = \phi(a)$, i.e., each element of the coset $\phi\mathscr{G}L$ sends the root a of $m(x)$ to the same root $\phi(a)$ of $m(x)$. Different cosets $\phi_1\mathscr{G}L$ and $\phi_2\mathscr{G}L$ give different roots $\phi_1(a)$ and $\phi_2(a)$, for otherwise $\phi_1^{-1}\phi_2(a) = a$, which would mean that $\phi_1^{-1}\phi_2 \in \mathscr{G}L$ since $L = E(a)$, contradicting the fact that $\phi_1\mathscr{G}L \neq \phi_2\mathscr{G}L$. Consequently $[\mathscr{G}E:\mathscr{G}L] \leq \deg m(x) = n = [L:E]$.

Proposition 2.4. Suppose $F \subseteq K$, $G = G(K:F)$, $1 \leq J \leq H \leq G$, and $[H:J]$ is finite. Then $[\mathscr{F}J:\mathscr{F}H] \leq [H:J]$.

Proof. Note that if $\phi \in H$, $\theta \in J$, and $a \in \mathscr{F}J$, then $\phi\theta(a) = \phi(a)$, so the image of a is constant as $\phi\theta$ ranges over the left coset ϕJ. Choose left coset representatives $\phi_1 = 1, \phi_2, \ldots, \phi_n$ for J in H. If $[\mathscr{F}J:\mathscr{F}H] > n = [H:J]$ choose $a_1, a_2, \ldots, a_{n+1} \in \mathscr{F}J$, linearly independent over $\mathscr{F}H$. Consider the homogeneous system of n linear equations in $n + 1$ unknowns $x_1, x_2, \ldots, x_{n+1}$ given in matrix form by $AX = 0$, where $X = (x_1, \ldots, x_{n+1})^t$ and the coefficient matrix A has ijth entry $\phi_i(a_j)$. The matrix A represents a linear transformation $T:K^{n+1} \to K^n$, so T has nontrivial kernel, i.e., there is a nontrivial solution to the system of equations. Choose a nontrivial solution with the least possible number of nonzero entries. By relabeling we may assume that the solution has the form $(b_1, b_2, \ldots, b_k, 0, \ldots, 0)$, $0 \neq b_i \in K$, and we may assume (by dividing) that $b_1 = 1$. If all b_i were in $\mathscr{F}H$, then the first equation in the system would be a dependence relation for $\{a_1, \ldots, a_k\}$, so we may assume further that $b_2 \notin \mathscr{F}H$. Choose $\phi \in H$ such that $\phi(b_2) \neq b_2$, and apply ϕ to each equation in the system $AX = 0$. By the remark at the beginning of the proof the equations are simply rearranged, since $\{\phi\phi_1, \phi\phi_2, \ldots, \phi\phi_n\}$ is a new set of coset representatives for J in H. Consequently

$$(\phi(b_1), \phi(b_2), \ldots, \phi(b_k), 0, \ldots, 0)$$

is also a solution vector for the system, with $\phi(b_1) = \phi(1) = 1$ and $\phi(b_2) \neq b_2$. If the new solution is subtracted from the old we obtain a third nontrivial solution having $k - 1$ or fewer nonzero entries, contradicting the minimality of k.

Theorem 2.5. (1) Suppose $F \subseteq E \subseteq L \subseteq K$, E is closed, and $[L:E]$ is finite. Then L is closed and $[\mathscr{G}E:\mathscr{G}L] = [L:E]$.

(2) Suppose $1 \leq J \leq H \leq G = G(K, F)$, J is closed, and $[H:J]$ is finite. Then H is closed and $[\mathscr{F}J:\mathscr{F}H] = [H:J]$.

Proof. (1) $[L:E] = [L:\mathscr{F}\mathscr{G}E] \leq [\mathscr{F}\mathscr{G}L:\mathscr{F}\mathscr{G}E] \leq [\mathscr{G}E:\mathscr{G}L] \leq [L:E]$ by Propositions 2.1, 2.3, and 2.4. Thus equality holds throughout.

The proof of (2) is similar to (actually it is "dual" to) the proof of (1).

Corollary. (1) All finite subgroups of $G = G(K:F)$ are closed.

(2) If K is Galois over F, $F \subseteq E \subseteq K$, and $[E:F]$ is finite, then E is closed, and consequently K is Galois over E.

Proof. (1) The subgroup 1 is closed.

(2) Since K is Galois F is closed, and hence E is closed by the proposition. Thus E is the fixed field of the Galois group $\mathscr{G}E = G(K:E)$, which says that K is Galois over E.

If $F \subseteq E \subseteq K$ we say that E is *stable* if $\phi(E) \subseteq E$ for all $\phi \in G = G(K:F)$. Note that if E is stable, then also $\phi^{-1}(E) \subseteq E$, all $\phi \in G$, so $E = \phi\phi^{-1}(E) \subseteq \phi(E) \subseteq E$, and in fact $\phi(E) = E$ for all $\phi \in G$.

Proposition 2.6. (1) If $F \subseteq E \subseteq K$, and E is stable, then $\mathscr{G}E \lhd G$.

(2) If $H \lhd G$, then $\mathscr{F}H$ is stable.

Exercise 2.2. Prove Proposition 2.6.

Corollary. (1) If $F \subseteq E \subseteq K$ and E is stable, then the closure $\mathscr{F}\mathscr{G}E$ is stable.

(2) If $H \lhd G$, then $\mathscr{G}\mathscr{F}H \lhd G$.

Proof. (1) Since E is stable $\mathscr{G}E \lhd G$, and hence $\mathscr{F}\mathscr{G}E$ is stable.
(2) Since $H \lhd G$ we have $\mathscr{F}H$ stable, and hence $\mathscr{G}\mathscr{F}H \lhd G$.

Proposition 2.7. If K is Galois over F, $F \subseteq L \subseteq K$, and L is stable, then L is Galois over F.

Proof. If $a \in L\backslash F$, then $\phi(a) \neq a$ for some $\phi \in G = G(K:F)$. But ϕ restricts to an F-automorphism θ of L since L is stable, so $\theta \in G(L:F)$ and $\theta(a) \neq a$. Thus L is Galois over F.

Theorem 2.8. Suppose K is Galois over F, $f(x) \in F[x]$ is irreducible, and $f(x)$ has a root $a \in K$. Then $f(x)$ splits over K and all its roots are distinct.

Proof. We may replace $f(x)$ by an associate if necessary and assume that $f(x)$ is monic, hence $f(x) = m_{a,F}(x)$. For each $\phi \in G$ observe that $\phi(f(x)) = f(x)$ and hence $\phi(a)$ is also a root of $f(x)$ in K. Let $a_1 = a, a_2, \ldots, a_k$ be all the distinct roots of $f(x)$ of the form $\phi(a)$, $\phi \in G$. Thus $k \leq n = \deg f(x)$. Set

$$g(x) = \prod_{i=1}^{k} (x - a_i).$$

Since each $\phi \in G$ merely permutes the roots a_i we see that $\phi(g(x)) = g(x)$ for all $\phi \in G$. It follows that $g(x) \in F[x]$ since F is the fixed field for G. But $g(x)$ is monic, $g(a) = 0$, $1 \leq \deg g(x) \leq \deg f(x)$, and $f(x) = m_a(x)$. Thus $g(x) = f(x)$ and the theorem follows.

Corollary 1. Suppose $F \subseteq L \subseteq K$ and L is both Galois and algebraic over F. Then L is stable.

Proof. If $\phi \in G(K:F)$ and $a \in L$, then $\phi(a)$ is a root of $m(x) = m_{a,F}(x)$ since $\phi(m(x)) = m(x)$. By the theorem $m(x)$ splits over L, so $\phi(a)$, being one of its roots, must lie in L.

Corollary 2. If K is Galois and algebraic over F and $F \subseteq L \subseteq K$, then L is stable if and only if L is Galois over F.

Proof. $\Leftarrow$: Corollary 1.
$\Rightarrow$: Proposition 2.7.

Proposition 2.9. Suppose $F \subseteq L \subseteq K$ and L is stable. Then the quotient group $G/\mathscr{G}L$ is isomorphic with the subgroup of $G(L:F)$ consisting of those $\theta \in G(L:F)$ that are restrictions to L of elements of G, i.e., those $\theta \in G(L:F)$ that may be extended to automorphisms of K.

Proof. If $\phi \in G$, then $\theta = \phi|L$ is in $G(L:F)$ since L is stable. The map $f: \phi \mapsto \phi|L$ is clearly a homomorphism from G into $G(L:F)$; its kernel comprises those $\phi \in G$ such that $\phi|L = 1$, i.e., $\ker f = \mathscr{G}L$. The proposition follows from the FHT.

Theorem 2.10 (The Fundamental Theorem of Galois Theory). Suppose K is a finite Galois extension of F, and $G = G(K:F)$ is its Galois group. Then the operation $\mathscr{F}$ of taking fixed fields is a 1–1 inclusion-reversing correspondence between the set of all subgroups of G and the set of all intermediate fields L between F and K. If $J \leq H \leq G$, then $[H:J] = [\mathscr{F}J:\mathscr{F}H]$, and in particular $|G| = [K:F]$. Furthermore, $H \lhd G$ if and only if $\mathscr{F}H = L$ is Galois over F, in which case $G(L:F) \cong G/H$.

Proof. Since K is a finite Galois extension of F all intermediate fields are closed and all subgroups of G are closed, by Theorem 2.5, and $[H:J] = [\mathscr{F}J:\mathscr{F}H]$ for $J \leq H \leq G$. The statement about normality is a consequence of Proposition 2.6 and Theorem 2.8, Corollary 2. If $H \lhd G$ and $L = \mathscr{F}H$, then by Proposition 2.9 we have $G/H = G/\mathscr{G}\mathscr{F}H = G/\mathscr{G}L$ isomorphic with a subgroup of $G(L:F)$. Note, however, that

$$\begin{aligned} |G/H| &= |G|/|H| = |G|/|\mathscr{G}L| = |G|/[\mathscr{G}L:\mathscr{G}K] \\ &= [K:F]/[K:L] = [L:F] = |G(L:F)|, \end{aligned}$$

so $G/H \cong G(L:F)$.

Corollary. If K is a finite extension of F and $|G(K:F)| = [K:F]$, then K is Galois over F.

Proof. It is clear from the definitions that $G = G(K:F) = G(K:\mathscr{F}G)$, so $\mathscr{F}G = \mathscr{F}G(K:\mathscr{F}G)$, which means that K is Galois over $\mathscr{F}G$, and hence $|G| = [K:\mathscr{F}G]$ by the theorem. Thus $[K:F] = [K:\mathscr{F}G]$, so $\mathscr{F}G = F$ and K is Galois over F.

3. NORMALITY AND SEPARABILITY

Over fields of characteristic $p > 0$ it is possible to have irreducible polynomials having multiple roots in an extension field. For example, let $F = \mathbb{Z}_2(t)$, t being an indeterminate. Let $f(x) = x^2 + t$; it is irreducible in $F[x]$. (Why?) By Proposition 1.8 $f(x)$ has a root, call it $\sqrt{t}$, in some extension field K of F. But then in $K[x]$ we have $(x - \sqrt{t})^2 = x^2 - 2\sqrt{t}x + t = x^2 + t = f(x)$, so $\sqrt{t}$ is a root of $f(x)$ with multiplicity 2.

Exercise 3.1. Suppose F is a field of characteristic $p > 0$.

(1) Show that $(a + b)^p = a^p + b^p$ and $(a - b)^p = a^p - b^p$ for all $a, b \in F$.

(2) Let $K = F(t)$ for an indeterminate t and let $f(x) = x^p - t \in K[x]$. Show that $f(x)$ is irreducible and that $f(x)$ has just one root with multiplicity p in any splitting field.

If F is a field, then an irreducible polynomial $f(x) \in F[x]$ is called *separable* if $f(x)$ has distinct roots in a splitting field (the definition is of course independent of the splitting field used since any two are F-isomorphic by the corollary to Theorem 1.10). In general a polynomial $f(x) \in F[x]$ is called *separable* if each of its irreducible factors is separable.

If F is a field and

$$f(x) = a_0 + a_1 x + \cdots + a_k x^k \in F[x]$$

the *derivative* of $f(x)$ can be defined formally as

$$f'(x) = a_1 + 2a_2 x + 3a_3 x^2 + \cdots + k a_k x^{k-1}.$$

Of course $f'(x) = 0$ if $f(x)$ is a constant.

Exercise 3.2. If $f(x), g(x) \in F[x]$ show that the derivative of $f(x) + g(x)$ is $f'(x) + g'(x)$ and the derivative of $f(x)g(x)$ is $f(x)g'(x) + f'(x)g(x)$.

Proposition 3.1. Suppose F is a field, $f(x) \in F[x]$, and $\deg f(x) > 0$. Then $f'(x) = 0$ if and only if $\operatorname{char} F = p > 0$ and there is a polynomial $g(x) \in F[x]$ such that $f(x) = g(x^p)$.

Proof. This becomes clear upon inspection of the coefficients of $f'(x)$.

Proposition 3.2. Suppose F is a field and $f(x)$ is a nonconstant polynomial in $F[x]$.

(1) If $f'(x) = 0$, then every root of $f(x)$ in any extension field has multiplicity 2 or greater.

(2) If $f'(x) \neq 0$ and $(f(x), f'(x)) = 1$, then $f(x)$ has no repeated roots in any extension field.

Proof. (1) Let a be a root of $f(x)$ in an extension field K of F, and write $f(x) = (x - a)g(x)$ in $K[x]$. Then

$$0 = f'(x) = g(x) + (x - a)g'(x)$$

and so $g(x) = -(x - a)g'(x)$. Setting $x = a$ we see that $g(a) = 0$ and hence $(x - a)^2 \mid f(x)$ in $K[x]$.

(2) Suppose to the contrary that $f(x)$ has a root a with multiplicity 2 (or more) in an extension K of F. Write $f(x) = (x - a)^2 g(x)$ in $K[x]$ and differentiate to see that a is also a root of $f'(x)$. By Proposition II.5.3 there are $h(x)$, $k(x)$ in $F[x]$ such that $h(x)f(x) + k(x)f'(x) = 1$. Substitute $x = a$ to obtain $0 = 1$, a contradiction.

Corollary 1. If $f(x) \in F[x]$ is irreducible, then $f(x)$ is separable if and only if $f'(x) \neq 0$.

Corollary 2. If $\operatorname{char} F = 0$, then every polynomial is separable.

Corollary 3. If $\operatorname{char} F = p > 0$, then an irreducible polynomial $f(x)$ is inseparable if and only if it has the form $f(x) = g(x^p)$ for some $g(x) \in F[x]$.

If F is a field, K is an extension of F, and $a \in K$ is algebraic over F, then a is called *separable* over F if its minimal polynomial $m_{a,F}(x)$ is separable. If K is algebraic over F, then K is called a *separable* extension of F if every $a \in K$ is separable over F. Thus for example every algebraic extension of a field F of characteristic 0 is a separable extension by Corollary 2 above.

Proposition 3.3. Suppose K is a splitting field over F for some set $\mathscr{F} \subseteq F[x]$, that $f(x) \in F[x]$ is separable and irreducible of degree $k > 0$, and that $f(x)$ splits over K. If $G = G(K:F)$ and if $a \in K$ is one root of $f(x)$, then $\{\phi(a):\phi \in G\}$ is the set of all roots of $f(x)$ in K. In fact, if $L = F(a)$ and $H = \mathscr{G}L \leq G$, then $[G:H] = k$, and if $\phi_1, \ldots, \phi_k$ are coset representatives for H in G, then $\phi_1 a, \ldots, \phi_k a$ are all the roots of $f(x)$.

Proof. Note that G acts as a permutation group on the roots of $f(x)$ in K, since $f(\phi c) = \phi f(c)$ for all $c \in K$, $\phi \in G$. In particular every ϕa, $\phi \in G$, is a root of $f(x)$. Let $b \in K$ be any root of $f(x)$. Then there is an F-isomorphism $\theta: F(a) \to F(b)$, with $\theta(a) = b$, by the corollary to Proposition 1.9. By Theorem 1.11 θ extends to an F-automorphism $\phi \in G$, and $\phi(a) = b$. Thus the set of roots in K of $f(x)$ is precisely $\{\phi(a):\phi \in G\}$. Clearly $\mathrm{Stab}_G(a) = \mathscr{G}L = H$, so there are $[G:H]$ distinct roots of $f(x)$, and $[G:H] = k$ since $f(x)$ is separable and irreducible of degree k. If $i \neq j$, then $\phi_i a \neq \phi_j a$, or else $\phi_j^{-1}\phi_i \in \mathrm{Stab}_G(a) = H$, and then $\phi_i H = \phi_j H$.

Theorem 3.4. If F is a field and K is an algebraic extension of F, then the following statements are equivalent.

(a) K is Galois over F.
(b) K is a separable splitting field for some set $\mathscr{F}_1 \subseteq F[x]$.
(c) K is a splitting field for some set $\mathscr{F}_2$ of separable polynomials in $F[x]$.

Proof. (a) $\Rightarrow$ (b): Set $\mathscr{F}_1 = \{m_a(x): a \in K\} \subseteq F[x]$. Each $m_a(x) \in \mathscr{F}_1$ splits over K and has distinct roots by Theorem 2.8. Thus K is a separable splitting field for $\mathscr{F}_1$.

(b) $\Rightarrow$ (c): Take $\mathscr{F}_2 = \mathscr{F}_1$.

(c) $\Rightarrow$ (a): Assume first that $[K:F]$ is finite and let $G = G(K:F)$. Choose an irreducible factor $f(x)$ of some $g(x) \in \mathscr{F}_2$, with $\deg f(x) = k > 1$. Let $a \in K$ be a root of $f(x)$, set $L = F(a)$, and set $H = \mathscr{G}L$. Then $[G:H] = k = [L:F]$ by Propositions 3.3 and 1.3. Now use induction on $[K:F]$, and consider the setting $L \subseteq K$. Since K is a splitting field over L for $\mathscr{F}_2$, and $[K:L] < [K:F]$, we have by induction that K is Galois over L, and hence $[K:L] = |G(K:L)| = |\mathscr{G}L| = |H|$. Multiply by $[L:F] = [G:H]$ to see that $|G| = [K:F]$. Thus K is Galois over F by the corollary to the Fundamental Theorem of Galois Theory.

Suppose then that $[K:F]$ is infinite, and take any $a \in K \backslash F$. Then $[F(a):F]$ is finite, so a is contained in a splitting field $M \subseteq K$ over F for some finite subset of $\mathscr{F}_2$. But then $[M:F]$ is finite, and M is Galois over F by the first part of the proof, above. Thus $\phi(a) \neq a$ for some $\phi \in G(M:F)$. By Theorem 1.11 ϕ extends to an element $\theta \in G(K:F)$. Since $\theta(a) \neq a$ we have K Galois over F.

Corollary. Suppose K is an algebraic extension of F and char $F = 0$. Then K is Galois over F if and only if K is a splitting field over F for some set of polynomials in $F[x]$.

An extension field K of F is called *normal* over F if every irreducible polynomial in $F[x]$ that has a root in K splits over K. The next theorem shows that the notion of normality characterizes splitting fields, and also characterizes stability within an algebraic closure.

Theorem 3.5. Suppose K is an algebraic extension of F and $\bar{F}$ is an algebraic closure of F with $K \subseteq \bar{F}$. Then the following statements are equivalent.

(a) K is normal over F.
(b) K is a splitting field for some set $\mathscr{F} \subseteq F[x]$.
(c) If $\phi \in G(\bar{F}:F)$, then $\phi(K) \subseteq K$.

Proof. (a) $\Rightarrow$ (b): Let $\mathscr{F} = \{m_a(x): a \in K\}$. It is clear from the definition of normality that K is a splitting field for $\mathscr{F}$ over F.

(b) $\Rightarrow$ (c): Say K is a splitting field for $\mathscr{F} = \{f_\alpha(x): \alpha \in A\} \subseteq F[x]$. Take $\phi \in G(\bar{F}:F)$ and $f_\alpha(x) \in \mathscr{F}$, and suppose $a \in K$ is a root of $f_\alpha(x)$. Then $0 = \phi(f_\alpha(a)) = f_\alpha(\phi(a))$ since $f_\alpha(x) \in F[x]$, so $\phi(a)$ is also a root of $f_\alpha(x)$, and consequently $\phi(a) \in K$. Since K is generated over F by elements such as a we have $\phi(K) \subseteq K$.

(c) $\Rightarrow$ (a): Take $f(x)$ irreducible in $F[x]$, with $f(a) = 0$ for some $a \in K$. Let b be any other root of $f(x)$ in $\bar{F}$. Then there is an F-isomorphism ϕ: $F(a) \to F(b)$ with $\phi(a) = b$ by the corollary to Proposition 1.9, and ϕ extends to an automorphism $\theta \in G(\bar{F}:F)$ by Theorem 1.11. But then $\theta(K) \subseteq K$, so $\theta(a) = b \in K$ and $f(x)$ splits over K.

Corollary. If K is an algebraic extension of F, then K is Galois over F if and only if it is both normal and separable over F.

Proof. Theorems 3.4 and 3.5.

Let K be an extension of F. A *normal closure* of K over F is a field $L \supseteq K$ that is normal over F and minimal in that respect. A *Galois* closure of K over F is a field $L \supseteq K$ that is Galois over F and minimal in that respect.

Theorem 3.6. If K is an algebraic extension of F, then K has a normal closure L over F which is unique up to K-isomorphism. If $[K:F]$ is finite then $[L:F]$ is finite. If K is separable over F, then L is a Galois closure for K over F.

Proof. Let $\{a_\beta : \beta \in B\}$ be a set of generators for K over F. Set $\mathscr{F} = \{m_{a_\beta,F}(x) : \beta \in B\}$, viewed as a subset of $K[x]$, and let L be a splitting field for $\mathscr{F}$ over K. Since some of the roots of elements of $\mathscr{F}$ generate K over F and the remaining roots generate L over K it is clear that L is also a splitting field for $\mathscr{F}$ over F. Thus L is normal over F by Theorem 3.5. If K is also separable over F, then L is Galois over F by Theorem 3.4. If $K \subseteq M \subseteq L$ and M is normal over F, then one root (viz., a_β) of each $m_{a_\beta}(x) \in \mathscr{F}$ is in M (in fact in K), and hence each $m_{a_\beta}(x) \in \mathscr{F}$ splits completely in M since M is normal over F. Thus $M = L$ and L is a normal closure for K over F. If L' is another normal closure for K over F, then L' is also a splitting field for $\mathscr{F}$ over K, so L' and L are K-isomorphic by Theorem 1.11. If $[K:F]$ is finite we may assume that B is finite and hence $[L:F]$ is finite.

Proposition 3.7. If G is a finite subgroup of the multiplicative group F^* of a field F, then G is cyclic.

Proof. Since G is the direct product of its Sylow subgroups it will suffice to show that each Sylow subgroup P of G is cyclic. Set $m = \max\{|a| : a \in P\}$, and choose $b \in P$ with $|b| = m$. Then $1, b, b^2, \ldots, b^{m-1}$ are m distinct roots of $f(x) = x^m - 1 \in F[x]$, so they are *all* the roots by Proposition 1.7. If $c \in P$, then $c^m = 1$, so c is a root of $f(x)$, and therefore $c = b^k$ for some k. Thus $P = \langle b \rangle$.

Recall that an extension K of F is called *simple*, with *primitive element* a, if $K = F(a)$.

Theorem 3.8. If K is a finite extension of F, then K is simple over F if and only if there are only finitely many intermediate fields L between F and K.

Proof. $\Rightarrow$: Say $K = F(a)$ and write $m(x)$ for the minimal polynomial $m_{a,F}(X)$. If $F \subseteq L \subseteq K$ set $f(x) = m_{a,L}(x)$, so $f(x) \mid m(x)$ in $K[x]$ by Proposition 1.2. If $f(x) = b_0 + b_1 x + \cdots + x^k$ set $M = F(b_0, b_1, \ldots, b_{k-1}) \subseteq L$. Then $m_{a,M}(x) = f(x)$ and $K = M(a)$, so $[K:M] = \deg f(x) = [K:L]$ by Proposition 1.3 and $M = L$. Thus L is completely determined by $f(x)$, and the number of intermediate fields can be no larger than the number of monic factors of $m(x)$ in $K[x]$, which is finite since $K[x]$ is a UFD.

$\Leftarrow$: We may assume that F is infinite since K^* is a cyclic group if K is finite, by Proposition 3.7. If $a, b \in K$, with $b \neq 0$, set $L = F(a, b)$. Consider all elements $c \in L$ obtained by forming $c = a + bd$, $d \in F$. Since F is infinite but there are only finitely many intermediate fields there are elements $c_1 = a + bd_1$ and $c_2 = a + bd_2 \neq c_1$ such that $F(c_1) = F(c_2) = E \subseteq L$. But then $c_1 - c_2 = b(d_1 - d_2) \in E$, and $0 \neq d_1 - d_2 \in F$, so $b \in E$. Also, $a = c_1 - bd_1 \in E$, so $E = L$, i.e., $F(a, b) = F(c_1)$. Now choose $a \in K$ so that $[F(a):F]$ is maximal. If $F(a) \neq K$ we could choose $b \in K \setminus F(a)$, and then write $F(a, b) = F(c)$ as above, contradicting the maximality of $[F(a):F]$. Thus $K = F(a)$ is simple.

Theorem 3.9. If K is a finite separable extension of F, then K is simple over F.

Proof. Let L be a Galois closure for K over F (Theorem 3.6). Then $G(L:F)$ is finite and has only finitely many subgroups. Thus there are only finitely many fields between F and K by the Fundamental Theorem of Galois Theory, and K is simple over F by Theorem 3.8.

Exercise 3.3. Find a primitive element over $\mathbb{Q}$ for a splitting field $K \subseteq \mathbb{C}$ for the polynomial $f(x) = x^4 - 5x^2 + 6$.

It is fashionable to assert that the Fundamental Theorem of Algebra is actually a theorem of analysis, presumably because all known proofs depend on "analytic" properties of $\mathbb{C}$ or $\mathbb{R}$. The proof that appears below, attributed by S. Lang [23] to E. Artin, provides interesting applications of some of the general field theory results obtained thus far. The only result needed from analysis is the fact that a polynomial of odd degree with real coefficients has a real root, a consequence of the Intermediate Value Theorem of calculus since such a polynomial clearly takes on both positive and negative values.

Theorem 3.10 (The Fundamental Theorem of Algebra). The field $\mathbb{C}$ of complex numbers is algebraically closed.

Proof. If $f(x) \in \mathbb{C}[x]$ let a be a root of $f(x)$ in some extension of $\mathbb{C}$. Let K be a Galois closure of $\mathbb{C}(a)$ over $\mathbb{R}$ (Theorem 3.6) and set $G = G(K:\mathbb{R})$. Let H be a 2-Sylow subgroup of G and let $L = \mathscr{F}H$. By the Fundamental Theorem of Galois Theory we have $[L:\mathbb{R}] = [G:H]$, an odd number. By Theorem 3.9 we may write $L = \mathbb{R}(b)$ for some $b \in L$, so the minimal polynomial $m_{b,\mathbb{R}}(x)$ is irreducible over $\mathbb{R}$ and of odd degree. By the remarks preceding the proof that degree must be 1, and hence $L = \mathbb{R}$, which means that $G = H$, a 2-group. Thus $G_1 = \mathscr{G}\mathbb{C} = G(K:\mathbb{C})$ is also a 2-group. If $G_1 \neq 1$ choose $G_2 \leq G_1$ such that $[G_1:G_2] = 2$ (see Exercise I. 2.5), and set $M = \mathscr{F}G_2$, so that $[M:\mathbb{C}] = [G_1:G_2] = 2$. But any polynomial of degree 2 over $\mathbb{C}$ has roots in $\mathbb{C}$ by the quadratic formula (verify that every $a \in \mathbb{C}$ has a square root in $\mathbb{C}$), so such a field M cannot exist! The contradiction shows that $G_1 = 1$, hence $K = \mathbb{C}$, and $a \in \mathbb{C}$, completing the proof.

As a final application in this section of the ideas above we describe all finite fields. Note that a finite field must have characteristic p for some prime $p \in \mathbb{Z}$. For any field F of characteristic p the mapping $\phi_p: a \mapsto a^p$ is a monomorphism; it is called the *Frobenius map* on F.

Theorem 3.11. Suppose F is a finite field with q elements, having prime field $F_p \cong \mathbb{Z}_p$. Then $q = p^n$, where $n = [F:F_p]$, and F is a splitting field over F_p for the polynomial $f(x) = x^q - x$. Conversely if $0 < n \in \mathbb{Z}$ and p is a prime, then there is a field F with $q = p^n$ elements, and F is uniquely determined up to isomorphism. The Galois group $G(F:F_p)$ is cyclic of order n, with the Frobenius map ϕ_p as a generator.

Proof. Note that $q = p^n$ is a special case of the fact that any vector space of dimension n over a field with p elements has p^n elements; it consists of all linear combinations of n basis elements so there are p choices for each of n coefficients. The multiplicative group F^* has order $q - 1$, so each $a \in F^*$ is a root of $x^{q-1} - 1$, and hence each $a \in F$ is a root of $f(x) = x^q - x$. Thus F is a splitting field for $f(x)$ over F_p. For the converse take $F_p = \mathbb{Z}_p$ and let $F \supseteq F_p$ be a splitting field for $f(x) = x^q - x \in F_p[x]$. Note that $f(x)$ has q distinct roots by Proposition 3.2, since $f'(x) = -1$. But if $a, b \in F$ are roots of $f(x)$, then $a \pm b$, ab, and a/b (if $b \neq 0$) are also roots of $f(x)$, so the set of q roots of $f(x)$ constitute a field over which $f(x)$ splits, and $|F| = q$. The uniqueness follows from Theorem 1.10. The Frobenius map ϕ_p is in $G(F:F_p)$ since F is finite, and $\phi_p^k(a) = a^{p^k}$ for all $a \in F$ if $0 \leq k \in \mathbb{Z}$, so clearly $\phi_p^n = 1$. If ϕ_p had order $k < n$, then all elements of F would be roots of $g(x) = x^{p^k} - x$, contradicting Proposition 1.7. Since $|G(F:F_p)| \leq [F:F_p] = n$ it follows that $G(F:F_p) = \langle \phi_p \rangle$.

Corollary. If F and K are both finite fields, with $F \subseteq K$, then K is a Galois extension of F.

Proof. If $|K| = q$ view $f(x) = x^q - x$ as a polynomial in $F[x]$. Since $f(x)$ has distinct roots it is separable, and K is a splitting field for $f(x)$ over F. Apply Theorem 3.4.

A field with $q = p^n$ elements is called a *Galois field* and will be denoted by F_q [another common notation is $GF(q)$].

4. THE GALOIS THEORY OF EQUATIONS

If F is a field and $f(x) \in F[x]$ let K be a splitting field for $f(x)$ over F. Then the *Galois group* of $f(x)$ over F is defined to be $G = G(K:F)$. The genesis of Galois theory was E. Galois's study in 1832 relating properties of $f(x)$ to the interplay between subgroups of G and fields L intermediate between F and K. In his work F and K were taken to be subfields of $\mathbb{C}$.

Note that the Galois group of $f(x)$ over F depends (up to isomorphism) only on $f(x)$ and F, since all splitting fields for $f(x)$ over F are F-isomorphic.

If $S \subseteq K$ is the set of (distinct) roots of $f(x)$, then G acts as a permutation group on S. The action is faithful since $F(S) = K$, so if $|S| = n$ we may (and often shall) view G as a subgroup of the symmetric group S_n.

Exercise 4.1. If $f(x) \in F[x]$ and K is a splitting field for $f(x)$ over F, denote by S the set of distinct roots of $f(x)$ in K and let $G = G(K:F)$. If $f(x)$ is irreducible over F show that G is transitive on S. If $f(x)$ has no repeated roots and G is transitive on S show that $f(x)$ is irreducible over F.

A *simple radical extension* of a field F is a field $K = F(a)$, where $a^n \in F$ for some positive integer n. Thus a is a root of a polynomial of the form $x^n - b$, where $b \in F$. For example, if char $F \neq 2$ and $[K:F] = 2$, then K is a simple radical extension of F, as is easily seen by the familiar process of completing the square.

If F is a field and $0 < n \in \mathbb{Z}$ consider $f(x) = x^n - 1 \in F[x]$. If char $F = p > 0$ we assume that $p \nmid n$. Let K be a splitting field for $f(x)$ over F. Since $f'(x) = nx^{n-1}$, and $p \nmid n$ if char $F = p$, we see by Proposition 3.2 that $f(x)$ has n distinct roots in K [$f(x)$ and $f'(x)$ are relatively prime since they have no roots in common]. The roots of $f(x)$ are called nth *roots of unity*. The nth roots of unity constitute a multiplicative subgroup of K^*. The subgroup is cyclic by Proposition 3.7, and its generators, which have multiplicative order n, are called the *primitive* nth roots of unity. By Exercise I.1.6 there are $\phi(n)$ primitive nth roots of unity, where ϕ is Euler's totient function. If ζ is a primitive nth root of unity in K, then $1, \zeta, \zeta^2, \ldots, \zeta^{n-1}$ are all of the nth roots of unity, and $K = F(\zeta)$, a simple radical extension.

Exercise 4.2. If $\eta \in K$ is an nth root of unity, $\eta \neq 1$, show that $1 + \eta + \eta^2 + \cdots + \eta^{n-1} = 0$.

Proposition 4.1. Suppose F is a field, $0 < n \in \mathbb{Z}$, and if char $F = p$, then $p \nmid n$. Then the Galois group G of $f(x) = x^n - 1$ over F is abelian.

Proof. By the remarks preceding the proposition we may take $G = G(K:F)$, where $K = F(\zeta)$ and ζ is a primitive nth root of unity. If $\phi, \theta \in G$, then $\phi(\zeta)$ and $\theta(\zeta)$ are roots of $f(x)$, so $\phi(\zeta) = \zeta^i$ and $\theta(\zeta) = \zeta^j$ for some integers i and j. Thus $\phi\theta(\zeta) = \zeta^{ij} = \theta\phi(\zeta)$. Since K is simple over F any element of G is completely determined by its effect on ζ, so $\phi\theta = \theta\phi$ and G is abelian.

If $f(x) = x^n - 1 \in F[x]$ as above let $\zeta_1, \zeta_2, \ldots, \zeta_{\phi(n)}$ be the primitive nth roots of unity in K and set

$$\Phi_n(x) = \prod\{(x - \zeta_i): 1 \leq i \leq \phi(n)\},$$

the nth *cyclotomic polynomial.* If η is any root of $f(x) = x^n - 1$, then η is a primitive dth root of unity, where d is the multiplicative order of η in K^*, so d is a divisor of n by Lagrange's Theorem. Consequently

$$x^n - 1 = \prod\{\Phi_d(x): 0 < d \in \mathbb{Z} \text{ and } d \mid n\}.$$

Thus

$$\Phi_n(x) = \frac{x^n - 1}{\prod\{\Phi_d(x): 0 < d < n \text{ and } d \mid n\}},$$

and $\Phi_n(x)$ can be determined recursively. Note that $\Phi_1(x) = x - 1$, so

$$\Phi_2(x) = \frac{x^2 - 1}{x - 1} = x + 1,$$

$$\Phi_3(x) = \frac{x^3 - 1}{x - 1} = x^2 + x + 1,$$

$$\Phi_6(x) = \frac{x^6 - 1}{(x - 1)(x + 1)(x^2 + x + 1)} = x^2 - x + 1,$$

$$\vdots$$

Since each successive cyclotomic polynomial is obtained by division we see inductively, by the division algorithm for $F[x]$, that $\Phi_n(x) \in F[x]$ for all n. In fact, if $F = \mathbb{Q}$ it follows that $\Phi_n(x)$ has coefficients in $\mathbb{Z}$.

Exercise 4.3. (1) Compute the first twelve cyclotomic polynomials.
(2) Find all n for which $\deg \Phi_n(x) = 2$.

Theorem 4.2. If p is a prime and $n = p^m$, then the cyclotomic polynomial $\Phi_n(x)$ is irreducible in $\mathbb{Q}[x]$.

Proof. Let us write $f(x)$ for $\Phi_n(x)$. Note that

$$\begin{aligned} f(x) &= (x^{p^m} - 1)/(x^{p^{m-1}} - 1) \\ &= x^{p^{m-1}(p-1)} + x^{p^{m-1}(p-2)} + \cdots + x^{p^{m-1}} + 1, \end{aligned}$$

since the roots of $x^{p^{m-1}} - 1$ are roots of unity of orders lower than $n = p^m$. If $f(x)$ is reducible in $\mathbb{Q}[x]$ it is reducible in $\mathbb{Z}[x]$ by Proposition II.5.15, so suppose $f(x)$ factors as $f(x) = g(x)h(x)$ in $\mathbb{Z}[x]$. Since $f(1) = g(1)h(1) = p$, a prime, we may assume that $g(1) = \pm 1$. If $n_1 = 1, n_2, \ldots, n_s$ are the integers between 1 and $n = p^m$ that are relatively prime to p set

$$k(x) = \prod\{g(x^{n_i}) : 1 \le i \le s\}.$$

If $\zeta \in \mathbb{C}$ is any primitive nth root of unity then $\{\zeta^{n_i} : 1 \le i \le s\}$ is the set of all primitive nth roots of unity, i.e., the set of all roots of $f(x)$. Some ζ^{n_j} is a root of $g(x)$, so $k(\zeta) = 0$. Since ζ was arbitrary every root of $f(x)$ is a root of $k(x)$ and hence $f(x) \mid k(x)$ in $\mathbb{Z}[x]$, say $k(x) = f(x) \cdot a(x)$ with $a(x) \in \mathbb{Z}[x]$. But then

$$k(1) = g(1)^s = \pm 1 = f(1)a(1) = p \cdot a(1),$$

a contradiction.

It should be remarked that in fact $\Phi_n(x)$ is irreducible in $\mathbb{Q}[x]$ for *all* n. For a proof see Van der Waerden [37].

Proposition 4.3. Suppose F is a field, $0 < n \in \mathbb{Z}$, and if char $F = p$, then $p \nmid n$. Suppose F contains a primitive nth root ζ of unity and $0 \neq b \in F$. Then a splitting field K for $f(x) = x^n - b$ over F is a simple radical extension of F and the Galois group G of $f(x)$ is abelian.

Proof. Let $a \in K$ be a root for $f(x)$. Then $a, \zeta a, \zeta^2 a, \ldots, \zeta^{n-1}a$ are n distinct roots of $f(x)$ in K, so $K = F(a)$, and $a^n = b \in F$ so K is a simple radical extension. The proof that G is abelian is very similar to the proof of Proposition 4.1 (in fact G is cyclic in this case).

A field K is called an *extension by radicals* of a field F if there is a sequence $L_0 \subseteq L_1 \subseteq L_2 \subseteq \cdots \subseteq L_k$ of fields, with $L_0 = F$ and $L_k = K$, such that L_i is a simple radical extension of L_{i-1}, $1 \leq i \leq n$. A polynomial $f(x) \in F[x]$ is said to be *solvable by radicals* over F if there is an extension K of F by radicals such that $f(x)$ splits over K.

Thus for example if char $F \neq 2$ and deg $f(x) = 2$, then $f(x)$ is solvable by radicals. For $F = \mathbb{Q}$ that fact goes back at least to Euclid, and we have the quadratic formula for expressing the roots of $f(x)$ in terms of its coefficients by means of "algebraic operations," i.e., the usual field operations together with the extraction of square roots (radicals). In the first half of the sixteenth century the Italian mathematicians del Ferro, Tartaglia, and Ferrari obtained formulas for the roots of polynomials of degrees 3 and 4, again in terms of the coefficients via field operations and extraction of roots (see Van der Waerden [37]; also see Section VI. 9). As a consequence polynomials of degree 3 or 4 over $\mathbb{Q}$ (and over many other fields) are solvable by radicals.

The search continued for formulas for the roots of polynomials of degrees higher than 4 until early in the nineteenth century, when P. Ruffini in Italy and N. H. Abel in Norway showed that not only could there be no general formulas for degrees 5 or greater, but that not all polynomials were even solvable by radicals. Galois subsequently set their explanations and examples in the general framework of Galois theory and gave necessary and sufficient conditions for solvability by radicals.

If $F \subseteq E \subseteq K$ and $F \subseteq L \subseteq K$ define the *join* of E and L to be $E \vee L = F(E \cup L)$, the smallest subfield of K that contains both E and L. If G is any group, $J \leq G$, and $H \leq G$, define the *join* of J and H to be $J \vee H = \langle J \cup H \rangle$, the smallest subgroup of G that contains both J and H.

Exercise 4.4. Suppose $F \subseteq E \subseteq K$, $F \subseteq L \subseteq K$, $G = G(K:F)$, $J \leq G$, and $H \leq G$.

(1) Show that $\mathcal{G}(E \vee L) = \mathcal{G}E \cap \mathcal{G}L$ and $\mathcal{F}(J \vee H) = \mathcal{F}J \cap \mathcal{F}H$.
(2) Show that $[E \vee L:F] \leq [E:F][L:F]$.

Proposition 4.4. Suppose F and L are fields, $F \subseteq K_1 \subseteq L$, $F \subseteq K_2 \subseteq L$, and K_1, K_2 are both extensions of F by radicals. Then the join $K_1 \vee K_2$ is also an extension of F by radicals.

Proof. Let $a_1, a_2, \ldots, a_m$ be the "radicals" that are successively adjoined to reach K_1, and let $b_1, b_2, \ldots, b_n$ be the corresponding elements for K_2. Then

$$K_1 \vee K_2 = F(a_1, \ldots, a_m, b_1, \ldots, b_n).$$

Proposition 4.5. If K is a separable extension of F by radicals and L is a Galois closure for K over F, then L is a separable extension of F by radicals.

Proof. Recall (Theorem 3.6) the construction of L as a splitting field over K of $\mathscr{F} = \{m_{a_i}(x)\}$, where $\{a_1, \ldots, a_n\}$ is an F-basis for K. Set $G = G(L:F)$. By Proposition 3.3 $\{\phi(a_i): \phi \in G, 1 \le i \le n\}$ spans L over F. If $G = \{\phi_1, \phi_2, \ldots, \phi_k\}$ set $K_i = \phi_i(K)$, $1 \le i \le k$. Then $L = K_1 \vee K_2 \vee \cdots \vee K_k$, and each K_i is F-isomorphic with K, so K_i is an extension of F by radicals. Thus L is an extension of F by radicals by Proposition 4.4.

Theorem 4.6 (Galois). Suppose char $F = 0$ and $f(x) \in F[x]$ is solvable by radicals. Then the Galois group of $f(x)$ is a solvable group.

Proof. Let $F = L_0 \subseteq L_1 \subseteq \cdots \subseteq L_k = K$ be a sequence of simple radical extensions, with $L_i = L_{i-1}(a_i)$, $a_i^{n_i} \in L_{i-1}$, such that there is a splitting field L for $f(x)$ over F, with $L \subseteq K$. By Proposition 4.5 we may assume that K is Galois over F. Let $G = G(L:F)$, the Galois group of $f(x)$. Since L is Galois over F we have by the Fundamental Theorem of Galois Theory that $G = G(L:F) \cong G(K:F)/G(K:L)$, so by Theorem I.5.4 it will suffice to show that $G(K:F)$ is solvable. Set $n = n_1 n_2 \cdots n_k$, let M be the splitting field for $x^n - 1$ over K, and choose a primitive nth root ζ of unity in M. Note that all n_ith roots of unity are in $F(\zeta)$, $1 \le i \le k$. Since K is Galois over F it is a splitting field for some $g(x) \in F[x]$. But then M is clearly a splitting field over F for $(x^n - 1)g(x)$, so M is Galois over F as well. As above $G(K:F)$ is a homomorphic image of $G(M:F)$, so it will suffice to show that $G(M:F)$ is solvable. To that end set

$$F = M_0, \quad F(\zeta) = M_1, \qquad M_2 = M_1(a_1), \ldots, M_{k+1} = M_k(a_k) = M.$$

Now each M_{i+1} is a splitting field over M_i, $0 \le i \le k$, because the appropriate roots of unity are present after $i = 0$, and hence M_{i+1} is Galois over M_i. Define subgroups $H_0, H_1, \ldots, H_{k+1}$ of $G(M:F)$ by setting $H_i = G(M:M_i)$, so that $H_0 = G(M:F)$ and $H_{k+1} = 1$. Since M is Galois over M_i and M_{i+1} is also Galois over M_i we apply the Fundamental Theorem of Galois Theory to see that

$$G(M:M_{i+1}) = H_{i+1} \lhd G(M:M_i) = H_i,$$

and that $H_i/H_{i+1} \cong G(M_{i+1}:M_i)$, which is abelian for each i by Propositions 4.1 and 4.3. Thus $G(M:F)$ is solvable by Theorem I.5.3, and the theorem is proved.

For a concrete example set $f(x) = x^5 + 5x^3 - 20x + 5 \in \mathbb{Q}[x]$. Then $f(x)$ is irreducible in $\mathbb{Z}[x]$ by the Eisenstein Criterion (Theorem II.5.18), and hence also in $\mathbb{Q}[x]$ by Proposition II.5.15. Since $f'(x) = 5(x^2 + 4)(x^2 - 1)$ we see by inspection of the graph of $f(x)$ that $f(x)$ has exactly three real roots a_1, a_2, and a_3. Thus it has two nonreal roots a_4 and a_5 in $\mathbb{C}$, which are necessarily complex conjugates of one another. Let $K \subseteq \mathbb{C}$ be a splitting field for $f(x)$ over $\mathbb{Q}$, and let $G = G(K:\mathbb{Q})$ be the Galois group of $f(x)$, viewed as a subgroup of S_5. Since $5 \mid [K:\mathbb{Q}]$, and hence $5 \mid |G|$, there must be a 5-cycle in G. By Theorem 3.5(c) the complex conjugation automorphism of $\mathbb{A}$ restricts to an automorphism of K. Its effect on the roots of $f(x)$ is to fix a_1, a_2, and a_3, and to interchange a_4 and a_5, so G also contains a 2-cycle. It follows (see Exercise I.12.21) that $G = S_5$. But S_5 is not solvable since $S_5' = A_5$, which is simple but not abelian. Thus $f(x)$ is not solvable by radicals.

Exercise 4.5. Show that $f(x) = x^5 - 2x^3 - 8x + 2$ is not solvable by radicals over $\mathbb{Q}$.

Suppose F is a field containing a primitive nth root of unity, $K = F(a)$ is a simple Galois extension with $[K:F] = n$, and the Galois group $G = G(K:F)$ is cyclic, say $G = \langle\phi\rangle$ of order n. If ζ is any nth root of unity in F define the *Lagrange resolvent* of ζ and a to be

$$L(\zeta, a) = a + \zeta\phi(a) + \zeta^2\phi^2(a) + \cdots + \zeta^{n-1}\phi^{n-1}(a).$$

Exercise 4.6. If F, K, and G are as above and $L(\zeta, a)$ is a Lagrange resolvent show that

$$\phi(L(\zeta, a)) = L(\zeta, \phi(a)) = \zeta^{-1}L(\zeta, a),$$

and conclude that the nth power of $L(\zeta, a)$ is in F.

Proposition 4.7. Suppose F is a field containing a primitive nth root ζ of unity, $K = F(a)$ is a Galois extension of F with $[K:F] = n$, and the Galois group $G = G(K:F)$ is cyclic. Then the Lagrange resolvent $L(\zeta^i, a)$ lies in $K\backslash F$ for some i.

Proof. Say $G = \langle\phi\rangle$. Then

$$\sum_{i=0}^{n-1} L(\zeta^i, a) = \sum_{i=0}^{n-1}\sum_{j=0}^{n-1} \zeta^{ij}\phi^j(a) = \sum_{j=0}^{n-1} \phi^j(a) \sum_{i=0}^{n-1} (\zeta^j)^i = na$$

(see Exercise 4.2). Since F has a primitive nth root of unity the characteristic of F cannot divide n, so $n1 \neq 0$ in F. Thus $na \in K\backslash F$ since $a \notin F$, and consequently at least one of the summands $L(\zeta^i, a)$ must lie in $K\backslash F$.

Proposition 4.8. Suppose $p \in \mathbb{Z}$ is a prime, F is a field containing a primitive pth root of unity, and K is a Galois extension of F, with $[K:F] = p$. Then K is a simple radical extension of F.

Proof. Choose $a \in K \backslash F$. Then $K = F(a)$ since $[K:F]$ is prime. The Galois group $G = G(K:F)$ is cyclic since $|G| = [K:F] = p$. By Proposition 4.7 there is a Lagrange resolvent $b \in K \backslash F$, so $K = F(b)$, and by Exercise 4.6 we have $b^p = c \in F$.

Proposition 4.9. Suppose $f(x) \in F[x]$ has Galois group G over F, and E is an extension field of F. Then the Galois group of $f(x)$ over E is isomorphic with a subgroup of G.

Proof. Say L is a splitting field for $f(x)$ over E, and that $f(x)$ has roots $a_1, a_2, \ldots, a_n$ in L. Then $K = F(a_1, \ldots, a_n)$ is a splitting field for $f(x)$ over F. If $\phi \in G(L:E)$, then ϕ permutes the roots $a_1, \ldots, a_n$, so $\phi(K) = K$, and ϕ acts as the identity on K only if ϕ fixes each root a_i, which is only if $\phi = 1 \in G(L:E)$. Thus $\phi \mapsto \phi | K$ is a monomorphism into $G(K:F)$.

Theorem 4.10 (Galois). Suppose F is a field of characteristic 0, $f(x) \in F[x]$, and the Galois group of $f(x)$ is solvable. Then $f(x)$ is solvable by radicals over F.

Proof. Let K be a splitting field for $f(x)$ over F, set $G = G(K:F)$, and say $[K:F] = n$. Let L be a splitting field over K for $x^n - 1$, and let $\zeta \in L$ be a primitive nth root of unity, so $L = K(\zeta)$. Set $E = F(\zeta)$, then clearly L is a splitting field for $f(x)$ over E. If we set $H = G(L:E)$, then H is isomorphic with a subgroup of G by Proposition 4.9, and hence H is also solvable. By Theorem I.5.3 H has a subnormal series $H = H_0 \geq H_1 \geq \cdots \geq H_k = 1$ with abelian factors, and by refining to a composition series we may assume that H_{i-1}/H_i is cyclic of prime order p_i, $1 \leq i \leq k$. In the setting of $E \subseteq L$ we set $L_i = \mathscr{F}H_i$, $0 \leq i \leq k$, so $E = L_0 \subseteq L_1 \subseteq \cdots \subseteq L_k = L$ and $[L_i:L_{i-1}] = p_i$. Since $G(L:L_i) = H_i \lhd H_{i-1} = G(L:L_{i-1})$ we have L_i Galois over L_{i-1}, and L_{i-1} contains a primitive p_ith root of unity (it is a power of ζ). By Proposition 4.8 L_i is a simple radical extension of L_{i-1}, $1 \leq i \leq k$. Thus L is an extension of E by radicals, and hence also of F since $E = F(\zeta)$.

5. SYMMETRIC FUNCTIONS

If F is any field let $x_1, x_2, \ldots, x_n$ be distinct indeterminates over F and set $K = F(x_1, x_2, \ldots, x_n)$, the field of rational functions.

$$\text{Set } f(x) = (x - x_1)(x - x_2)\cdots(x - x_n) \in K[x];$$

$f(x)$ is called the *general polynomial* of degree n. Multiply the linear factors of $f(x)$ together to obtain

$$f(x) = x^n - \sigma_1 x^{n-1} + \sigma_2 x^{n-2} - \cdots + (-1)^n \sigma_n,$$

so that

$$\begin{aligned}
\sigma_1 &= x_1 + x_2 + x_3 + \cdots + x_n, \\
\sigma_2 &= x_1x_2 + x_1x_3 + \cdots + x_{n-1}x_n = \sum\{x_ix_j : 1 \le i < j \le n\}, \\
&\vdots \\
\sigma_n &= x_1x_2x_3\cdots x_n.
\end{aligned}$$

In general

$$\sigma_k = \sum\{x_{i_1}x_{i_2}\cdots x_{i_k} : 1 \le i_1 < i_2 < \cdots < i_k \le n\},$$

a homogeneous polynomial of degree k in $x_1, x_2, \ldots, x_n$. The coefficients σ_k are called the *elementary symmetric polynomials* in $x_1, x_2, \ldots, x_n$.

Exercise 5.1. (1) Write out the elementary symmetric polynomials for $n = 3, 4$, and 5.
(2) How many monomial summands are there in σ_k?

Each element ϕ of the symmetric group S_n defines an F-automorphism of K by permuting the indeterminates $x_1, \ldots, x_n$ as follows: if $g(x_1, x_2, \ldots, x_n) \in K$, then

$$\phi g(x_1, x_2, \ldots, x_n) = g(x_{\phi(1)}, x_{\phi(2)}, \ldots, x_{\phi(n)}).$$

For example if $\phi = (132)$ and $g(x_1, x_2, x_3) = x_1^2 - x_2x_3 + x_3^3$, then

$$\phi g(x_1, x_2, x_3) = g(x_3, x_1, x_2) = x_3^2 - x_1x_2 + x_2^3.$$

Set $F_0 = F(\sigma_1, \sigma_2, \ldots, \sigma_n) \subseteq K$ and let F_1 be the fixed field in K of S_n. Clearly $F_0 \subseteq F_1$ since each σ_k is invariant under permutations of $x_1, x_2, \ldots, x_n$. The elements of F_1 are called the *symmetric* rational functions in $x_1, x_2, \ldots, x_n$ over F. Note that $f(x) \in F_0[x]$, and that $K = F_0(x_1, x_2, \ldots, x_n)$, so K is a splitting field for $f(x)$ over F_0. Thus $[K:F_0] \le n!$ by the corollary to Proposition 1.8. Since $f(x)$ has distinct roots, K is Galois over F_0 by Theorem 3.4. Since $G(K:F_1) \ge S_n$ we have $|G(K:F_1)| = [K:F_1] \ge n!$, and so $n! \ge [K:F_0] \ge [K:F_1] \ge n!$ and equality holds throughout. Consequently $F_0 = F_1$ and every symmetric rational function can be represented as a rational function in the elementary symmetric polynomials $\sigma_1, \sigma_2, \ldots, \sigma_n$ (this statement will be refined later, in Theorem 5.3). Also the Galois group $G = G(K:F_0)$ of the general polynomial $f(x)$ is the symmetric group S_n. The next theorem is an immediate consequence of this fact and Theorem 4.6.

Theorem 5.1 (Ruffini, Abel). If F is any field of characteristic 0 and $n \ge 5$, then the general polynomial of degree n is not solvable by radicals over the field of symmetric rational functions in $x_1, \ldots, x_n$ over F.

Exercise 5.2. Show that the general polynomial $f(x)$ is irreducible in $F_0[x]$.

Theorem 5.2. If H is any finite group then there exist a field L and a finite Galois extension K of L such that H is (isomorphic with) the Galois group $G(K:L)$.

Proof. By Cayley's Theorem (I.2.2) H is isomorphic with a subgroup H_1 of a symmetric group S_n (we may take $n = |H|$). Take F, F_0, and K as above so that $G(K:F_0) = S_n$. Set $L = \mathscr{F}H_1$. Then by the Fundamental Theorem of Galois Theory $H_1 = \mathscr{G}L = G(K:L)$.

It is a famous unsolved problem to determine whether every finite group H occurs as the Galois group of a finite Galois extension of the rational field $\mathbb{Q}$. Shafarevich [34] gave an affirmative solution in 1954 for solvable groups H, but the general question remains open.

We describe next a constructive procedure for expressing arbitrary polynomials in $x_1, x_2, \ldots, x_n$ in a particular form involving the elementary symmetric polynomials $\sigma_1, \sigma_2, \ldots, \sigma_n$. It will be convenient to replace the field F by an arbitrary commutative ring R with 1. Note that $\sigma_1, \ldots, \sigma_n$ can be viewed as elements of $R[x_1, \ldots, x_n]$. As usual each $\phi \in S_n$ acts on elements of $R[x_1, \ldots, x_n]$ by permuting the indeterminates, and we say that $g(x_1, \ldots, x_n)$ is a *symmetric* polynomial over R if $\phi g(x_1, \ldots, x_n) = g(x_1, \ldots, x_n)$ for all $\phi \in S_n$.

Set $f_n(x) = f(x)$, the general polynomial of degree n, with coefficients in $R[x_1, \ldots, x_n]$, and note that $f_n(x)$ is monic of degree n, has x_n as a root, and has $1, -\sigma_1, \sigma_2, \ldots$ as its coefficients. Next set

$$f_{n-1}(x) = f_n(x)/(x - x_n).$$

On the one hand, of course,

$$f_{n-1}(x) = (x - x_1)\cdots(x - x_{n-1}).$$

On the other hand if we view $f_n(x)$ as $x^n - \sigma_1 x^{n-1} + \cdots$ we may obtain $f_{n-1}(x)$ by the division algorithm for polynomials with coefficients in the ring $R[\sigma_1, \ldots, \sigma_n, x_n]$. Observe that $f_{n-1}(x)$ is monic of degree $n - 1$, has x_{n-1} as a root, and has coefficients that are various polynomials over R in $\sigma_1, \sigma_2, \ldots, \sigma_n$, and x_n.

$$\begin{array}{r|l} & x^{n-1} + (x_n - \sigma_1)x^{n-2} + (\sigma_2 + (x_n - \sigma_1)x_n)x^{n-3} + \cdots \\ \hline x - x_n & x^n - \sigma_1 x^{n-1} + \sigma_2 x^{n-2} - \sigma_3 x^{n-3} + \cdots \end{array}$$

Next set

$$f_{n-2}(x) = f_{n-1}(x)/(x - x_{n-1}),$$

and in general set

$$f_k(x) = f_{k+1}(x)/(x - x_{k+1}), \qquad 1 \le k \le n - 1.$$

Just as for $f_n(x)$ and for $f_{n-1}(x)$ we see in general that $f_k(x)$ is monic of degree k, has x_k as a root, and has coefficients that are polynomials over R in $\sigma_1, \ldots, \sigma_n, x_n, x_{n-1}, \ldots, x_{k+1}$.

For example, if $n = 3$ we have

$$f_2(x) = (x - x_1)(x - x_2) = x^2 + (-\sigma_1 + x_3)x + (\sigma_2 - x_3\sigma_1 + x_3^2)$$

and

$$f_1(x) = x - x_1 = x + (-\sigma_1 + x_2 + x_3).$$

If we now set $x = x_k$ in $f_k(x)$ we may solve for x_k^k as a polynomial of degree $k - 1$ or less evaluated at x_k, the coefficients being polynomials over R in $\sigma_1, \ldots, \sigma_n, x_n, \ldots, x_{k+1}$. Taking $n = 3$ again for an example we have

$$\begin{aligned} x_1 &= \sigma_1 - x_2 - x_3, \\ x_2^2 &= (\sigma_1 - x_3)x_2 - (\sigma_2 - x_3\sigma_1 + x_3^2), \\ x_3^3 &= \sigma_1 x_3^2 - \sigma_2 x_3 + \sigma_3. \end{aligned}$$

Given any polynomial

$$g(x_1, \ldots, x_n) \in R[x_1, x_2, \ldots, x_n]$$

we may replace each x_1 by $\sigma_1 - x_2 - \cdots - x_n$, so that x_1 no longer appears. Then we may replace each x_2^2 by its expression in terms of x_2 (to the first power only), $\sigma_1, \ldots, \sigma_n, x_n, \ldots, x_3$ so that x_2 finally appears only to the first power, if at all. Continuing, we may make replacements so that x_k appears only to the power $k - 1$ or less for each k. Ultimately we will have expressed the original polynomial $g(x_1, \ldots, x_n)$ as a sum of terms, each having the form

$$h(\sigma_1, \ldots, \sigma_n)x_2^{k_2}x_3^{k_3} \cdots x_n^{k_n},$$

where $0 \le k_2 < 2, 0 \le k_3 < 3, \ldots, 0 \le k_n < n$, and $h(\sigma_1, \ldots, \sigma_n)$ is a polynomial in n variables over R evaluated at $\sigma_1, \ldots, \sigma_n$.

For example, with $n = 3$ again, if

$$g(x_1, x_2, x_3) = x_1^2 - x_2x_3 + x_3^3,$$

then the process above gives

$$g(x_1, x_2, x_3) = \sigma_1 x_3^2 - (\sigma_1 + \sigma_2)x_3 - \sigma_1 x_2 + (\sigma_1^2 - \sigma_2 + \sigma_3).$$

Exercise 5.3. Apply the above process to

(1) $g(x_1, x_2, x_3) = x_1^2 + x_2^2 + x_3^2$, and

(2) $g(x_1, x_2, x_3) = x_1 - x_2^2 + x_3^3$.

Theorem 5.3 (The Fundamental Theorem on Symmetric Polynomials). Suppose R is a commutative ring with 1 and

$$g(x_1, \ldots, x_n) \in R[x_1, \ldots, x_n]$$

is a symmetric polynomial. Then there is a polynomial $f(x_1,\dots,x_n)$ in $R[x_1,\dots,x_n]$ such that $g(x_1,\dots,x_n) = f(\sigma_1,\dots,\sigma_n)$, i.e., $g(x_1,\dots,x_n)$ is a polynomial over R in the elementary symmetric polynomials.

Proof. By the procedure above we may write $g(x_1,\dots,x_n)$ as a sum of monomials $h(\sigma_1,\dots,\sigma_n)x_2^{k_2}\cdots x_n^{k_n}$. Since $g(x_1,\dots,x_n)$ is symmetric it is invariant under the various transpositions $(12),(13),\dots,(1n)$, from which it follows that $k_2 = k_3 = \cdots = k_n = 0$ since x_1 does not appear in any of the monomials and each coefficient $h(\sigma_1,\dots,\sigma_n)$ is symmetric. The theorem follows.

Theorem 5.4. Suppose

$$f(x) = b_0 + b_1 x + \cdots + b_n x^n \in \mathbb{Z}[x]$$

has roots $a_1, a_2, \dots, a_n$ in $\mathbb{C}$, and suppose $g(x_1, x_2, \dots, x_n)$ is a symmetric polynomial over $\mathbb{Q}$ of degree d. Then $g(a_1, a_2, \dots, a_n) \in \mathbb{Q}$, and if $g(x_1,\dots,x_n)$ has coefficients in $\mathbb{Z}$ then $b_n^d g(a_1,\dots,a_n) \in \mathbb{Z}$.

Proof. Note that

$$f(x) = b_n \cdot \prod\{(x - a_i): 1 \le i \le n\},$$

so $b_i = \pm b_n \sigma_i(a_1,\dots,a_n)$ for each i, and hence $\sigma_i(a_1,\dots,a_n) = \pm b_i/b_n \in \mathbb{Q}$. Thus $g(a_1,\dots,a_n) \in \mathbb{Q}$ by Theorem 5.3. If we set $h(x) = f(x/b_n)$, then the roots of $h(x)$ are $b_n a_1, b_n a_2, \dots, b_n a_n$. Note that $b_n^{n-1}h(x)$ is monic in $\mathbb{Z}[x]$, say

$$\begin{aligned} b_n^{n-1}h(x) &= c_0 + c_1 x + \cdots + c_{n-1}x^{n-1} + x^n \\ &= \prod\{(x - b_n a_i): 1 \le i \le n\} \in \mathbb{Z}[x], \end{aligned}$$

and so $\sigma_i(b_n a_1,\dots,b_n a_n) = \pm c_i \in \mathbb{Z}$ for each i. If

$$g(x_1,\dots,x_n) \in \mathbb{Z}[x_1,\dots,x_n],$$

then it is a sum of homogeneous symmetric polynomials over $\mathbb{Z}$ of various degrees $r \le d$. If $k(x_1,\dots,x_n)$ is one of those homogeneous summands, then

$$b_n^d k(a_1,\dots,a_n) = b_n^{d-r} k(b_n a_1,\dots,b_n a_n) \in \mathbb{Z},$$

again by Theorem 5.3, and so $b_n^d g(a_1,\dots,a_n) \in \mathbb{Z}$.

Recall that an *algebraic number* is a complex number that is algebraic over $\mathbb{Q}$. All other complex numbers are called *transcendental*.

Theorem 5.5 (Lindemann, 1882). The real number π is transcendental.

Proof (Niven [28]). If not, then $i\pi$ is also algebraic (where $i^2 = -1$) and there is an irreducible polynomial $f_1(x) \in \mathbb{Z}[x]$ such that $f_1(i\pi) = 0$. Let $a_1 = i\pi, a_2, \dots, a_n$ be all the roots of $f_1(x)$ in $\mathbb{C}$. Since $e^{i\pi} = -1$ we have

$$\prod\{(1 + e^{a_j}): 1 \le j \le n\} = 0.$$

Multiply to see that

$$0 = 1 + e^{a_1} + \cdots + e^{a_n} + \sum_{i<j} e^{a_i + a_j} + \cdots + e^{a_1 + a_2 + \cdots + a_n}. \qquad (*)$$

Set

$$g_2(x) = \prod\{(x - (a_i + a_j)) : 1 \le i < j \le n\}.$$

If $\phi \in S_n$ apply ϕ to $g_2(x)$ in its factored form by permuting the subscripts of the a_is, and note that each $\phi \in S_n$ leaves $g_2(x)$ invariant. Thus the coefficients of $g_2(x)$ are symmetric polynomials (over $\mathbb{Z}$) in $a_1, \ldots, a_n$, and hence $g_2(x) \in \mathbb{Q}[x]$ by Theorem 5.4. Multiply by a suitable integer to obtain $f_2(x) \in \mathbb{Z}[x]$ having all $a_i + a_j$ as its roots, $1 \le i < j \le n$. Similarly we may obtain $f_3(x) \in \mathbb{Z}[x]$ having all $a_i + a_j + a_k$ as its roots, $1 \le i < j < k \le n$, etc. Thus

$$\prod\{f_j(x) : 1 \le j \le n\} \in \mathbb{Z}[x]$$

has as its roots all of the exponents $a_i, a_i + a_j$, etc., in (*) above. Suppose $\prod\{f_j(x) : 1 \le j \le n\}$ has 0 as a root with multiplicity $k - 1$, and set

$$\begin{aligned} f(x) &= \left(\prod\{f_j(x) : 1 \le j \le n\}\right)/x^{k-1} \\ &= c_0 + c_1 x + \cdots + c_r x^r \in \mathbb{Z}[x], \end{aligned}$$

$c_0 \ne 0$, $c_r \ne 0$. Thus the roots $b_1, b_2, \ldots, b_r \in \mathbb{C}$ of $f(x)$ are just the nonzero exponents in (*), and (*) becomes

$$e^{b_1} + e^{b_2} + \cdots + e^{b_r} + k = 0,$$

with $0 < k \in \mathbb{Z}$.

Let $p \in \mathbb{Z}$ be some prime and set

$$g(x) = (c_r^{rp-1}/(p-1)!)x^{p-1}(f(x))^p.$$

Then set

$$h(x) = \sum\{g^{(j)}(x) : 0 \le j \le p(r+1)\},$$

where $g^{(0)}(x) = g(x)$ and $g^{(j)}(x)$ is the jth derivative of $g(x)$.

Exercise 5.4. (1) Show that

$$\frac{d}{dx}(e^{-x}h(x)) = -e^{-x}g(x).$$

(2) Show that

$$\begin{aligned} g(0) = g'(0) = \cdots = g^{(p-2)}(0) &= 0, \\ g^{(p-1)}(0) &= c_r^{rp-1}c_0^p, \end{aligned}$$

and

$$g^{(j)}(0) \in p\mathbb{Z} \qquad \text{if} \quad p \leq j \leq p + rp - 1.$$

(3) If $p > \max\{k, c_0, c_r\}$ conclude that $k \cdot h(0) \in \mathbb{Z}$ and $(k \cdot h(0), p) = 1$.

Continuing the proof we apply Exercise 5.4.1 and the Fundamental Theorem of Calculus to see that

$$-\int_0^x e^{-t} g(t)\, dt = e^{-x} h(x) - h(0).$$

Change variables, $t = sx$, and get

$$-x \int_0^1 e^{-sx} g(sx)\, ds = e^{-x} h(x) - h(0)$$

or

$$h(x) - e^x h(0) = -x \int_0^1 e^{(1-s)x} g(sx)\, ds.$$

Substitute $x = b_1, b_2, \ldots, b_r$ in succession and add to obtain

$$\begin{aligned} \sum_{j=1}^{r} (h(b_j) - e^{b_j} h(0)) &= \sum_{j=1}^{r} h(b_j) + k \cdot h(0) \\ &= -\sum_{j=1}^{r} b_j \int_0^1 e^{(1-s)b_j} g(sb_j)\, ds. \end{aligned}$$

Note next that

$$g(b_j) = g'(b_j) = \cdots = g^{(p-1)}(b_j) = 0 \qquad \text{for all} \quad j,$$

since b_j is a root of multiplicity p, and that $(p-1)!g(x) \in \mathbb{Z}[x]$. The pth and higher derivatives of $(p-1)!g(x)$ are all in $\mathbb{Z}[x]$, and each has all its coefficients divisible by p or more consecutive integers. It follows (Why?) that each has all its coefficients divisible by $p!$, and so all $g^{(j)}(x)$, $j \geq p$, are in $p\mathbb{Z}[x]$.

For each $j \geq p$, $\sum_{m=1}^{r} g^{(j)}(b_m)$ is a symmetric polynomial over $\mathbb{Z}$ in $b_1, \ldots, b_r$. By definition of $g(x)$ each coefficient of $g^{(j)}(x)$ is divisible by c_r^{rp-1}, and $\deg g^{(j)}(x) \leq rp - 1$. By Theorem 5.4 and the observations above

$$\sum_{m=1}^{r} g^{(j)}(b_m) = pk_j \in p\mathbb{Z} \qquad \text{for all} \quad j \geq p.$$

Thus

$$\sum_{m=1}^{r} h(b_m) = \sum_{m=1}^{r} \sum_{j=0}^{p(r+1)} g^{(j)}(b_m) = \sum_{j=p}^{p(r+1)} pk_j \in p\mathbb{Z}.$$

Since $k \cdot h(0) \in \mathbb{Z}$ and $p \nmid k \cdot h(0)$ we conclude that

$$0 \neq \sum_{j=1}^{r} h(b_j) + k \cdot h(0) \in \mathbb{Z}.$$

But

$$\int_0^1 e^{(1-s)b_j} g(sb_j)\, ds = \int_0^1 e^{(1-s)b_j} \frac{c_r^{rp-1}}{(p-1)!} b_j^{p-1} s^{p-1} f(sb_j)^p\, ds. \qquad (**)$$

Note that

$$\left|e^{(1-s)b_j}\right| \leq e^{|b_j|} \qquad \text{for} \quad 0 \leq s \leq 1,$$

and that

$$\left|\frac{c_r^{rp-1}}{(p-1)!} b_j^{p-1} s^{p-1} f(sb_j)^p\right| \leq \frac{1}{|b_j c_r|} \cdot \frac{|c_r^r b_j \cdot f(sb_j)|^p}{(p-1)!}.$$

If M is an upper bound for $|c_r^r b_j f(sb_j)|$, $0 \leq s \leq 1$, then $\lim_{p \to \infty} M^p/(p-1)! = 0$ (e.g., since the series $\sum M^{n+1}/n!$ converges by the ratio test). It follows that each integrand in (**) converges uniformly to 0 as $p \to \infty$, and hence the integral can be made as small as we like by a suitable choice of p. That, of course, is incompatible with the sum of the integrals being a nonzero integer, and the theorem is proved.

6. GEOMETRICAL CONSTRUCTIONS

The ancient Greeks left unsolved several problems involving geometrical constructions. Their point of view was that the only tools allowed for the construction of geometrical figures are compasses and (unmarked) straightedges. It will be convenient to take the point of view of analytic geometry in discussing what can and what can not be constructed.

If we choose two points in the plane we may label one as the origin $(0, 0)$ and the other as $(1, 0)$, thus choosing a unit distance and determining an x-axis. With a straightedge we may draw line segments joining previously constructed points. With a compass we may draw circles having previously constructed centers and radii. We say that a point is *constructible* if it can be obtained as a point of intersection of such lines and/or circles. Recall from high school geometry that a perpendicular to a line can be drawn at a given point on that line. Clearly then any point (a, b), with $a, b \in \mathbb{Z}$, is constructible.

A real number is called *constructible* if it appears as a coordinate for a constructible point in the plane. Denote by K the set of all constructible real numbers. Thus for example $\mathbb{Z} \subseteq K$.

Proposition 6.1. The set K of constructible real numbers is a field.

Proof. Take $a, b \in K$, with $b \neq 0$. The following pictures (Fig. 1) indicate how to construct $a + b$, $-b$, ab, and a/b, and hence to conclude that K is a field.

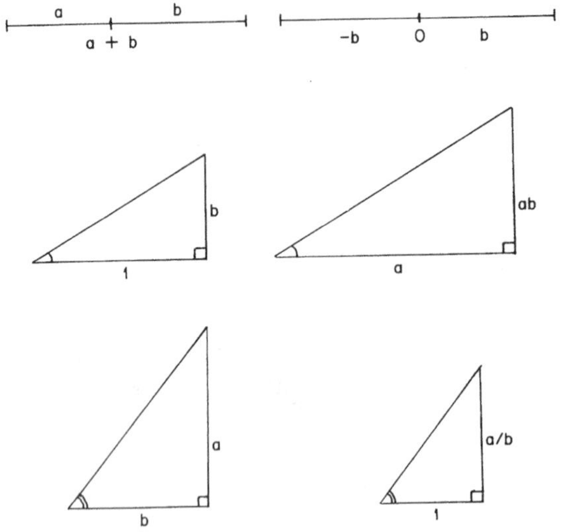

Figure 1

In order to describe the field K somewhat more explicitly let us suppose we have a field L, with $\mathbb{Q} \subseteq L \subseteq K$, and ask what sort of numbers in K can be obtained by constructions using points with coordinates in L (think of the elements in L as having been constructed previously). Intersecting two "L-lines" amounts to solving two linear equations with coefficients in L, so the solutions will be in L and no new numbers are obtained. Suppose we intersect two "L-circles", i.e., solve the following equations simultaneously:

$$(x - a)^2 + (y - b)^2 = u,$$
$$(x - c)^2 + (y - d)^2 = v,$$

where $a, b, c, d, u, v \in L$. If the squares are expanded and the second equation is subtracted from the first we obtain

$$2(c - a)x + 2(d - b)y = u - v - a^2 - b^2 + c^2 + d^2,$$

the equation of an L-line whose intersection with either circle is the same as the intersection of the two circles. Thus it suffices to consider the result of intersecting a line and a circle, say

$$(x-a)^2+(y-b)^2=u \qquad \text{and} \qquad cx+dy=v,$$

with $a, b, c, d, u, v \in L$ and either $c \neq 0$ or $d \neq 0$. We may solve for x or y in the linear equation, substitute into the quadratic equation, and then solve the quadratic equation. Ultimately then the new numbers obtained are roots of a quadratic equation and hence are either in L or in an extension L' having degree 2 over L.

We have established half of the next theorem.

Theorem 6.2. A real number a is constructible if and only if there is a sequence $L_0 \subseteq L_1 \subseteq L_2 \subseteq \cdots \subseteq L_k$ of subfields of $\mathbb{R}$ such that $L_0 = \mathbb{Q}$, $[L_i : L_{i-1}] = 2$ for $1 \leq i \leq k$, and $a \in L_k$.

Proof. It will suffice to show that a square root can be constructed for any constructible number, and we appeal again to a picture (Fig. 2) to indicate the construction.

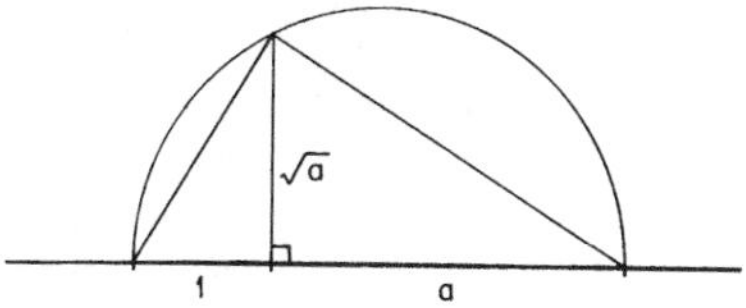

Figure 2

Exercise 6.1. (1) Verify the constructions above for ab, a/b, and $\sqrt{a}$.

(2) Construct the number $1+(2+\sqrt{3})^{1/2}$ with straightedge and compass.

Perhaps the best known construction problem left by the Greeks was that of trisecting an arbitrary angle. Note that an angle ϕ can be constructed if and only if $\cos\phi$ is a constructible number (see Fig. 3).

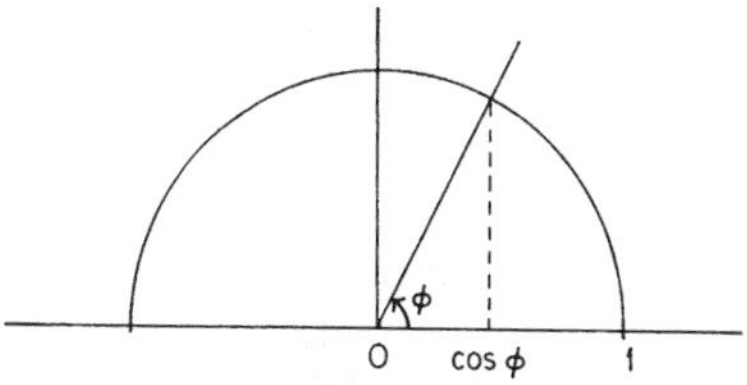

Figure 3

For the trisection problem we suppose that we are given an angle θ, and we wish to construct the angle $\phi = \theta/3$, so we assume $\cos\theta = \cos 3\phi$ as known (and constructible), and ask whether $\cos\phi$ is constructible. By elementary trigonometric identities we have $\cos\theta = 4\cos^3\phi - 3\cos\phi$, so if we set $a = \cos\theta$, then $x = 2\cos\phi$ is a root of $f(x) = x^3 - 3x - 2a$. If $f(x)$ is reducible [over $\mathbb{Q}(a)$], then its roots lie either in $\mathbb{Q}(a) \subseteq K$ or in an extension of degree 2 over $\mathbb{Q}(a)$, and hence are constructible by Theorem 6.2. However, if $f(x)$ is irreducible over $\mathbb{Q}(a)$, then each of its roots lies in an extension of degree 3 over $\mathbb{Q}(a)$, so the roots cannot be constructible, by Theorem 6.2 and Proposition 1.1.

For example, if $\theta = \pi/2$, then $a = \cos(\pi/2) = 0$, so $f(x) = x(x^2 - 3)$ and θ can be trisected. However, if $\theta = \pi/3$, then $a = \cos(\pi/3) = \frac{1}{2}$, and $f(x) = x^3 - 3x - 1$, which is irreducible over $\mathbb{Q}$, so $\pi/3$ can not be trisected. It follows in particular that there can be no general procedure for trisecting angles with compass and straightedge.

It is of some interest to observe that with a very slight relaxation of the construction rules any given angle can be trisected. All that is needed is a straightedge that has a point marked at unit distance from one end. Suppose the angle θ has its vertex at the origin and one side along the x axis. Draw a unit circle centered at the origin and slide the marked straightedge so that its end is on the x axis, the unit mark is on the circle, and the edge passes through the point where the other side of the angle meets the circle, as pictured in Fig. 4. If the resulting angles are labeled as in Fig. 4, then $\alpha + \alpha + (\pi - \beta) = \pi$, so $\beta = 2\alpha$, and also $\beta + \beta + (\pi - \alpha - \theta) = \pi$, and hence $\theta = 2\beta - \alpha = 3\alpha$.

The problem of duplication (i.e., doubling the volume) of the cube is very ancient and has an extensive history. An oracle on the island of Delos apparently suggested in about 430 B.C. that a plague might be lifted if a new altar for Apollo were constructed having the same cubical shape as the old altar but of twice the size (i.e., volume). As a result the problem is often referred to as the *Delian problem*. If the old altar is assumed to have had an edge of length 1, then the assignment was to construct a new edge of length a such that $a^3 = 2$, or $a = \sqrt[3]{2}$. Since $x^3 - 2$ is irreducible in $\mathbb{Q}[x]$ we have $[\mathbb{Q}(\sqrt[3]{2}):\mathbb{Q}] = 3$, and hence $\sqrt[3]{2}$ is not constructible by Theorem 6.2 and Proposition 1.1.

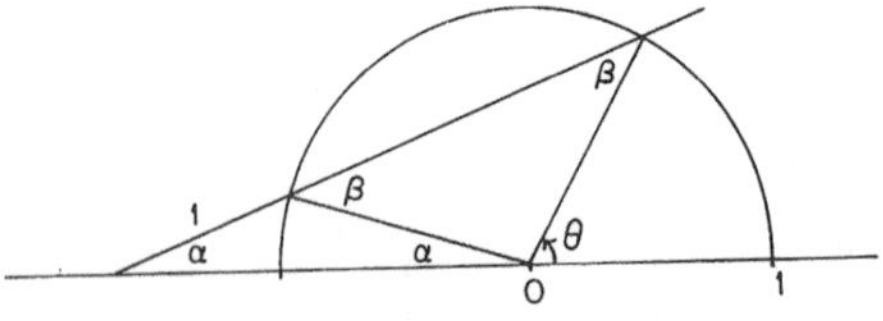

Figure 4

Exercise 6.2. A straightedge with a point marked at unit distance from one end can also be used to solve the Delian problem. Let AB be a segment of unit length. Drop a perpendicular ray from A and another ray from A inclined at angle 120° to AB. Slide the straightedge with its end on the latter ray and passing through B until the unit mark lies on the perpendicular to AB, say at point P. Show that then PB has length $\sqrt[3]{2}$ (see Fig. 5).

Each side of a regular polygon with n sides subtends a central angle of $2\pi/n$, so a regular n-gon can be constructed with straightedge and compass if and only if $\cos(2\pi/n)$ is a constructible number. Note for example that a regular 18-gon can not be constructed since its central angle would give a trisection of $\pi/3$.

If $n = 2^k, k \geq 3$, the regular n-gon is easily constructed by bisecting angles after a square has been constructed. If $n = ab$, with $(a, b) = 1$, and if an n-gon has been constructed we need only join every bth vertex to construct an a-gon. On the other hand if both the a-gon and the b-gon are constructible write $1 = ra + sb$, with $r, s \in \mathbb{Z}$, so that $2\pi/n = 2\pi r/b + 2\pi s/a$, to see that the n-gon is also constructible. As a result we need only consider the constructibility of a regular n-gon in case $n = p^k$, p an odd prime. Most of the following discussion is valid for arbitrary n, however.

Let $\zeta = e^{2\pi i/n}$, a primitive nth root of unity in $\mathbb{C}$. Then $\zeta = \cos(2\pi/n) + i\sin(2\pi/n)$ and $\bar{\zeta} = \zeta^{-1} = \cos(2\pi/n) - i\sin(2\pi/n)$, so $\cos(2\pi/n) = (\zeta + \bar{\zeta})/2$. We conclude that a regular n-gon is constructible if and only if $[\mathbb{Q}(\zeta + \bar{\zeta}):\mathbb{Q}]$ is a power of 2.

First consider $\mathbb{Q}(\zeta)$. It is a Galois extension of $\mathbb{Q}$, being a splitting field for $x^n - 1$, and we know that $[\mathbb{Q}(\zeta):\mathbb{Q}] = \phi(n)$ by Theorem 4.2. Set $L = \mathbb{Q}(\zeta + \bar{\zeta}) \subseteq \mathbb{Q}(\zeta)$ and $H = \mathcal{G}L$. If $\theta \in H$, say with $\theta(\zeta) = \zeta^r$, $1 \leq r < n$, then $\theta(\zeta + \bar{\zeta}) = \zeta^r + \zeta^{-r} = \zeta + \bar{\zeta}$ and so $\cos(2\pi r/n) = \cos(2\pi/n)$, which holds only for $r = 1$ and $r = n - 1$ (look at the graph of the cosine function). Thus either

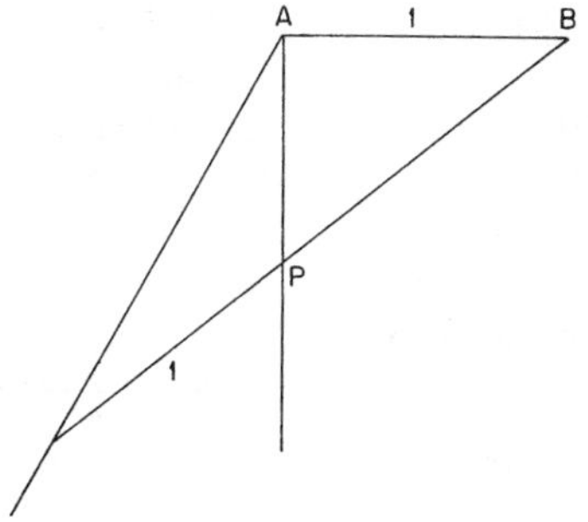

Figure 5

$\theta(\zeta) = \zeta$ (so $\theta = 1$) or $\theta(\zeta) = \bar{\zeta}$, and $|H| = 2$. Consequently $[\mathbb{Q}(\zeta + \bar{\zeta}):\mathbb{Q}] = \phi(n)/2$, and the regular n-gon is constructible if and only if $\phi(n)$ is a power of 2.

Assume that $n = p^k$, p an odd prime. Then $\phi(n) = \phi(p^k) = p^{k-1}(p-1)$, which is a power of 2 if and only if $k = 1$ and $p = 2^c + 1$ for some positive $c \in \mathbb{Z}$. If c has an odd prime factor, then $2^c + 1$ is easily factored algebraically, so $c = 2^m$, and hence $p = 2^{2^m} + 1$, a so-called *Fermat prime*.

We have proved the following theorem, which was first proved by Gauss in 1796 when he was 19 years old.

Theorem 6.3. A regular n-gon, $n \geq 3$, can be constructed by straightedge and compass if and only if $n = 2^k p_1 p_2 \cdots p_s$, where $0 \leq k \in \mathbb{Z}$ and the p_i are distinct Fermat primes of the form $2^{2^{m_i}} + 1, 0 \leq m_i \in \mathbb{Z}$.

In general the numbers $2^{2^m} + 1 = F_m$, $0 \leq m \in \mathbb{Z}$, are called *Fermat numbers*, since Fermat conjectured in 1650 that every F_m is prime. The first five are indeed primes; $F_0 = 3$, $F_1 = 5$, $F_2 = 17$, $F_3 = 257$, and $F_4 = 65537$. However, Euler observed in 1732 that $F_5 = 641 \cdot 6700417$. Subsequently at least 30 more Fermat numbers have been proved composite, and to date no further Fermat primes have been found.

Constructions for a regular triangle, square, and pentagon were given in Euclid's "*Elements*." The first explicit construction of the 17-gon was given by Erchinger in about 1800. Richelot and Schwendenwein found constructions for the 257-gon in 1892, and Hermes spent 10 years on the construction of the 65537-gon at Göttingen around the turn of the century. The construction processes seem now to have been computerized (see the article by Bishop [5]).

Exercise 6.3. (1) Observe that $641 = 5^4 + 2^4 = 5 \cdot 2^7 + 1$. Conclude that $641 \mid 5^4 \cdot 2^{28} + 2^{32}$ and $641 \mid 5^4 \cdot 2^{28} - 1$, then show that $641 \mid F_5$ without finding the other factor.

(2) If $0 \leq n < m \in \mathbb{Z}$ show that $F_n \mid F_m - 2$. Conclude that $(F_n, F_m) = 1$, and then conclude that there are infinitely many primes in $\mathbb{Z}$.

As a final application of Theorem 6.2 we discuss the quadrature, or squaring, of a circle, which is the construction of a square having the same area as the given circle. An ancient Egyptian solution was to take the side of the square to be $\frac{16}{9}$ times the radius of the circle, which amounts to assuming that $\pi = \frac{256}{81}$, a considerably better approximation than that which appears in I Kings 7:23, or in II Chronicles 4:2.

If the circle is taken to have radius 1, then its area is π, and the problem is to construct a segment of length $\sqrt{\pi}$. But $\pi \in \mathbb{Q}(\sqrt{\pi})$, and $[\mathbb{Q}(\pi):\mathbb{Q}]$ is infinite by Theorem 5.5, so $\sqrt{\pi}$ can not be constructible by Theorem 6.2.

Exercise 6.4. The *spiral of Archimedes* has equation $r = \theta$ in polar coordinates. Sketch the curve and observe that it first crosses the y axis (for $r > 0$) at height $\pi/2$. If the spiral is given show how to square a circle.

7. NORM AND TRACE

The concepts introduced in this section will not be needed in the remainder of the book except for a brief encounter in Section 4 of Chapter VI. The concepts are of considerable importance in the study of algebraic number theory.

Suppose F is a field and K is a finite extension of F, say with $[K:F] = n$. If we view K as a vector space over F, then each $a \in K$ determines an F-linear transformation $t_a: K \to K$, defined by $t_a(u) = au$, all $u \in K$. Define the *norm* $N_{K/F}$ and the *trace* $\mathrm{Tr}_{K/F}$, both maps from K to F, by means of

$$N_{K/F}(a) = \det(t_a), \qquad \mathrm{Tr}_{K/F}(a) = \mathrm{trace}\,(t_a).$$

When confusion is unlikely we write simply N for $N_{K/F}$ and Tr for $\mathrm{Tr}_{K/F}$.

The most readily apparent means for computing values of N and Tr is to choose a basis for K over F, represent each t_a, $a \in K$, by a matrix, then compute the determinant and trace as usual in linear algebra. For the usual reasons N and Tr are independent of the basis chosen.

For example, take $F = \mathbb{Q}$ and $K = \mathbb{Q}(\sqrt{m})$, where $1 \neq m \in \mathbb{Z}^*$ and m is square-free. Then $\{1, \sqrt{m}\}$ is a basis for K over F. If $a = r + s\sqrt{m} \in K$, then t_a is represented by the matrix $\left[\begin{smallmatrix} r & ms \\ s & r \end{smallmatrix}\right]$, and consequently $N(a) = r^2 - ms^2$ (as in Section 5 of Chapter II), and $\mathrm{Tr}(a) = 2r$.

Elementary properties of the determinant and trace functions for linear transformations translate immediately to corresponding properties of N and Tr, yielding the following proposition.

Proposition 7.1. Suppose $[K:F] = n$, $a, b \in K$, and $u, v \in F$. Then

(1) $N(ab) = N(a)N(b)$,
(2) $\mathrm{Tr}(ua + vb) = u\,\mathrm{Tr}(a) + v\,\mathrm{Tr}(b)$,
(3) $N(u) = u^n$, $\mathrm{Tr}(u) = nu$.

Proposition 7.2. Suppose K is a finite extension of F and $F \subseteq L \subseteq K$, with $[K:L] = s$. If $a \in L$, then $\mathrm{Tr}_{K/F}(a) = s \cdot \mathrm{Tr}_{L/F}(a)$ and $N_{K/F}(a) = (N_{L/F}(a))^s$.

Proof. Define $t_a: K \to K$ as above. Choose a basis $\{b_1, \ldots, b_r\}$ for L over F and let $t_a | L$ be represented by the matrix $A = [a_{ij}]$ relative to $\{b_i\}$, i.e., $ab_i = \sum_{j=1}^r a_{ji}b_j$ for each i. Choose a basis $\{c_1, \ldots, c_s\}$ for K over L. By Proposition 1.1 $\{b_ic_j\}$ is a basis for K over F, which we order as $b_1c_1, b_2c_1, \ldots, b_rc_1; b_1c_2, \ldots, b_rc_2; \ldots$. Then we have

$$t_a(b_ic_j) = a(b_ic_j) = (ab_i)c_j = \textstyle\sum_{k=1}^s a_{ki}(b_kc_j) \qquad \text{for all } i \text{ and } j.$$

Consequently the matrix representing t_a relative to $\{b_ic_j\}$ is block diagonal of the form

$$\begin{bmatrix} A & \cdot & & 0 \\ & & \cdot & \\ 0 & & \cdot & A \end{bmatrix},$$

with r As down the diagonal. Thus $\det(t_a) = (\det A)^r$ and $\operatorname{trace}(t_a) = r \cdot \operatorname{trace} A$, and the proposition follows.

Theorem 7.3. Suppose K is a finite Galois extension of F, and $F \subseteq L \subseteq K$. Set $G = G(K:F)$ and $H = \mathscr{F}L$, and let $\{\phi_1, \phi_2, \ldots, \phi_k\}$ be a set of coset representatives for H in G. If $a \in L$, then

$$\mathrm{Tr}_{L/F}(a) = \sum\{\phi_i(a) : 1 \leq i \leq k\},$$

and

$$N_{L/F}(a) = \prod\{\phi_i(a) : 1 \leq i \leq k\}.$$

Proof. Let

$$m(x) = b_0 + b_1 x + \cdots + b_{s-1}x^{r-1} + x^r$$

be the minimal polynomial for a over F. We assume first that $L = F(a)$, so $r = k$. Then $\{1, a, \ldots, a^{k-1}\}$ is a basis for L over F, and the matrix representing t_a on L is

$$A = \begin{bmatrix} 0 & 0 & \cdots & 0 & -b_0 \\ 1 & 0 & \cdots & 0 & -b_1 \\ 0 & 1 & \cdots & 0 & -b_2 \\ \vdots & \vdots & & \vdots & \vdots \\ 0 & 0 & \cdots & 1 & -b_{k-1} \end{bmatrix},$$

sometimes called the *companion matrix* of $m(x)$. Thus $\mathrm{Tr}_{L/F}(a) = -b_{k-1}$ and $N_{L/F}(a) = (-1)^k b_0$ (expand the determinant along the first row of A). But also $-b_{k-1}$ is the sum of the roots (in K) of $m(x)$, and $(-1)^k b_0$ is the product of the roots (see the beginning of Section 5, above). Those roots are precisely $\{\phi_i(a) : 1 \leq i \leq k\}$, by Proposition 3.3, and the theorem follows in this special case.

In general say $[L:F(a)] = s$, so $rs = [L:F] = k$. Set $J = \mathscr{G}F(a) = \mathrm{Stab}_G(a)$, so $[J:H] = s$. Each left coset of J in G is then a union of s distinct left cosets of H. We may relabel if necessary and assume that $\phi_1, \ldots, \phi_r$ are coset representatives for J in G. The remaining H-coset representatives ϕ_i are distributed evenly among the cosets $\phi_1 J, \ldots, \phi_r J$. Then $\phi_1(a), \ldots, \phi_r(a)$ are the distinct roots of $m(x)$. In the list $\phi_1(a), \ldots, \phi_r(a), \phi_{r+1}(a), \ldots, \phi_k(a)$ each of those r distinct roots appears s times due to the s cosets of H lying within each coset of $J = \mathrm{Stab}_G(a)$. Thus by the first part of the proof and by Proposition 7.2

we have

$$\sum_{i=1}^{k} \phi_i(a) = s\sum_{i=1}^{r} \phi_i(a) = s \cdot \mathrm{Tr}_{F(a)/F}(a) = \mathrm{Tr}_{L/F}(a),$$

$$\prod_{i=1}^{k} \phi_i(a) = \left(\prod_{i=1}^{r} \phi_i(a)\right)^s = \left(N_{F(a)/F}(a)\right)^s = N_{L/F}(a).$$

Corollary. If K is a finite Galois extension of F and $a \in K$, then

$$N_{K/F}(a) = \prod\{\phi(a) : \phi \in G(K:F)\}$$

and

$$\mathrm{Tr}_{K/F}(a) = \sum\{\phi(a) : \phi \in G(K:F)\}.$$

Proposition 7.4. If K is any field and S is any set of field automorphisms of K, then S is K-linearly independent in the vector space of all functions from K to K.

Proof. We may clearly assume that $|S| \geq 2$. Suppose by way of contradiction that S is dependent, and take a dependence relation

$$a_1\phi_1 + a_2\phi_2 + \cdots + a_k\phi_k = 0,$$

with all $\phi_i \in S$, all $a_i \in K^*$, and $k(\geq 2)$ minimal. Choose $b \in K$ with $\phi_1(b) \neq \phi_2(b)$. Then we have

$$a_1\phi_1(bc) + a_2\phi_2(bc) + \cdots + a_k\phi_k(bc) = 0$$

or

$$a_1\phi_1(b)\phi_1(c) + a_2\phi_2(b)\phi_2(c) + \cdots + a_k\phi_k(b)\phi_k(c) = 0$$

for all $c \in K$, and hence another dependence relation

$$a_1\phi_1(b)\phi_1 + a_2\phi_2(b)\phi_2 + \cdots + a_k\phi_k(b)\phi_k = 0.$$

Multiply the original dependence relation by $\phi_1(b)$ and subtract from the new relation to obtain

$$a_2(\phi_2(b) - \phi_1(b))\phi_2 + \cdots + a_k(\phi_k(b) - \phi_1(b))\phi_k = 0,$$

contradicting the minimality of k.

Corollary. If L is a finite separable extension of F, then $\mathrm{Tr}_{L/F}$ maps L onto F.

Proof. Since Tr is an F-linear transformation from L to the one-dimensional F-space F, its image is either F or 0. If K is a Galois closure of L over F, then by Theorem 7.3 $\mathrm{Tr}_{L/F}$ is a sum of automorphisms of K. Those automorphisms are linearly independent, so $\mathrm{Tr}_{L/F}$ is not the zero function, and hence $\mathrm{Tr}(L) = F$.

The next theorem appeared in D. Hilbert's famous "Zahlbericht" of 1897.

Theorem 7.5 (Hilbert's Satz 90). Suppose K is a finite Galois extension of F with $G = G(K:F)$ cyclic, say $G = \langle \sigma \rangle$ of order n. If $a \in K$, then $N_{K/F}(a) = 1$ if and only if $a = b/\sigma(b)$ for some $b \in K$.

Proof. $\Leftarrow$: If $a = b/\sigma(b)$, then by the corollary to Theorem 7.3 we see that

$$N(a) = [b/\sigma b][\sigma b/\sigma^2 b] \cdots [\sigma^{n-1} b/\sigma^n b] = 1$$

since $\sigma^n = 1 \in G$.

$\Rightarrow$: Suppose $N(a) = 1$, $a \in K$. By Proposition 7.4 the automorphisms $1, \sigma, \ldots, \sigma^{n-1}$ are K-linearly independent, so the function

$$\phi = a \cdot 1 + (a\sigma(a))\sigma + (a\sigma(a)\sigma^2(a))\sigma^2 + \cdots + \left(\prod_{i=0}^{n-1} \sigma^i(a)\right)\sigma^{n-1}$$

from K to K is not 0. Note that the coefficient of σ^{n-1} in ϕ is $N(a) = 1$, by the corollary to Proposition 7.4. Thus we have $b = \phi(c) \neq 0$ for some $c \in K$. But then (verify)

$$\sigma b = (1/a)[b - ac] + \sigma^n c = (1/a)[b - ac] + c = b/a,$$

and hence $a = b/\sigma b$.

Theorem 7.5 affords an alternative proof of Proposition 4.8. See Exercise 8.57 below.

The map $N_{K/F}$ restricts to a homomorphism of multiplicative groups, $K^* \to F^*$, and Theorem 7.5 describes its kernel when K is Galois over F with cyclic Galois group. We conclude this section with an analog of Theorem 7.5 for the homomorphism $\mathrm{Tr}_{K/F}$ of additive groups. The proofs are also analogous.

Theorem 7.6. Suppose K is a finite Galois extension of F with $G = G(K:F)$ cyclic, say $G = \langle \sigma \rangle$ of order n. If $a \in K$, then $\mathrm{Tr}(a) = 0$ if and only if $a = b - \sigma b$ for some $b \in K$.

Proof. $\Leftarrow$: Compute, using Theorem 7.3.

$\Rightarrow$: Suppose $\mathrm{Tr}(a) = 0$. Choose $c \in K$ with $\mathrm{Tr}(c) = 1$ (Corollary, Proposition 7.4). Set

$$b = \sum_{i=0}^{n-2} \left(\sum_{j=0}^{i} \sigma^j(a) \right) \sigma^i(c).$$

Then

$$\sigma b = \sum_{i=0}^{n-2} \left(\sum_{j=0}^{i} \sigma^{j+1}(a) \right) \sigma^{i+1}(c) = \sum_{i=1}^{n-1} \left(\sum_{j=1}^{i} \sigma^j(a) \right) \sigma^i(c),$$

and so

$$\begin{aligned} b - \sigma b &= ac + \sum_{i=1}^{n-2} a\sigma^i(c) - \sum_{j=1}^{n-1} \sigma^j(a)\sigma^{n-1}(c) \\ &= a \sum_{i=0}^{n-2} \sigma^i(c) - (\mathrm{Tr}(a) - a)\sigma^{n-1}(c) \\ &= a \sum_{i=0}^{n-1} \sigma^i(c) = a\,\mathrm{Tr}(c) = a \end{aligned}$$

(Theorem 7.3 was used twice).

An application of Theorem 7.6 appears in Exercise 8.58.

8. FURTHER EXERCISES

1. Find a splitting field $K \subseteq \mathbb{C}$ for $x^3 - 2 \in \mathbb{Q}[x]$, and determine $[K:\mathbb{Q}]$.
2. Find a splitting field $K \subseteq \mathbb{C}$ for $x^5 - 1 \in \mathbb{Q}[x]$, and determine $[K:\mathbb{Q}]$.
3. Suppose K is an algebraic extension of F and R is a ring, with $F \subseteq R \subseteq K$. Show that R is a field.
4. Find all solutions to $x^2 + 1 = 0$ in the ring $\mathbb{H}$ of quaternions. Why is your answer not at odds with Proposition 1.7?
5. If $f:\mathbb{C} \to \mathbb{R}$ is a (ring) homomorphism show that $f(a) = 0$ for all $a \in \mathbb{C}$.
6. If $S = \{\sqrt{p}: p \in \mathbb{Z}, p \text{ prime}\}$ show that $[\mathbb{Q}(S):\mathbb{Q}]$ is infinite.
7. Suppose K is an extension of F, $a \in K$ is algebraic over F, and $\deg m_a(x)$ is odd. Show that $F(a^2) = F(a)$.
8. Show that the field $\mathbb{A} \subseteq \mathbb{C}$ of algebraic numbers is algebraically closed.
9. (S. Raffer). Let F be a field that is not algebraically closed, and suppose that the class $\mathcal{S}$ of all algebraic extensions K of F is a set. Say $|\mathcal{S}| = \alpha$. Construct different algebraic extensions of F corresponding to each element of the power set 2^α and obtain a contradiction.
10. If $K \subseteq \mathbb{C}$ is a splitting field over $\mathbb{Q}$ for $x^3 - 2$ find all subfields of K.
11. Determine the Galois groups (over $\mathbb{Q}$) of the following polynomials:

(a) $x^3 - 1$, (b) $x^5 - 1$, (c) $x^6 - 1$, (d) $x^3 - 2$,
(e) $x^4 - 2$, (f) $x^4 + 1$, (g) $x^4 - 7x^2 + 10$, (h) $x^6 - 3x^3 + 2$.

12. Suppose $f(x)$ is irreducible in $F[x]$ and K is a Galois extension of F. Show that all irreducible factors of $f(x)$ in $K[x]$ have the same degree. [*Hint*: Show that $G(K:F)$ acts as a permutation group on the factors.]
13. Let $G = G(\mathbb{R}:\mathbb{Q})$.

(1) If $\phi \in G$ and $a \leq b$ in $\mathbb{R}$ show that $\phi(a) \leq \phi(b)$. [*Hint*: $b - a$ is a square in $\mathbb{R}$.]

(2) Show that $G = 1$. [*Hint*: If not choose $\phi \in G$ and $a \in \mathbb{R}$ such that $\phi(a) \neq a$. Choose $b \in \mathbb{Q}$ between a and $\phi(a)$.]

14. Suppose K is a finite Galois extension of F, and $F \subseteq E \subseteq K$, $F \subseteq L \subseteq K$. Show that the join $E \vee L$ is Galois over E, that L is Galois over $E \cap L$, and that $G(E \vee L : E) \cong G(L : E \cap L)$.

15. Let $F = \mathbb{C}$, let K be the field $\mathbb{C}(t)$ of rational functions, and let $G = G(K:F)$. If ϕ and θ in G are determined by $\phi(t) = \zeta t$ and $\theta(t) = 1/t$, where $\zeta \in \mathbb{C}$ is a primitive nth root of unity, show that $H = \langle \phi, \theta \rangle \leq G$ is isomorphic with the dihedral group D_n. Show that $\mathscr{F}H = \mathbb{C}(t^n + t^{-n})$.

16. For any $f(x) \in F[x]$ set $f^{(0)}(x) = f(x), f^{(1)}(x) = f'(x)$, and in general let $f^{(n)}(x)$ be the derivative of $f^{(n-1)}(x)$, $1 \leq n \in \mathbb{Z}$. If $f(x)$, $g(x) \in F[x]$ set $h(x) = f(x)g(x)$ and show that

$$h^{(n)}(x) = \sum_{k=0}^{n} \binom{n}{k} f^{(n-k)}(x) g^{(k)}(x)$$

(this is *Leibniz's rule*).

17. If char $F = 0$, $a \in F$, and $f(x)$ has degree n in $F[x]$ show that

$$f(x) = \sum_{k=0}^{n} (f^{(k)}(a)/k!)(x - a)^k.$$

18. Give an example of fields $F \subseteq E \subseteq K$ such that K is normal over E and E is normal over F but K is not normal over F (think about $\sqrt{2}$ and $\sqrt[4]{2}$).

19. Find a primitive element (over $\mathbb{Q}$) for $K = \mathbb{Q}(\sqrt{3}, \sqrt[3]{2}) \subseteq \mathbb{C}$.

20. Let s and t be distinct indeterminates over $\mathbb{Z}_p$, and let $F = \mathbb{Z}_p(s, t)$, the field of rational functions in s and t. Let a and b be roots of $x^p - s$ and $x^p - t$, respectively, in some extension of F, and let $K = F(a, b)$. Show that K is not a simple extension of F.

21. Suppose char $F = p \neq 0$ and K is an extension of F. An element $a \in K$ is called *purely inseparable* over F if it is a root of a polynomial of the form $x^{p^k} - b \in F[x]$, $0 \leq k \in \mathbb{Z}$.

(1) Show that if $a \in K$ is both separable and purely inseparable over F, then $a \in F$.

(2) Show that the set of all elements of K that are purely inseparable over F constitute a field. Conclude that there is a unique largest "purely inseparable" extension of F within K.

22. A field F is called *perfect* if either char $F = 0$ or else char $F = p$ and $F = F^p = \{a^p : a \in F\}$.

(1) If F is finite show that the map $a \mapsto a^p$ is a monomorphism and conclude that F is perfect.

(2) Show that the field $\mathbb{Z}_p(t)$ of rational functions in the indeterminate t is not perfect.

(3) Show that a field F is perfect if and only if every finite extension K of F is separable over F, and hence every $f(x) \in F[x]$ is separable.

23. Suppose F_q and F_r are Galois fields, with $q = p^m$ and $r = p^n$, p prime. Show that F_q has a subfield (isomorphic with) F_r if and only if $n \mid m$.

24. List all subfields of F_q if $q = 2^{20}$; if $q = p^{30}$, p prime.

25. Suppose the Galois fields F_q and F_r are both subfields of a field K. Determine $F_q \cap F_r$ and $F_q \vee F_r$.

26. If $0 < k \in \mathbb{Z}$ show that there exists an irreducible polynomial of degree k over any Galois field F_q.

27. If F is a finite field of characteristic p show that every element of F has a unique pth root.

28. If P is a prime and $1 \leq n \in \mathbb{Z}$ show that $f(x) = x^{p^n} - x$ is the product of all the monic irreducible polynomials over $\mathbb{Z}_p$ whose degrees are divisors of n. (*Hint*: Use Exercise 23.)

29. Let F be a finite field.

(1) Show that the product of all elements in F^* is -1.

(2) Conclude *Wilson's Theorem*: If $p \in \mathbb{Z}$ is a prime, then $(p-1)! \equiv -1 \pmod{p}$.

30. If F is a finite field show that every element of F is a sum of two squares in F. (*Hint*: See Exercise I.12.8, and let S be the set of squares in the additive group F.)

31. Solve $x^5 - 1 = 0$ explicitly in $\mathbb{R}$ and show that $\cos(2\pi/5) = (-1 + \sqrt{5})/4$. [*Hint*: Write

$$x^4 + x^3 + x^2 + 1 = x^2(x^2 + x + 1 + 1/x + 1/x^2)$$

and make the substitution $y = x + 1/x$.]

32. (Artin). Let F be a field, $\mathbb{Q} \subseteq F \subseteq \mathbb{A}$, maximal with respect to $\sqrt{2} \notin F$ (Why does F exist?).

(1) If $F \subseteq K \subseteq \mathbb{A}$, with K normal and finite over F, and $K \neq F$, show that $G = G(K:F)$ is a 2-group having a unique subgroup of index 2. Conclude that G is cyclic.

(2) If $F \subseteq L \subseteq \mathbb{A}$ and $[L:F]$ is finite show that L is normal over F and $G(L:F)$ is cyclic. Conclude that the set of finite extensions of F (in $\mathbb{A}$) is an ascending chain.

33. Suppose F is a field of characteristic $p \neq 0$, $n = p^k m$ with $(p, m) = 1$ and $k > 0$, and K is a splitting field for $x^n - 1$ over F. Show that K has roots of unity of order m but not of order n, and in particular K has no primitive nth roots of unity.

34. Show (by factoring) that $\Phi_{12}(x)$ is reducible in $\mathbb{Z}_{11}[x]$.

35. If $n > 1$ is odd show that $\Phi_{2n}(x) = \Phi_n(-x)$.

36. Let p be a prime and set

$$f(x) = 1 + x + x^2 + \cdots + x^{p-1}.$$

Show that for any field F the irreducible factors of $f(x)$ in $F[x]$ all have the same degree.

37. Show that

$$f(x) = x^7 - 7x^6 - 189x^4 + 1701x^2 - 7 \in \mathbb{Q}[x]$$

is not solvable by radicals.

38. Factor $x^n - 1$ into a product of polynomials of degrees 1 or 2 in $\mathbb{R}[x]$. (*Hint*: The nth roots of unity in $\mathbb{C}$ have the form

$$\eta = e^{2k\pi i/n} = \cos(2k\pi/n) + i\sin(2k\pi/n).$$

Except for ± 1 they occur in complex conjugate pairs.)

39. Suppose F is a field and $K = F(x)$, the field of rational functions in the indeterminate x over F. If $u \in K \backslash F$ show that u is transcendental over F.

40. If F is a field show that the set of monomials

$$\{x_2^{k_2} x_3^{k_3} \cdots x_n^{k_n} : 0 \leq k_i < i, 2 \leq i \leq n\}$$

is a basis for $K = F(x_1, x_2, \ldots, x_n)$ over $F_0 = F(\sigma_1, \sigma_2, \ldots, \sigma_n)$.

41. Write $x_1^2 + x_2^2 + \cdots + x_n^2$ as a polynomial in the elementary symmetric polynomials $\sigma_1, \sigma_2, \ldots, \sigma_n$.

42. Find the sums of the squares of the roots of each of the polynomials $f(x) = x^3 - 3x^2 + 7$ and $g(x) = 3x^4 - x^3 + 5x + 7$.

43. Show that the elementary symmetric polynomials $\sigma_1, \ldots, \sigma_n$ are algebraically independent over any commutative ring R with 1, i.e., if

$$f(x_1, \ldots, x_n) \in R[x_1, \ldots, x_n] \qquad \text{and} \qquad f(\sigma_1, \ldots, \sigma_n) = 0,$$

then $f(x_1, \ldots, x_n) = 0$. (*Hint*: Take a counterexample of minimal degree and view it as a polynomial in x_n with coefficients in $R[x_1, \ldots, x_{n-1}]$. Show that the "constant term" is not 0, substitute 0 for x_n, and use induction on n.)

44. The *discriminant* of the general polynomial

$$f(x) = \prod\{(x - x_i): 1 \le i \le n\}$$

is

$$D = \prod\{(x_i - x_j)^2: 1 \le i < j \le n\}.$$

Write D as a polynomial in $\sigma_1, \ldots, \sigma_n$ in the cases $n = 2$ and $n = 3$.

45. If char $F \neq 3$ show that a monic cubic polynomial

$$f(x) = a_0 + a_1 x + a_2 x^2 + x^3 \in F[x]$$

can be written as

$$f(x) = g(y) = y^3 + ay + b$$

by means of the substitution $x = y - a_2/3$. Observe that $f(x)$ and $g(y)$ have the same Galois group over F.

46. Suppose F is a field, $f(x) = x^3 + ax + b \in F[x]$, and $K \supseteq F$ is a splitting field for $f(x)$ over F. Write D for the discriminant of $f(x)$ (see Exercise 44, above) and let d be a square root for D.

(1) Show that $D = -4a^3 - 27b^2$ and that $d \in K$.

(2) If $f(x)$ is irreducible show that $G = G(K:F)$ is the alternating group A_3 if and only if $d \in F$.

(3) Conclude, if $f(x)$ is irreducible, that G is A_3 if $-4a^3 - 27b^2$ is a square in F, and otherwise G is S_3.

47. Use Exercise 46 (and also 44 if necessary) to compute the Galois groups over $\mathbb{Q}$ of the following cubics:

(1) $x^3 - 4x + 2$, (2) $x^3 - 12x + 8$,
(3) $x^3 + 6x^2 + 9x + 3$, (4) $x^3 - 6x^2 + 3x - 11$,
(5) $x^3 - 39x + 26$, (6) $x^3 - 84x + 56$.

48. Determine whether or not the angles $\pi/4$, $\pi/6$, $5\pi/6$, $2\pi/3$, and $2\pi/5$ can be trisected with straightedge and compass.

49. Show that an angle of n degrees, $n \in \mathbb{Z}$, is constructible if and only if $3 \mid n$.

50. The spiral of Archimedes (see Exercise 6.4) can be used to trisect any angle. Place the angle with vertex at the origin O and initial side on the x axis, and say the other side first cuts the spiral at P. Trisect the segment OP and draw circles about O through the trisection points. Show that lines from O through the points of intersection of those circles with the spiral trisect the original angle.

51. Solve the Delian problem by intersecting the two parabolas $x = y^2$ and $2y = x^2$.

52. If $\cos\theta = a/b$, with $a, b \in \mathbb{Z}$, $(a, b) = 1$, $b > 1$, and b not divisible by a cube (> 1) in $\mathbb{Z}$, show that θ can not be trisected by straightedge and compass.

53. Verify the construction shown in Fig. 6 (Richmond, 1893) of a regular pentagon inscribed in a circle. Let AB be a diameter and $OC \perp AB$ a radius. Bisect AO at D and choose E on AB such that $\overline{DE} = \overline{DC}$. Let FG be the perpendicular bisector of OE. Then BG is an edge of the pentagon (see Exercise 31).

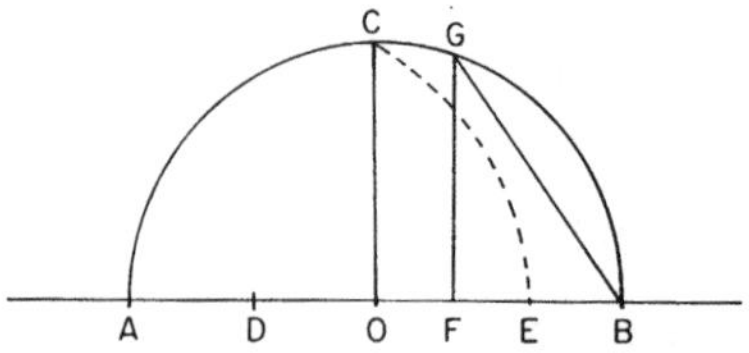

Figure 6

54. A curve called a *cissoid* was studied by Diocles (ca. 180 B.C.) in connection with the Delian problem. In terms of analytic geometry the cissoid can be taken to have the equation $y^2 = x^3/(1 - x)$. Sketch the curve. If the cissoid meets the line $y = -2x + 2$ at the point P show that the line joining the origin to P meets the line $x = 1$ at height $y = \sqrt[3]{2}$.

55. If F is a finite field and K is a finite extension of F show that both $\mathrm{Tr}_{K/F}$ and $N_{K/F}$ map K onto F.

56. If K is finite over F but not separable show that $\mathrm{Tr}_{K/F} \equiv 0$.

57. Use Hilbert's Satz 90 (Theorem 7.5) to prove Proposition 4.8. (*Hint*: Apply Satz 90 to a primitive pth root of unity, and show that the resulting element of K is a primitive element.)

58. Suppose char $F = p$, K is a Galois extension of F, and $[K:F] = p$. Show that $K = F(a)$ for some a with $m_{a,F}(x)$ of the form $x^p - x - b$, $b \in F$. [*Hint*: Let $G(K:F) = \langle\sigma\rangle$, use Theorem 7.6 to write $1 = \sigma c - c$, and set $a = c^p - c$.]

Chapter IV | Modules

1. PRELIMINARIES

Suppose R is a ring. An additive abelian group M is called a *(left) R-module* if there is defined a *scalar multiplication* $(r, x) \mapsto rx \in M$ satisfying the requirements

(i) $r(x + y) = rx + ry$,
(ii) $(r + s)x = rx + sx$, and
(iii) $(rs)x = r(sx)$

for all $r, s \in R$, $x, y \in M$. It follows easily that $r \cdot 0 = 0 \cdot x = 0$ and that $(-r)x = r(-x) = -rx$.

Right R-modules are defined similarly.

Exercise 1.1. Recall that the set $\mathrm{End}(M)$ of all endomorphisms of the abelian group M is a ring (Example 5, p. 48). If M is an R-module show that there is a ring homomorphism $\phi: r \mapsto \phi_r$ from R to $\mathrm{End}(M)$, with $\phi_r(x) = rx$, all $r \in R$, $x \in M$. Conversely, if M is an abelian group and $\phi: R \to \mathrm{End}(M)$ is a homomorphism show that M becomes an R-module if we define $rx = \phi_r(x)$.

This provides an alternative definition of R-modules.

Perhaps the most obvious source of examples of modules is to take $R = F$, a field, and $M = V$, any vector space over F. In fact the theory of modules can reasonably be viewed as an attempt to generalize linear algebra by allowing more general scalars. Further examples abound; we list a few of them.

EXAMPLES

1. Take $R = \mathbb{Z}$ and let M be any additive abelian group. Then rx already has a meaning for any $r \in \mathbb{Z}$, $x \in M$, and it serves to define a module action. Thus the theory of abelian groups is subsumed by the theory of modules.

2. If S is a ring and R a subring of S, then S is an R-module with rx for $r \in R$, $x \in S$, the usual product in S.

3. If R is a ring and J is a left ideal in R, then J is an R-module with rx for $r \in R$, $x \in J$, the product in R.

4. For a very important example let $R = F[x]$, F a field, let V be a vector space over F, and let $T: V \to V$ be a linear transformation. If

$$f(x) = a_0 + a_1 x + \cdots + a_n x^n \in F[x]$$

define $f(T)$ to be $a_0 I + a_1 T + \cdots + a_n T^n$, also a linear transformation on V. It is easily verified that "substitution" of T for x effects a homomorphism $f(x) \mapsto f(T)$ from the polynomial ring R into the ring of all linear transformations of V (see Theorem II.3.4). If we define $f(x) \cdot v = f(T)(v)$, the result of applying the linear transformation $f(T)$ to the vector $v \in V$, then V becomes an $F[x]$-module, which we will usually denote by V_T.

5. For a rather trivial class of examples let R be any ring, M any additive abelian group, and define $rx = 0$ for all $r \in R$, $x \in M$.

Exercise 1.2. (1) Verify that V_T, in Example 4 above, is an R-module.
(2) If $F = \mathbb{Q}$, V is the $\mathbb{Q}$-space of all column vectors with 2 entries from $\mathbb{Q}$, $T: V \to V$ is the result of multiplication by the matrix $A = \left[\begin{smallmatrix} 1 & 1 \\ 0 & 1 \end{smallmatrix}\right]$, and $f(x) = x^m - x$, determine the module action $f(x)v$ on an arbitrary vector $v = \left[\begin{smallmatrix} a \\ b \end{smallmatrix}\right]$ in V_T.

If R is a ring with 1, M is an R-module, and $1 \cdot x = x$ for all $x \in M$, then M is called a *unitary* R-module (some authors prefer to make that requirement part of the definition of a module). Any vector space (as an F-module) and Examples 1, 3, and 4 above provide examples of unitary modules. Example 2 may not, even when both R and S are rings with 1, as an example on page 48 shows.

A subgroup N of an R-module M is called a *submodule* if $rx \in N$ for all $r \in R$ and $x \in N$. If N is a submodule of M, then the quotient group M/N is also an R-module if we define $r(x + N) = rx + N$, all $r \in R$, $x \in M$.

Proposition 1.1. If M is an R-module and $\{M_\alpha\}$ is any nonempty collection of submodules of M, then $\bigcap_\alpha M_\alpha$ is also a submodule.

If M is an R-module and $S \subseteq M$ is any subset, then by Proposition 1.1 there is a unique smallest submodule of M that contains S as a subset, viz., $R\langle S\rangle = \bigcap\{N : S \subseteq N \text{ and } N \text{ is a submodule of } M\}$. We call $R\langle S\rangle$ the submodule of M *generated by* S. If $M = R\langle S\rangle$ for some finite set S we say that M is *finitely generated* over R, and if $M = R\langle a\rangle$ for some single element $a \in M$ we say that M is *cyclic* over R.

Exercise 1.3. Let V_T be the module of Exercise 1.2(2). Show that V_T is a cyclic module and that $N = \left\{\left[\begin{smallmatrix} a \\ 0 \end{smallmatrix}\right] : a \in \mathbb{Q}\right\}$ is a cyclic submodule (neither, of course, is a cyclic group).

If M and N are both R-modules, then a *homomorphism* (or an *R-homomorphism*, if we wish to stress the ring R) from M to N is a (group) homomorphism $f: M \to N$ such that $f(rx) = rf(x)$ for all $r \in R$, $x \in M$. The usual language of *R-isomorphism*, *R-automorphism*, etc., will be used without further explanation. The *kernel* and *image* of an R-homomorphism are defined as usual; they are clearly submodules of M and N, respectively. Denote by $\mathrm{Hom}_R(M, N)$ the set of all R-homomorphisms from M to N.

Exercise 1.4. Show that $\mathrm{Hom}_R(M, N)$ is an abelian group if we define $(f + g)(x) = f(x) + g(x)$. If R is a commutative ring show that $\mathrm{Hom}_R(M, N)$ is an R-module if we define $(rf)(x) = r \cdot f(x)$. If R is a commutative ring with 1 and N is a unitary R-module show that $\mathrm{Hom}_R(M, N)$ is a unitary R-module.

It is a very simple matter to extend the basic homomorphism and isomorphism theorems from abelian groups to arbitrary R-modules. We state the results.

Theorem 1.2 (The Fundamental Homomorphism Theorem for Modules). Suppose M and N are R-modules and $f \in \mathrm{Hom}_R(M, N)$ has kernel K. If $\eta: M \to M/K$ denotes the canonical quotient R-homomorphism, then there is a unique R-isomorphism $g: M/K \to \mathrm{Im}\, f \subseteq N$ such that the diagram

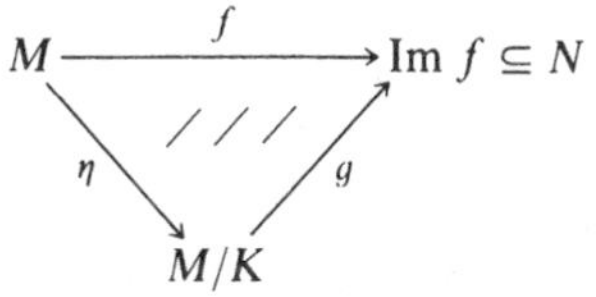

is commutative.

Proposition 1.3. Suppose M and N are R-modules and $f: M \to N$ is an epimorphism, with kernel K. Then there is a 1–1 correspondence between the set of all submodules L of N and the set of all those submodules P of M that contain K, given by $L \leftrightarrow f^{-1}(L) = P$. In particular each submodule of a quotient module M/K has the form P/K for some submodule P, $K \subseteq P \subseteq M$.

Theorem 1.4 (The Freshman Theorem for Modules). Suppose K and N are submodules of an R-module M, with $K \subseteq N$. Then N/K is a submodule of M/K and $(M/K)/(N/K) \cong M/N$.

Theorem 1.5 (The Isomorphism Theorem for Modules). If K and N are submodules of an R-module M, then $K + N$ and $K \cap N$ are submodules of M and $(K + N)/K \cong N/(K \cap N)$.

An R-module $M \neq 0$ is called *simple* if its only submodules are M and 0. A sequence

$$M = M_0 \supseteq M_1 \supseteq M_2 \supseteq \cdots \supseteq M_k = 0$$

of submodules of an R-module M is called a *composition series* if each M_{i+1} is a maximal (proper) submodule of M_i, or equivalently (see Proposition 1.3) if every factor M_i/M_{i+1} is simple.

Theorem 1.6 (Jordan–Hölder). Suppose an R-module M has composition series

$$M = M_0 \supseteq M_1 \supseteq \cdots \supseteq M_k = 0$$

and

$$M = N_0 \supseteq N_1 \supseteq \cdots \supseteq N_m = 0.$$

Then $k = m$ and there is a 1–1 correspondence between the sets of factors so that corresponding factors are R-isomorphic.

Proof. Essentially the same as the proof of Theorem I.5.5.

An R-module M is called *Noetherian* if it satisfies the ascending chain condition (ACC) for submodules, i.e., if $M_1 \subseteq M_2 \subseteq \cdots$ is any ascending chain of submodules of M, then either the chain is finite or there is some k such that $M_m = M_k$ for all $m \geq k$. In either case we say that the chain *terminates*. (Recall the corresponding notion for commutative rings; see p. 66.)

Proposition 1.7. The following statements regarding an R-module are equivalent.

(i) M is Noetherian.
(ii) Every submodule of M is finitely generated.
(iii) Every nonempty set $\{M_\alpha\}$ of submodules of M has a maximal element with respect to set inclusion.

Proof. i $\Rightarrow$ ii: If a submodule N is not finitely generated choose $x_1 \in N$ and set $M_1 = R\langle x_1 \rangle$. Then choose $x_2 \in N \backslash M_1$ and set $M_2 = R\langle x_1, x_2 \rangle$, etc. The process does not terminate since N is not finitely generated, and the sequence $M_1 \subset M_2 \subset \cdots$ violates the ACC for M.

ii $\Rightarrow$ iii: If there is a set of submodules without a maximal element we may choose a strictly increasing infinite chain $M_1 \subset M_2 \subset \cdots$ from the set. Set $N = \bigcup M_i$, a submodule of M. Then $N = R\langle x_1, \ldots, x_k \rangle$ for some $x_1, \ldots, x_k$ in N, and hence x_i is in some M_{j_i} for each i. If j_m is the largest of the j_i, then all

x_i are in M_{j_m} and hence $N = M_{j_m}$, contradicting the fact that the sequence is strictly increasing.

iii $\Rightarrow$ i: Obvious.

Proposition 1.8. Let K be a submodule of the R-module M and set $N = M/K$. Then M is Noetherian if and only if both K and N are Noetherian.

Proof. $\Rightarrow$: All submodules of K are submodules of M, so they are all finitely generated, and hence K is Noetherian by Proposition 1.7. Any ascending chain $N_1 \subseteq N_2 \subseteq \cdots$ of submodules of N has the form $M_1/K \subseteq M_2/K \subseteq \cdots$, with each M_i a submodule of M and $M_1 \subseteq M_2 \subseteq \cdots$, so the chain must terminate and N is Noetherian.

$\Leftarrow$: Let $M_1 \subseteq M_2 \subseteq \cdots$ be an ascending chain of submodules of M. Then the chain $M_1 \cap K \subseteq M_2 \cap K \subseteq \cdots$ terminates, say at $M_j \cap K$, and likewise the chain $(M_1 + K)/K \subseteq (M_2 + K)/K \subseteq \cdots$ terminates, say at $(M_m + K)/K$. Set $n = \max\{j, m\}$. If $k \geq n$ and $x \in M_k$, then $(M_k + K)/K = (M_n + K)/K$, so $x + K = y + K$ for some $y \in M_n$. But then $x - y \in M_k \cap K = M_n \cap K$, and hence $x \in M_n$. Thus the chain $M_1 \subseteq M_2 \subseteq \cdots$ terminates at M_n and M is Noetherian.

2. DIRECT SUMS, FREE MODULES

The definition and construction of a (direct) product for an arbitrary nonempty family $\{M_\alpha\}$ of R-modules is completely analogous to the corresponding definition and construction for the direct product of groups, and will not be repeated. We will write $\prod\{M_\alpha : \alpha \in A\}$, or just $\prod M_\alpha$, for the product of a nonempty family of R-modules.

The notion of *direct sum*, in a sense dual to the notion ofdirect product, is also of considerable importance for modules. Suppose $\{M_\alpha : \alpha \in A\}$, $A \neq \varnothing$, is a family of R-modules. A *direct sum* of the modules M_α is an R-module M together with a family $i_\alpha: M_\alpha \to M$, $\alpha \in A$, of R-homomorphisms with the following universal property. Given any R-module N and R-homomorphisms $f_\alpha: M_\alpha \to N$, $\alpha \in A$, there is a unique $f \in \operatorname{Hom}_R(M, N)$ such that $f i_\alpha = f_\alpha$ for all $\alpha \in A$, i.e., the diagrams

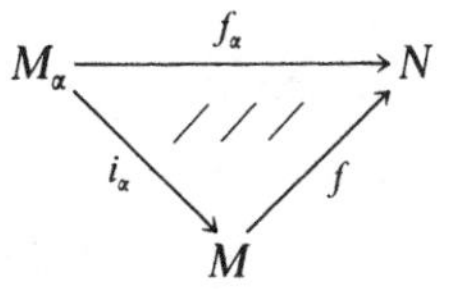

are all commutative. Note that the definition results from the definition of a product by simply reversing all the arrows. In the language of category theory the direct sum is a *coproduct* in the category of R-modules.

Proposition 2.1. If a direct sum exists for a family $\{M_\alpha\}$ of R-modules, then it is unique up to R-isomorphism, and each i_α is an R-monomorphism.

Proof. As usual (see the proof of Proposition I.6.1).

It is perhaps reasonable by now to be a bit brief in discussing the construction of a direct sum.

Theorem 2.2. Any nonempty family $\{M_\alpha : \alpha \in A\}$ of R-modules has a direct sum M.

Proof. Take M to be the submodule of $\prod M_\alpha$ consisting of those m such that $m_\alpha = 0$ for all but finitely many $\alpha \in A$ (in particular, if A is finite, then $M = \prod M_\alpha$). Define $i_\alpha : M_\alpha \to M$ by agreeing that $i_\alpha(x) = m$ if and only if $m_\alpha = x$ and $m_\beta = 0$ for all $\beta \neq \alpha$. It is easy to check that $i_\alpha \in \operatorname{Hom}_R(M_\alpha, M)$. Given any R-module N and family $f_\alpha : M_\alpha \to N$ of R-homomorphisms define $f : M \to N$ by setting $f(m) = \sum\{f_\alpha(m_\alpha) : \alpha \in A\}$ (there are only finitely many nonzero summands). Then $f \in \operatorname{Hom}_R(M, N)$, $fi_\alpha = f_\alpha$ for all α, and f is uniquely determined.

Exercise 2.1. Prove the last statement in the proof above, and show that $i_\alpha \in \operatorname{Hom}_R(M_\alpha, M)$.

Let us agree to write $\bigoplus\{M_\alpha : \alpha \in A\}$, or simply $\bigoplus M_\alpha$, for the direct sum of the M_α. For a finite collection $M_1, M_2, \ldots, M_n$ of R-modules we write $M_1 \oplus M_2 \oplus \cdots \oplus M_n$ for the direct sum.

In general, if $\{M_\alpha : \alpha \in A\}$ is any collection of submodules of an R-module M, we shall write $\sum\{M_\alpha : \alpha \in A\}$ for the submodule consisting of all sums $\sum\{x_\alpha : \alpha \in A\}$, where $x_\alpha \in M_\alpha$ and $x_\alpha = 0$ for all but finitely many $\alpha \in A$.

The next theorem is an analog for modules of Theorem I.6.3, and describes an R-module as the *internal direct sum* of submodules.

Theorem 2.3. Suppose M is an R-module, $\{M_\alpha\}$ is a family of submodules, $R\langle \bigcup M_\alpha \rangle = M$, and $\sum_\alpha M_\alpha = M$, and $M_\alpha \cap \sum\{M_\beta : \beta \neq \alpha, \beta \in A\} = 0$ for all α. Then M is R-isomorphic with $\bigoplus M_\alpha$.

When M is the internal direct sum of a family $\{M_\alpha\}$ of submodules we will often write $M = \bigoplus M_\alpha$. It will generally be clear from the context whether $\bigoplus M_\alpha$ represents an internal or an external direct sum.

Exercise 2.2. Show that $M = \bigoplus M_\alpha$, an internal direct sum of submodules, if and only if each $x \in M$ has a unique expression of the form $x = x_1 + x_2 + \cdots + x_k$ for some k, with $x_i \in M_{\alpha_i}$.

Proposition 2.4. If $M_1, M_2, \ldots, M_n$ are Noetherian R-modules, then $M = M_1 \oplus M_2 \oplus \cdots \oplus M_n$ is Noetherian.

Proof. It will suffice, by induction, to assume that $M = M_1 \oplus M_2$. Then $N_1 = \{(x, 0): x \in M_1\}$ is a submodule of M that is R-isomorphic with M_1, so N_1 is Noetherian, and M/N_1 is R-isomorphic with M_2 [via $(x, y) + N_1 \mapsto y$], so M/N_1 is Noetherian. Thus M is Noetherian by Proposition 1.8.

A ring R is called *left* (*right*) Noetherian if it satisfies the ascending chain condition for left (right) ideals, or equivalently if R is Noetherian when viewed as a left (right) R-module.

Proposition 2.5. Suppose R is a left Noetherian ring with 1 and M is a finitely generated unitary R-module. Then M is a Noetherian R-module.

Proof. Say $M = R\langle a_1, \ldots, a_n \rangle$. Let N be the R-module $R^n = R \oplus R \oplus \cdots \oplus R$. Then N is Noetherian by Proposition 2.4. Define $f: N \to M$ by setting

$$f(r_1, \ldots, r_n) = r_1 a_1 + \cdots + r_n a_n.$$

Clearly $f \in \mathrm{Hom}_R(N, M)$ and f is onto since M is unitary. Thus $M \cong N/\ker f$ and M is Noetherian by Proposition 1.8.

Let R be a ring with 1 and B an arbitrary set. A *free* R-module based on B is a unitary R-module F together with a function $\phi: B \to F$ such that given any unitary R-module M and any function $\theta: B \to M$ there is a unique $f \in \mathrm{Hom}_R(F, M)$ such that $f\phi = \theta$, i.e., the diagram

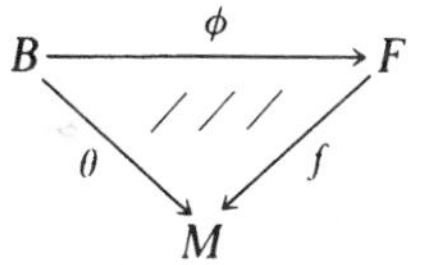

is commutative.

Exercise 2.3. (1) If $B = \varnothing$ show that $F = 0$ is a free R-module based on B.

(2) If $B = \{b\}$, a singleton, show that $F = R$ is a free R-module based on B, with $\phi(b) = 1 \in R$.

As usual we may check that a free R-module F based on a set B is unique (to within R-isomorphism) if it exists, and that ϕ is 1–1. In order to construct a free R-module (if $B \neq \varnothing$) let $M_\beta = R$, viewed as a left R-module, for each $\beta \in B$ and set $F = \oplus M_\beta = \bigoplus\{R: \beta \in B\}$, a direct sum of $|B|$ copies of the R-module R. For each $\beta \in B$ define $\phi(\beta) = m \in F$ by means of $m_\beta = 1 \in M_\beta = R$ and $m_\alpha = 0$ if $\beta \neq \alpha \in B$. It is then a routine matter to verify that F is a free R-module based on B.

Exercise 2.4. Do so.

Since $\phi: B \to F$ is 1–1 we may identify β with $\phi(\beta)$ for each $\beta \in B$, and hence assume that $B \subseteq F$. If $b_1, \ldots, b_k \in B$, $r_1, \ldots, r_k \in R$, and $\sum_{j=1}^{k} r_j b_j = 0$, then it is immediate from the construction that $r_1 = \cdots = r_k = 0$. In general we say that a subset B of an R-module M is *R-linearly independent* if given $b_1, \ldots, b_k \in B$ and $r_1, \ldots, r_k \in R$ with $\sum r_j b_j = 0$ we may conclude that all $r_j = 0$. A *basis* for an R-module M is an R-linearly independent subset of M such that $M = R\langle B \rangle$.

Theorem 2.6. If R is a ring with 1, then a unitary R-module is free if and only if it has a basis.

Proof. $\Rightarrow$: See the construction and discussion above. If $M = 0$, then $B = \varnothing$ is a basis.

$\Leftarrow$: Let B be a basis for M. If $B = \varnothing$, then $M = 0$ and M is free. Suppose then that $B \neq \varnothing$. Let $\phi: B \to M$ be the inclusion map, i.e., $\phi(b) = b$ for all $b \in B$. Set $M_b = Rb$ for each $b \in B$; then $r \mapsto rb$ is an R-isomorphism between R and M_b. The fact that B is a basis allows us to apply Theorem 2.3 and conclude that $M = \bigoplus\{M_b : b \in B\}$, which is isomorphic with $\bigoplus\{R : b \in B\}$, and we may conclude that M is free just as above.

As a particular case we observe that a *free abelian group* is (to within isomorphism) just a direct sum of some number (possibly transfinite) of copies of the additive group $\mathbb{Z}$ of integers. Note also that if $R = K$, a field, and V is a vector space over K, then V has a basis (see the Appendix), and hence V is a free K-module. We know in that case that any two bases have the same cardinality, and we shall see shortly that the same result holds for free abelian groups, but it does not hold true in general (see Exercise 7.6).

Theorem 2.7. Suppose R is a ring with 1 having an ideal I such that R/I is a division ring, and F is a free R-module. Then any two bases of F have the same cardinality.

Proof. Set $E = IF = \{\sum_i r_i x_i : r_i \in I, x_i \in F\}$, and note that E is a submodule of F. Then F/E is a vector space over $K = R/I$, with scalar multiplication given by $(r + I)(x + E) = rx + E$ (Verify!). If B is a basis for F set $\bar{b} = b + E$ for each $b \in B$, and set $\bar{B} = \{\bar{b} : b \in B\}$. Each $x \in F/E$ can be written as

$$x = \sum_{i=1}^{k} r_i b_i + E = \sum_i (r_i + I)\bar{b}_i, \qquad r_i \in R, \quad b_i \in B,$$

so $\bar{B}$ spans F/E over K. If

$$\sum_{i=1}^{k} (r_i + I)\bar{b}_i = \sum_i r_i b_i + E = E$$

(i.e., $= 0$ in F/E), then $\sum_i r_i b_i \in E$, so there are $s_1, \ldots, s_k \in I$ such that $\sum_i r_i b_i = \sum_i s_i b_i$ (there is no harm in using the same set $\{b_i\}$ of basis elements, some coefficients may be 0). Thus $\sum_i(r_i - s_i)b_i = 0$, so $r_i = s_i$, all i, and hence each $r_i + I = I$. Thus $\bar{B}$ is a K-basis for F/E (in particular the map $b \mapsto \bar{b}$ is 1–1), and F/E has K-dimension $|B|$. The same argument applies to any other basis for F, so all bases must have the same cardinality.

Corollary. If R is a commutative ring with 1 and F is a free R-module, then any two bases of F have the same cardinality.

Proof. Use Proposition II.1.8 and the corollary to Proposition II.1.9.

Because of Theorem 2.7 we may define the *rank* of a free module F over a ring R with 1 having a division ring as a homomorphic image to be the cardinality of any basis for F. In particular the notion of rank is defined for free abelian groups.

Theorem 2.8. If R is any ring with 1 and M is a unitary R-module, then M is a homomorphic image of a free R-module F.

Proof. See the proof of Theorem I.10.1.

Corollary. If R is a commutative ring with 1 and M is a unitary R-module, with $M = R\langle S\rangle$ for some set $S \subseteq M$, then M is a homomorphic image of a free R-module F of rank $|S|$.

The ring $R = \mathbb{Z}_4$ has an ideal $M = 2R$, so M is a submodule of the free R-module R. Note, however, that M is not a free R-module, since $|M| = 2$ and $|R| = 4$. Thus submodules of free modules are not necessarily free.

Theorem 2.9. Suppose R is a PID, F is a free R-module, and E is a submodule of F. Then E is also free, and $\text{rank}(E) \le \text{rank}(F)$.

Proof. We may assume $E \neq 0$. Let B be a basis for F. For any $C \subseteq B$ set $F_C = R\langle C\rangle$ and $E_C = E \cap F_C$. Let $\mathscr{S}$ be the set of all triples (C, C', f), where $C' \subseteq C \subseteq B$, E_C is free, and $f: C' \to E_C$ is a function such that $f(C')$ is a basis for E_C. Then $\mathscr{S} \neq \varnothing$ since $(\varnothing, \varnothing, \varnothing) \in \mathscr{S}$. Partially order $\mathscr{S}$ by agreeing that $(C, C', f) \le (D, D', g)$ if $C \subseteq D$, $C' \subseteq D'$, and $g|C' = f$. Zorn's Lemma applies and we may choose a maximal element (A, A', h) in $\mathscr{S}$. The proof will be complete if we can show that $A = B$, since $E = E_B$. If $A \neq B$ choose $b \in B\backslash A$ and set $D = A \bigcup \{b\}$. If $E_D = E_A$, then $(A, A', h) < (D, A', h)$, contradicting maximality. If $E_D \neq E_A$, then there are various elements $y + rb \in E_D$, with $y \in F_A$, $r \in R$, and in some cases $r \neq 0$. Let I be the set of all $r \in R$ such that $y + rb \in E$ for some $y \in F_A$. Then I is an ideal in R, say $I = (s)$, and so $w = x + sb \in E$ for some $x \in F_A$ (note: $s \neq 0$). Set $D' = A' \bigcup \{b\}$ and extend $h': D' \to E_D$ by setting $h'(b) = w$. If $z \in E_D$, then $z = y + rb$ for some $y \in F_A$ and $r = r's \in I$, so

$$z = (y - r'x) + r'w, \qquad \text{and} \qquad z - r'w = y - r'x \in E \cap F_A = E_A.$$

It follows that $R\langle h'(D')\rangle = E_D$. A dependence relation on $h'(D')$ translates easily into a dependence relation on D, so $h'(D')$ is a basis for E_D. Thus $(A, A', h) < (D, D', h')$, maximality is violated again, and the theorem is proved.

Corollary. Suppose R is a PID, M is a finitely generated unitary R-module, and N is a submodule of M. Then N is finitely generated.

Proof. As shown in the proof of Proposition 2.5 there is an R-homomorphism $f: R^n \to M$ for some integer n. Thus $f^{-1}(N)$ is a submodule of R^n, so it is free of rank $m \leq n$ by the theorem. Thus $N = ff^{-1}(N)$ has a set of m generators.

Proposition 2.10. Suppose R is a ring with 1, M is a unitary R-module, F is a free R-module, and $f \in \mathrm{Hom}_R(M, F)$ is onto. Then M has a free submodule E that is R-isomorphic with F such that $M = E \oplus \ker f$.

Proof. If B is a basis for F choose $x_b \in M$ such that $f(x_b) = b$ for each $b \in B$. Set $E = R\langle\{x_b : b \in B\}\rangle$. If $\sum_{i=1}^{k} r_i x_{b_i} = 0$, then $0 = f(\sum r_i x_{b_i}) = \sum r_i b_i$, so all $r_i = 0$ and $\{x_b : b \in B\}$ is linearly independent. Thus $\{x_b : b \in B\}$ is a basis for E, so E is free and is clearly R-isomorphic with F. If $x \in M$ write $f(x) = \sum r_i b_i$, $b_i \in B$, and note that then $x - \sum r_i x_{b_i} \in \ker f$. Thus $M = E + \ker f$. Since $E \cap \ker f = 0$ we conclude that $M = E \oplus \ker f$ by Theorem 2.3.

3. FINITELY GENERATED MODULES OVER A PID

Assume throughout this section that R is a PID. The most important examples will be $R = F$, a field, $R = \mathbb{Z}$, and $R = F[x]$. *All R-modules considered will be assumed unitary.*

If M is an R-module, then an element $x \in M$ is called a *torsion* element if $rx = 0$ for some $r \neq 0$ in R. This generalizes the idea of an element of finite order in an abelian group.

If $x \in M$ is a torsion element we define its *annihilator*, or *order ideal*, to be $A(x) = \{r \in R : rx = 0\}$. Then $A(x)$ is an ideal in R, so $A(x) = (s_x)$ is principal; the generator s_x, which is determined up to associates, is called the *order* of x. We write $|x| = s_x$ for the order of x.

Exercise 3.1. Let $R = \mathbb{Q}[x]$ and let $M = V_T$ as in Exercise 1.2(2). If $u = \left[\begin{smallmatrix}1\\0\end{smallmatrix}\right]$ and $v = \left[\begin{smallmatrix}0\\1\end{smallmatrix}\right]$ find $|u|$ and $|v|$.

An R-module M is a *torsion module* if every element of M is a torsion element. At the other extreme M is *torsion-free* if 0 is the only torsion element in M.

Proposition 3.1. Let T be the set of all torsion elements in an R-module M. Then T is a submodule and M/T is torsion free.

Proof. If $x, y \in T$ say $|x| = a$ and $|y| = b$. Then $ab(x - y) = b \cdot ax - a \cdot by = 0$, and if $r \in R$, then $a(rx) = r(ax) = 0$, so T is a submodule of M. Suppose $r(x + T) = rx + T = T$ in M/T. Then $rx \in T$, so $A(rx) \neq 0$. If $0 \neq s \in A(rx)$, then $sr \cdot x = 0$, so $0 \neq sr \in A(x)$, and $x \in T$. Thus M/T is torsion free.

The submodule T of torsion elements in M is called its *torsion submodule*. In particular every abelian group has a *torsion subgroup* consisting of all its elements of finite order.

Note that the proof of Proposition 3.1 did not actually require that R be a PID, but only an integral domain.

Proposition 3.2. If M is a finitely generated torsion-free R-module, then M is free of finite rank.

Proof. Let $\{x_1, \ldots, x_n\}$ be a generating set for M and let $\{y_1, \ldots, y_m\}$ be a maximal linearly independent subset of $\{x_1, \ldots, x_n\}$. Let $N = R\langle y_1, \ldots, y_m\rangle$, a free submodule of M. For each x_i there exists some $s_i \neq 0$ in R and elements $r_{ij} \in R$ such that $s_i x_i + \sum_j r_{ij} y_j = 0$, by the maximality of $\{y_j\}$, and so $s_i x_i \in N$. Set $s = \prod\{s_i : 1 \leq i \leq n\}$, so that $sx \in N$ for all $x \in M$. Define $f: M \to N$ by setting $f(x) = sx$; then $f \in \operatorname{Hom}_R(M, N)$ and $\ker f = 0$ since M is torsion-free. Thus M is R-isomorphic with a submodule of N, so M is free and of finite rank by Theorem 2.9.

Theorem 3.3. If M is a finitely generated R-module with torsion submodule T, then M has a free submodule F of finite rank such that $M = T \oplus F$. The rank of F is uniquely determined by M.

Proof. Since M/T is torsion-free and the quotient map $\eta: M \to M/T$ is onto we conclude from Propositions 2.10 and 3.2 that $M = T \oplus F$ for some free submodule F of finite rank that is R-isomorphic with M/T. Any other free submodule F' such that $M = T \oplus F'$ is also R-isomorphic with M/T so rank $F' = \operatorname{rank} F$.

As a consequence of Theorem 3.3 we may define the *rank* of a finitely generated R-module M to be rank (F), where $M = T \oplus F$, T is the torsion submodule of M, and F is free.

Suppose that M is a torsion module and that there is some $r \in R^*$ with $rx = 0$ for all $x \in M$. Set $I = \{s \in R : sx = 0, \text{ all } x \in M\}$. Then I is an ideal in R; say $I = (a)$, so $a \mid r$. The generator a of I, which is determined up to associates, is called the *exponent* of M. It is the least common multiple of the orders of all elements of M.

Of course not every torsion module has an exponent. For example, take $M = \oplus\{\mathbb{Z}_n : n \in \mathbb{Z}\}$. However, if $M = R\langle x_1, \ldots, x_n\rangle$ is finitely generated and torsion, and if $|x_i| = r_i$, then $r = \prod\{r_i : 1 \leq i \leq n\}$ annihilates every $x \in M$, and M has an exponent.

Exercise 3.2. Suppose M is an R-module, that x and y are torsion elements with orders r and s, respectively, and that r and s are relatively prime in R. Show that $x + y$ has order rs.

Proposition 3.4. Suppose M is a torsion module having exponent r. Then M has an element of order r.

Proof. Recall that R, being a PID, is a UFD. Write $r = \prod\{p_i^{e_i}: 1 \le i \le k\}$, a product of distinct prime powers, with each $e_i > 0$, and set $r_i = r/p_i$. Then $r \nmid r_i$ and so for each i there is some $x_i \in M$ with $r_i x_i \neq 0$. Set $y_i = (r/p_i^{e_i})x_i$ and note that $p_i^{e_i} y_i = 0$ but $p_i^{e_i - 1} y_i = r_i x_i \neq 0$, and hence $|y_i| = p_i^{e_i}$ for each i. Set $x = y_1 + \cdots + y_k$. Then $|x| = r$ by Exercise 3.2.

If M is an R-module and $s \in R$ define $M[s] = \{x \in M: sx = 0\}$, the set of all elements in M having order a divisor of s.

Exercise 3.3. Show that $M[s]$ and $sM = \{sx: x \in M\}$ are both submodules of M.

Recall that if $r, s \in R$, not both 0, then the symbol (r, s) is used both to denote the ideal generated by $\{r, s\}$ and to denote a GCD for r and s, which is in fact a generator for that ideal. Both uses appear in the next proposition.

Proposition 3.5. Suppose $M = R\langle y\rangle$ is cyclic of order r, and $s \in R$. Then

(i) $M[s] = R\langle (r/(r,s))y\rangle \cong R/(r,s)$, and
(ii) $sM = R\langle sy\rangle \cong R/(r/(r,s))$,

so $M[s]$ is cyclic of order (r, s) and sM is cyclic of order $r/(r, s)$.

Proof. Note that $M[s] = \{uy: u \in R \text{ and } suy = 0\}$, and that $suy = 0$ if and only if u is a multiple of $r/(r, s)$. Thus $M[s] = R\langle (r/(r,s))y\rangle$. Define a map $\phi: R \to M[s]$ by setting $\phi(v) = (vr/(r,s))y$. Then ϕ is an epimorphism and $\ker\phi = (r, s)$, so (i) is proved. The proof of (ii) is similar and is left as an exercise.

Corollary. If $(r, s) = 1$, then $M[s] = 0$ and $sM = M$.

Proposition 3.6. Suppose $C = R\langle z\rangle$ is a cyclic R-module of order r_0, M is a finitely generated torsion R-module whose exponent divides r_0, N is a submodule of M, and $f \in \operatorname{Hom}_R(N, C)$. Then f can be extended to $g \in \operatorname{Hom}_R(M, C)$.

Proof. If $M = R\langle x_1, \ldots, x_k\rangle$ set

$$N_1 = N + R\langle x_1\rangle, \quad N_2 = N_1 + R\langle x_2\rangle, \quad \ldots, \quad N_k = N_{k-1} + R\langle x_k\rangle = M.$$

By induction it will suffice to show that f can be extended to an R-homomorphism from N_1 to C. If s is the order of $x_1 + N$ in N_1/N, then $s \mid r_0$ (since $r_0 x_1 + N = N$), say $r_0 = st$. Thus $sx_1 \in N$, so $f(sx_1) \in C$, and $tf(sx_1) = f(tsx_1) = f(r_0 x_1) = 0$, so $f(sx_1)$ has order dividing t in C. On the other hand $f(sx_1) = uz$ for some $u \in R$, so $tuz = 0$ and $r_0 | tu$, or $ts | tu$. Thus $s | u$, say $u = sv$, and so $f(sx_1) = s(vz)$. Set $z_0 = vz \in C$ and define $f' \in \operatorname{Hom}_R(N \oplus R, C)$ by setting $f'(y,r) = f(y) + rz_0$. Next define $h \in \operatorname{Hom}_R(N \oplus R, N_1)$ by setting $h(y,r) = y + rx_1$. Let us see that $\ker h \subseteq \ker f'$. If $(y,r) \in \ker h$, then $rx_1 = -y \in N$, so $r(x_1 + N) = 0$ in N_1/N, and hence $s | r$ in R, say $r = -ws$. Then $y = swx_1$, $(y,r) = (swx_1, -sw)$, and we have

$$f'(y,r) = f'(swx_1, -sw) = f(swx_1) - swz_0 = f(swx_1) - wf(sx_1) = 0.$$

Since h maps $N \oplus R$ onto N_1 we may define f_1 on N_1 by setting $f_1(h(y,r)) = f'(y,r)$ (f_1 is well defined since $\ker h \subseteq \ker f'$). Note that if $y \in N$, then

$$f(y) = f'(y,0) = f_1(h(y,0)) = f_1(y) + 0 \cdot x_1 = f_1(y),$$

so f_1 extends f to N_1 and the proof is complete.

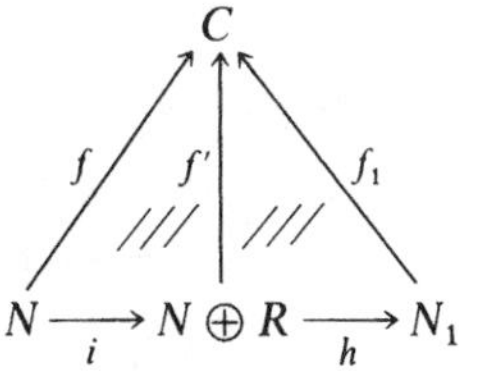

Theorem 3.7. If M is a finitely generated torsion R-module, then $M = M_1 \oplus M_2 \oplus \cdots \oplus M_k$, where each M_i is cyclic of order r_i, with $r_i | r_{i-1}$, $2 \le i \le k$, and r_1 the exponent of M.

Proof. By Proposition 3.4 we may choose an element $x_1 \in M$ whose order is the exponent r_1 of M. Set $M_1 = R\langle x_1 \rangle$. By Proposition 3.6 the identity map $i: M_1 \to M_1$ extends to an R-homomorphism $g_1: M \to M_1$. Set $N_2 = \ker g_1$. If $x \in M$, then

$$g_1(x - g_1(x)) = g_1(x) - g_1(x) = 0$$

since g_1 extends the identity map, so $x - g_1(x) \in N_2$ and

$$x = g_1(x) + (x - g_1(x)) \in M_1 + N_2.$$

If $x \in M_1 \cap N_2$, then $x = g_1(x) = 0$, so $M = M_1 \oplus N_2$ by Theorem 2.3. If N_2 has exponent r_2, then $r_2 | r_1$; we may choose $x_2 \in N_2$ of order r_2 and set $M_2 = R\langle x_2 \rangle$. Then $N_2 = M_2 \oplus N_3$ as above, hence $M = M_1 \oplus M_2 \oplus N_3$, and we may continue. The process must terminate in a finite number of steps with some $N_k = 0$ since M is Noetherian (Proposition 2.5).

Corollary. $M \cong R/(r_1) \oplus R/(r_2) \oplus \cdots \oplus R/(r_k)$.

Proof. The map $r \mapsto rx_i$ from R to M_i in the proof above has kernel (r_i), so $M_i \cong R/(r_i)$ by the FHTM.

Theorem 3.8 (Uniqueness). Suppose M is a finitely generated torsion R-module and that

$$M = M_1 \oplus \cdots \oplus M_k = N_1 \oplus \cdots \oplus N_m.$$

with M_i cyclic of order r_i, N_j cyclic of order s_j, $r_i | r_{i-1}$ for each i, $s_j | s_{j-1}$ for each j, and neither r_k nor s_m is a unit in R. Then $k = m$, r_i and s_i are associates in R, and $M_i \cong N_i$ for each i.

Proof. We may assume $k \geq m$. Choose a prime $p \in R$ such that $p | r_k$, and hence $p | r_i$, $1 \leq i \leq k$. Since M has both r_1 and s_1 as exponent r_1 and s_1 must be associates ($r_1 \sim s_1$), so $p | s_1$. Suppose $p | s_j$ for $1 \leq j \leq n$, where $n \leq m$, but $p \nmid s_{n+1}$ (if it exists). Since p is a prime $R/(p) = K$ is a field. By Proposition 3.5 we see that

$$M[p] = M_1[p] \oplus \cdots \oplus M_k[p] \cong \oplus\{K : 1 \leq i \leq k\},$$

a vector space of dimension k over K, and also

$$\begin{aligned} M[p] = N_1[p] \oplus \cdots \oplus N_m[p] &= N_1[p] \oplus \cdots \oplus N_n[p] \\ &= \oplus\{K : 1 \leq j \leq n\}, \end{aligned}$$

a vector space of dimension n over K. It follows, since $n \leq m \leq k$, that $m = k$. We have already observed that $r_1 \sim s_1$, and we wish to show that $r_i \sim s_i$ for all i. If not suppose that $r_i \sim s_i$ for $1 \leq i \leq n - 1$, but that $r_n \not\sim s_n$, say because $s_n \nmid r_n$. Set $M' = r_n M$, then

$$M' = r_n M_1 \oplus \cdots \oplus r_n M_{n-1} = r_n N_1 \oplus \cdots \oplus r_n N_{n-1} \oplus r_n N_n \oplus \cdots,$$

with $r_n N_n \neq 0$, violating the conclusion drawn above. Thus $r_i \sim s_i$ for all i, and furthermore we have

$$M_i \cong R/(r_i) = R/(s_i) \cong N_i \qquad \text{for all } i.$$

The orders $r_1, \ldots, r_k$ of the cyclic submodules $M_1, \ldots, M_k$ are called the *invariant factors* of the torsion module M. In case $R = \mathbb{Z}$, so $M = A$ is an abelian group, we take the orders $r_i = n_i$ to be positive integers, as usual. In that case we see by Theorem 3.7 that $|A| = n_1 n_2 \cdots n_k$, so A is finite.

Theorems 3.7 and 3.8 thus tell us that all finite abelian groups are determined up to isomorphism by their invariant factors. For example, if $n = |A| = 100$ we may write $100 = 50 \cdot 2 = 20 \cdot 5 = 10 \cdot 10$, and $\mathbb{Z}_{100}, \mathbb{Z}_{50} \oplus \mathbb{Z}_2, \mathbb{Z}_{20} \oplus \mathbb{Z}_5, \mathbb{Z}_{10} \oplus \mathbb{Z}_{10}$ are all the abelian groups of order 100. If $n = 48$ we have $48 = 24 \cdot 2 = 12 \cdot 4 = 12 \cdot 2 \cdot 2 = 6 \cdot 2 \cdot 2 \cdot 2$, and a corresponding list of five groups.

We may combine Theorems 3.3, 3.7, and 3.8 for a complete description of the structure of finitely generated modules over a PID.

Theorem 3.9. If M is a finitely generated R-module, then there exist a nonnegative integer m and nonzero nonunits $r_1, \ldots, r_k$ in R, with $r_i | r_{i-1}$ for $2 \leq i \leq k$, such that $M = M_1 \oplus \cdots \oplus M_k \oplus F$, where M_i is cyclic of order r_i and F is free of rank m. Equivalently

$$M \cong R/(r_1) \oplus \cdots \oplus R/(r_k) \oplus R^m.$$

The module M is determined up to R-isomorphism by m and the sequence $r_1, \ldots, r_k$ of invariant factors.

Corollary. If A is a finitely generated abelian group, then there exist a nonnegative integer m and integers $n_1, \ldots, n_k$, all larger than 1, with $n_i | n_{i-1}$ for $2 \leq i \leq k$, such that

$$A \cong \mathbb{Z}_{n_1} \oplus \cdots \oplus \mathbb{Z}_{n_k} \oplus \mathbb{Z}^m.$$

The group A is determined up to isomorphism by m and the sequence $n_1, \ldots, n_k$ of invariant factors, and the order of the torsion subgroup of A is $n_1 n_2 \cdots n_k$.

A torsion R-module M is called *primary*, or *p-primary*, if its exponent is some power p^e of a prime $p \in R$.

Proposition 3.10. Suppose M is a torsion module with exponent $r \in R$. If $r = st$, with $(s, t) = 1$, then $M = M[s] \oplus M[t]$. Consequently if $r = p_1^{e_1} p_2^{e_2} \cdots p_k^{e_k}$, where the p_i are distinct primes in R, then $M = M[p_1^{e_1}] \oplus \cdots \oplus M[p_k^{e_k}]$, a direct sum of primary submodules.

Proof. We may write $1 = as + bt$ for some $a, b \in R$ by Proposition II.5.2. For any $x \in M$ we have

$$x = t \cdot bx + s \cdot ax \in M[s] + M[t].$$

If $x \in M[s] \cap M[t]$, then $x = asx + btx = 0$. Thus $M = M[s] \oplus M[t]$ by Theorem 2.3.

Note that the primary submodules in the direct sum decomposition in Proposition 3.10 are unique: $M[p_i^{e_i}]$ is the submodule consisting of all elements of M having order a power of the prime divisor p_i of r.

We may now apply the Invariant Factor Theorem 3.7 to each primary submodule of a torsion module M, the result being the next theorem.

Theorem 3.11. Suppose M is a finitely generated torsion module of exponent $r = \prod\{p_i^{e_i}: 1 \leq i \leq k\}$, where the p_i are distinct primes in R and each $e_i \geq 1$. Then $M = \bigoplus\{M_i: 1 \leq i \leq k\}$, where M_i is p_i-primary, and $M_i = \bigoplus\{M_{ij}: 1 \leq j \leq k_i\}$, where M_{ij} is cyclic of order $p_i^{e_{ij}}$, with $1 \leq e_{ij} \leq e_{i(j-1)} \leq e_i$ for all i and j. M is determined up to isomorphism by the set $\{p_i^{e_{ij}}: 1 \leq j \leq k_i, 1 \leq i \leq k\}$ of prime powers, which are called the *elementary divisors* of M.

The statement of Theorem 3.11 applies directly to finite abelian groups; it will not be restated as a corollary.

As an application of Theorem 3.11 let us describe all abelian groups A of order $72 = 2^3 \cdot 3^2$ by means of elementary divisors. If $p_1 = 2$ and $p_2 = 3$ we require that $\sum_j e_{1j} = 3$ and $\sum_j e_{2j} = 2$. Thus the possible sets of elementary divisors are

$$\{2^3, 3^2\}, \quad \{2^3, 3, 3\}, \quad \{2^2, 2, 3^2\}, \quad \{2^2, 2, 3, 3\},$$
$$\{2, 2, 2, 3^2\}, \quad \text{and} \quad \{2, 2, 2, 3, 3\}.$$

Thus up to isomorphism an abelian group of order 72 is one of

$$\mathbb{Z}_8 \oplus \mathbb{Z}_9, \qquad \mathbb{Z}_8 \oplus \mathbb{Z}_3 \oplus \mathbb{Z}_3, \qquad \mathbb{Z}_4 \oplus \mathbb{Z}_2 \oplus \mathbb{Z}_9, \qquad \mathbb{Z}_4 \oplus \mathbb{Z}_2 \oplus \mathbb{Z}_3 \oplus \mathbb{Z}_3,$$
$$\mathbb{Z}_2 \oplus \mathbb{Z}_2 \oplus \mathbb{Z}_2 \oplus \mathbb{Z}_9, \qquad \text{and} \qquad \mathbb{Z}_2 \oplus \mathbb{Z}_2 \oplus \mathbb{Z}_2 \oplus \mathbb{Z}_3 \oplus \mathbb{Z}_3.$$

4. APPLICATIONS TO LINEAR ALGEBRA

We concentrate now on the modules $M = V_T$, where V is a finite-dimensional vector space over a field F and $T: V \to V$ is a linear transformation (see Example 4, p. 126).

Since the set of all linear transformations from V to V is itself a finite-dimensional vector space over F the set $\{I, T, T^2, \ldots\}$ is linearly dependent, and hence

$$a_0 I + a_1 T + \cdots + a_k T^k = f(T) = 0$$

for some nonzero $f(x) \in F[x]$. Just as in the proof of Proposition III.1.2 we see that there is a unique monic polynomial $m_T(x)$ of minimal positive degree such that $m_T(T) = 0$, and that if $f(x)$ is any polynomial such that $f(T) = 0$, then $m_T(x) \mid f(x)$ in $F[x]$. We call $m_T(x)$ the *minimal polynomial* of the linear transformation T.

For example, if $F = \mathbb{Q}$ and T is multiplication by $\left[\begin{smallmatrix} 1 & 1 \\ 0 & 1 \end{smallmatrix}\right]$, then $m_T(x) = x^2 - 2x + 1$; if T is multiplication by $\left[\begin{smallmatrix} -10 & 18 \\ -6 & 11 \end{smallmatrix}\right]$, then $m_T(x) = x^2 - x - 2$.

We recall from linear algebra that if $T: V \to V$ is represented by the matrix A, then the *characteristic polynomial* of T is $f(x) = \det(A - xI)$.

Exercise 4.1. Suppose $F = \mathbb{Q}$ and T is multiplication by

$$A = \begin{bmatrix} 4 & -6 & 3 \\ 4 & -7 & 4 \\ 6 & -12 & 7 \end{bmatrix}.$$

Show that T has minimal polynomial $m_T(x) = x^2 - 3x + 2$ and characteristic polynomial $f(x) = -x^3 + 4x^2 - 5x + 2$.

It is clear from the very definition of the minimal polynomial $m_T(x)$ that V_T is a torsion module and that $m_T(x)$ is its exponent. Any basis for V as an F-

vector space also generates V_T as a module, so V_T is finitely generated. Note that if W is any subspace of V, then $TW \subseteq W$ if and only if $f(T)W \subseteq W$ for all $f(x) \in F[x]$, so the submodules of V_T are just the T-*invariant* subspaces of V.

Exercise 4.2. Suppose $W = R\langle v\rangle$ is a cyclic submodule of V_T, and that W has order $f(x) \in F[x]$, where $\deg f(x) = k > 0$. Show that the set $\{v, Tv, T^2v, \ldots, T^{k-1}v\}$ is a (vector space) basis for W. We call v a *cyclic* vector for W.

Suppose F is a field and

$$f(x) = a_0 + a_1x + \cdots + a_{n-1}x^{n-1} + x^n$$

is a monic polynomial in $F[x]$. Define the *companion matrix* of $f(x)$ to be the $n \times n$ matrix

$$C(f) = \begin{bmatrix} 0 & 0 & 0 & 0 & \cdots & 0 & -a_0 \\ 1 & 0 & 0 & 0 & \cdots & 0 & -a_1 \\ 0 & 1 & 0 & 0 & \cdots & 0 & -a_2 \\ 0 & 0 & 1 & 0 & \cdots & 0 & -a_3 \\ \vdots & \vdots & \vdots & \vdots & & \vdots & \vdots \\ 0 & 0 & 0 & 0 & \cdots & 1 & -a_{n-1} \end{bmatrix}$$

For example, the companion matrix of $f(x) = 2 - 3x + x^3$ is

$$C(f) = \begin{bmatrix} 0 & 0 & -2 \\ 1 & 0 & 3 \\ 0 & 1 & 0 \end{bmatrix}.$$

Proposition 4.1. If F is a field, $f(x) \in F[x]$ is monic of degree n, and T is the linear transformation of F^n represented by the companion matrix $C(f)$, then the minimal polynomial $m_T(x)$ is $f(x)$.

Proof. Let $e_1, e_2, \ldots, e_n$ be the usual basis elements for F^n, i.e.,

$$e_1 = \begin{bmatrix} 1 \\ 0 \\ \vdots \\ 0 \end{bmatrix},$$

etc. Then $Te_i = e_{i+1}$ for $1 \le i \le n-1$, and

$$Te_n = -a_0e_1 - a_1e_2 - \cdots - a_{n-1}e_n.$$

Thus $e_2 = Te_1, e_3 = T^2e_1, \ldots, e_n = T^{n-1}e_1$, and hence

$$Te_n = T^ne_1 = -a_0Ie_1 - a_1Te_1 - \cdots - a_{n-1}T^{n-1}e_1,$$

which says that $f(T)e_1 = 0$. But then also

$$f(T)e_i = f(T)T^{i-1}e_1 = T^{i-1}f(T)e_1 = 0 \qquad \text{for all } i,$$

and hence $f(T)v = 0$ for all $v \in V$. If

$$g(x) = b_0 + b_1 x + \cdots + b_k x^k \in F[x],$$

with $k < n$ and $g(T) = 0$, then

$$g(T)e_1 = \sum_{j=0}^{k} b_j T^j e_1 = \sum_{j=0}^{k} b_j e_{j+1} = 0,$$

so all $b_j = 0$ and $g(x) = 0$. Thus $m_T(x) = f(x)$.

Exercise 4.3. Show that the characteristic polynomial of a companion matrix $C(f)$ is $\pm f(x)$.

Proposition 4.2. Suppose $W = R\langle v\rangle$ is a cyclic submodule of V_T, and that the order of W is the monic polynomial $f(x) \in F[x]$ of degree k. Then the restriction of T to W is represented by the companion matrix $C(f)$ with respect to the basis $\{v, Tv, \ldots, T^{k-1}v\}$ for W.

Proof. We saw in Exercise 4.2 that $\{v, Tv, \ldots, T^{k-1}v\}$ is indeed a basis for W. If $0 \le i \le k-2$, then $T(T^i v) = T^{i+1}v$, and if

$$f(x) = a_0 + a_1 x + \cdots + a_{k-1}x^{k-1} + x^k,$$

then

$$T(T^{k-1}v) = T^k v = -a_0 Iv - a_1 Tv - \cdots - a_{k-1}T^{k-1}v$$

since v has order $f(x)$. The proposition follows.

Theorem 4.3. If V is a finite-dimensional vector space over a field F and $T: V \to V$ is a linear transformation, then there is a basis for V with respect to which the matrix that represents T has the block diagonal form

$$A = \begin{bmatrix} C(f_1) & & & \\ & C(f_2) & & 0 \\ 0 & & \ddots & \\ & & & C(f_k) \end{bmatrix},$$

where the $f_i(x)$ are the (monic) invariant factors of V_T, so that $f_i(x) \mid f_{i-1}(x)$, $2 \le i \le k$. Furthermore $m_T(x) = f_1(x)$ and the characteristic polynomial of T is

$$f(x) = \pm\prod\{f_i(x): 1 \le i \le k\}.$$

Proof. By Theorem 3.7 we know that $V_T = \oplus\{W_i: 1 \le i \le k\}$, where $W_i = F[x]\langle v_i\rangle$ is cyclic of order $f_i(x)$, the ith invariant factor of V_T, chosen monic. That T has the matrix A is thus a consequence of Proposition 4.2. We have $m_T(x) = f_1(x)$ since both are the exponent of V_T, and the statement about the characteristic polynomial follows from Exercise 4.3.

Corollary (The Cayley–Hamilton Theorem). The minimal polynomial $m_T(x)$ is a divisor of the characteristic polynomial $f(x)$, and hence $f(T) = 0$.

The matrix A in Theorem 4.3 is called a *rational canonical* matrix for T. The word "rational" refers to the fact that the entries of A lie in F itself and no extension field is required. Recall from linear algebra that two matrices A and B represent the same linear transformation $T: V \to V$ (with respect to different bases for V) if and only if there is an invertible matrix C with entries in F (the *change-of-basis* matrix) such that $B = C^{-1}AC$. In that case we say that A and B are *similar* matrices over F. In fact, suppose A represents T with respect to the basis $\{v_i\}$ and B represents T with respect to the basis $\{w_i\}$, and let C be the matrix representing the identity transformation $I: V \to V$ with respect to $\{w_i\}$ in the domain and $\{v_i\}$ in the range, i.e., $w_i = \sum_j c_{ji}v_j$ for all i. Then we have

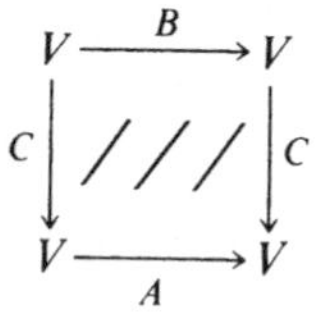

and so A and B are similar over F.

The next theorem is thus a direct consequence of Theorem 4.3 and the uniqueness Theorem 3.8.

Theorem 4.4. If F is a field and A is an $n \times n$ matrix with entries in F, then A is similar over F to a unique rational canonical matrix, called its *rational canonical form*. Two matrices A and B with entries in F are similar over F if and only if they have the same rational canonical form.

Corollary. Suppose K is an extension field of F, A and B are $n \times n$ matrices with entries in F, and A and B are similar over K. Then A and B are similar over F.

Proof. A and B are similar over F to rational canonical matrices P and Q, respectively, say $C^{-1}AC = P$ and $D^{-1}BD = Q$, all the matrices having entries in F. But A and B have entries in K, so we must have $P = Q$ by the theorem, and hence $A = (DC^{-1})^{-1}B(DC^{-1})$.

Let us next interpret primary decomposition and elementary divisors in the case $M = V_T$ (see Proposition 3.10 and Theorem 3.11).

If the minimal polynomial $m_T(x)$ of the linear transformation $T: V \to V$ factors as

$$m_T(x) = \prod\{p_i(x)^{e_i} : 1 \leq i \leq k\},$$

where the $p_i(x)$ are distinct, monic, and irreducible, and each $e_i \geq 1$, then V_T is the direct sum of the primary submodules $V_T[p_i(x)^{e_i}]$. Each primary submodule decomposes according to its (monic) invariant factors $p_i(x)^{e_{ij}}$. If the appropriate basis is chosen (via a cyclic vector) in each cyclic submodule, then the union of those bases is a basis for V_T, and with respect to that basis the matrix representing T is the block diagonal matrix with the companion matrices $C(p_i^{e_{ij}})$ as the diagonal blocks.

The uniquely determined set $\{p_i(x)^{e_{ij}}\}$ of monic polynomials is called the set of *elementary divisors* of T, or of any matrix A that represents T. If follows that two $n \times n$ matrices A and B with entries in a field F are similar over F if and only if they have the same set of elementary divisors.

Suppose now that one of the irreducible factors $p_i(x)$ of $m_T(x)$ is linear, say $p_i(x) = x - a_i$. For example, that is true of *all* $p_i(x)$ if F is algebraically closed. Suppose $p_i(x)^{e_{ij}} = (x - a_i)^{e_{ij}}$ is one of the elementary divisors of T and let V_{ij} be a cyclic submodule of order $(x - a_i)^{e_{ij}}$ with cyclic vector v. Note that T restricts to a linear transformation on V_{ij} and that the minimal polynomial of that restriction is $p_i(x)^{e_{ij}} = (x - a_i)^{e_{ij}}$. It follows (Why?) that the set

$$\{v, (T - a_iI)v, (T - a_iI)^2v, \ldots, (T - a_iI)^{e_{ij}-1}v\}$$

is a basis for V_{ij}. We then have

$$\begin{aligned}
Tv &= a_iv + (T - a_iI)v,\\
T(T - a_iI)v &= a_i(T - a_iI)v + (T - a_iI)^2v,\\
&\;\;\vdots\\
T(T - a_iI)^{e_{ij}-2}v &= a_i(T - a_iI)^{e_{ij}-2}v + (T - a_iI)^{e_{ij}-1}v,\\
T(T - a_iI)^{e_{ij}-1}v &= a_i(T - a_iI)^{e_{ij}-1}v
\end{aligned}$$

(verify these equations). Thus the matrix of the restriction of T to V_{ij} is the $e_{ij} \times e_{ij}$ matrix

$$A_{ij} = \begin{bmatrix}
a_i & 0 & 0 & \cdots & 0 & 0\\
1 & a_i & 0 & \cdots & 0 & 0\\
0 & 1 & a_i & \cdots & 0 & 0\\
\vdots & \vdots & \vdots & \ddots & \vdots & \vdots\\
0 & 0 & 0 & \cdots & a_i & 0\\
0 & 0 & 0 & \cdots & 1 & a_i
\end{bmatrix}.$$

The matrix A_{ij} is called a *Jordan block*; a block diagonal matrix whose diagonal blocks are Jordan blocks is called a *Jordan matrix*. If $T: V \to V$ can be represented by a Jordan matrix J, then J is called the *Jordan canonical form* for T, or for any matrix that represents T. Since the Jordan matrix (when it exists) is completely determined by the elementary divisors of T it is unique, except that the linear factors $p_i(x)$ may be relabeled and the diagonal blocks rearranged accordingly. It is customary to say that the matrices with the blocks rearranged in that way are the "same" Jordan canonical form.

Theorem 3.11 and the discussion above yield the next theorem.

Theorem 4.5. If $T: V \to V$ is a linear transformation such that all irreducible factors of $m_T(x)$ are linear, then V has a basis relative to which T is represented by a Jordan matrix. If F is algebraically closed and A, B are $n \times n$ matrices with entries in F, then A and B are similar over F if and only if they have the same Jordan canonical form.

Theorem 4.6. If F is any field, V is a finite-dimensional vector space over F, and $T: V \to V$ is a linear transformation, then T can be represented by a diagonal matrix if and only if the minimal polynomial $m_T(x)$ splits as a product of distinct linear factors in $F[x]$.

Proof. It is necessary and sufficient that T be represented by a Jordan matrix all of whose Jordan blocks are 1×1, which holds if and only if every elementary divisor is of degree 1. Thus the exponent of each primary submodule must be linear and the exponent of V_T, which is $m_T(x)$, is just the product of those distinct linear factors.

5. COMPUTATIONS

Suppose again that R is a PID, and assume that all R-modules discussed are unitary.

Suppose M is a free R-module of rank n, with $\{x_1, x_2, \ldots, x_n\}$ as a basis, N is a free R-module of rank m, with $\{y_1, y_2, \ldots, y_m\}$ as a basis, and $f \in \mathrm{Hom}_R(M, N)$. For each x_i we may write $f(x_i) = \sum_{j=1}^{m} a_{ji} y_j$, with $a_{ji} \in R$. Just as in linear algebra we say that f is *represented by* the $m \times n$ matrix $A = (a_{ij})$ relative to the bases $\{x_i\}$ for M and $\{y_j\}$ for N; the map $f \mapsto A$ is an R-isomorphism from $\mathrm{Hom}_R(M, N)$ onto the R-module of all $m \times n$ matrices with entries from R.

It should be stressed that in discussing the matrix representing an R-homomorphism it is important that the bases $\{x_i\}$ and $\{y_j\}$ are *ordered*; clearly a different matrix might result if basis elements are permuted.

Theorem 5.1. Suppose R is a PID, M and N are free R-modules of finite rank, and $f \in \mathrm{Hom}_R(M, N)$. Set $E = \mathrm{Im}(f)$. Suppose there are bases $\{x_1, \ldots, x_n\}$ for M and $\{y_1, \ldots, y_m\}$ for N such that the matrix representing f

relative to $\{x_i\}$ and $\{y_j\}$ has the block diagonal form

$$B = \begin{bmatrix} U & 0 & 0 \\ 0 & D & 0 \\ 0 & 0 & 0 \end{bmatrix},$$

where

$$U = \begin{bmatrix} u_1 & & 0 \\ & \ddots & \\ 0 & & u_s \end{bmatrix} \quad \text{and} \quad D = \begin{bmatrix} d_1 & & 0 \\ & \ddots & \\ 0 & & d_k \end{bmatrix},$$

with $u_i \in U(R)$, all i, $0 \neq d_j \notin U(R)$, all j, and $d_j | d_{j+1}$ for $1 \leq j \leq k-1$. Then the quotient module N/E is the direct sum of the cyclic submodules $R\langle y_i + E\rangle$, $s+1 \leq i \leq m$, its torsion submodule has invariant factors $d_k, d_{k-1}, \ldots, d_1$, and its rank is $m - s - k$.

Proof. Note that

$$f(x_i) = \begin{cases} u_i y_i & \text{if } 1 \leq i \leq s, \\ d_{i-s} y_i & \text{if } s+1 \leq i \leq s+k, \\ 0 & \text{if } s+k+1 \leq i \leq n, \end{cases}$$

and that $y_i \in E$ if $1 \leq i \leq s$ since $u_i \in U(R)$. Thus $\{y_i + E : s+1 \leq i \leq m\}$ generates N/E as an R-module. If we write $W_i = R\langle y_i + E\rangle$, $s+1 \leq i \leq m$, we have then that $N/E = W_{s+1} + \cdots + W_m$. Suppose $x \in W_i \cap \sum\{W_j : j \neq i\}$, say

$$x = -r_i y_i + E = \sum\{r_j y_j + E : j \neq i\}.$$

Then $\sum_{s+1}^{m} r_j y_j \in E$ and we may write

$$\sum_{s+1}^{m} r_j y_j = \sum_{j=1}^{s} s_j u_j y_j + \sum_{j=s+1}^{s+k} s_j d_{j-s} y_j,$$

with all $s_j \in R$. But then $s_j = 0$ if $1 \leq j \leq s$, $r_j = 0$ if $s+k+1 \leq j \leq m$, and $r_j = s_j d_{j-s}$ if $s+1 \leq j \leq s+k$ since $\{y_j\}$ is a basis for N. Thus $r_j y_j \in E$ for $s+1 \leq j \leq s+k$ and we conclude that $x = 0$. Thus $N/E = \bigoplus\{W_i : s+1 \leq i \leq m\}$.

Suppose now that $\sum_{i=s+1}^{m} r_i(y_i + E) = E$ in N/E, $r_i \in R$. As above we have

$$\sum_{s+1}^{m} r_j y_j = \sum_{j=s+1}^{s+k} s_j d_{j-s} y_j,$$

where $s_j \in R$, so $r_j = s_j d_{j-s}$ if $s+1 \leq j \leq s+k$ and $r_j = 0$ if $s+k+1 \leq j \leq m$. It follows that each $y_i + E$, $s+1 \leq i \leq s+k$, has order d_{i-s} in N/E and the set

$$\{y_i + E : s+k+1 \leq i \leq m\}$$

is R-linearly independent. Thus

$$\bigoplus\{W_i : s+1 \le i \le s+k\}$$

is the torsion submodule of N/E, and the invariant factors are $d_k, \ldots, d_1$ (see Theorem 3.9).

If $f \in \mathrm{Hom}_R(M, N)$ is represented by the matrix A relative to (ordered) bases $\{x_i\}$ and $\{y_j\}$, then changes in the bases result in changes in A. Let us consider some very specific sorts of changes and their results.

If x_i and x_j are interchanged then the ith and jth columns of A are interchanged. The same change in A can be effected by a matrix multiplication $A \mapsto AC_{ij}$, where C_{ij} is the matrix that results from interchanging the ith and jth columns in the $n \times n$ identity matrix I. For example

$$\begin{bmatrix} 1 & 2 & 3 \\ 4 & 5 & 6 \end{bmatrix} \begin{bmatrix} 0 & 1 & 0 \\ 1 & 0 & 0 \\ 0 & 0 & 1 \end{bmatrix} = \begin{bmatrix} 2 & 1 & 3 \\ 5 & 4 & 6 \end{bmatrix}.$$

Note that the matrix C_{ij} is invertible; in fact $C_{ij}^2 = I$.

Likewise if y_i and y_j are interchanged the effect on A is to interchange the ith and jth rows. The same change is effected by the matrix multiplication $A \mapsto R_{ij}A$, where R_{ij} results from interchanging the ith and jth rows of I. Note that $R_{ij}^2 = I$.

If x_i is replaced by $x_i + rx_j$ for some $r \in R$ and $j \neq i$ the resulting set is still a basis, and the effect on A is to add r times the jth column to the ith column (verify this). Again the same effect can be achieved by a matrix multiplication $A \mapsto AC_{ijr}$, where C_{ijr} is the result of adding r times the jth column of I to the ith column. The matrix C_{ijr} is also invertible, with $C_{ijr}^{-1} = C_{ij(-r)}$. Similar remarks (with a slight twist) apply to the invertible matrix R_{ijr} obtained by adding r times the jth row of I to the ith row. The result of multiplying $R_{ijr} \cdot A$ is to add r times the jth row of A to the ith row, but that row operation corresponds to replacing y_j by $y_j - ry_i$ in the basis for N.

The changes in A discussed above are called elementary *row* and *column operations*. The net effect of several row and/or column operations in succession is to replace A by $B = PAQ$, where P and Q are invertible matrices with entries in R, obtained as products of various R_{ij}s and R_{ijr}s (for P) and various C_{ij}s and C_{ijr}s (for Q). In general we say that $m \times n$ matrices B and A with entries in R are *equivalent* (over R) if $B = PAQ$ for invertible matrices P and Q with entries in R.

It is sometimes convenient, although we will not find it strictly necessary, to admit a third type of elementary operation, viz., multiplying a row or column by a unit u from R. This reduces to multiplication by R_{ijr} or C_{ijr} if we allow $i = j$ and take $r = u - 1$.

Theorem 5.2. Suppose R is a Euclidean ring with function $d: R^* \to \mathbb{Z}$, and $A = (a_{ij})$ is an $m \times n$ matrix with entries from R. Then there are invertible matrices P and Q with entries in R such that $PAQ = B$ has the block diagonal form

$$B = \begin{bmatrix} U & 0 & 0 \\ 0 & D & 0 \\ 0 & 0 & 0 \end{bmatrix},$$

as in Theorem 5.1. The diagonal entries d_i are unique up to associates.

Proof. By the discussion above it will suffice to show that A can be transformed to B by a succession of elementary row and column operations. We may of course assume that $A \neq 0$. Note first that we may apply elementary operations until the nonzero matrix entry with minimal d-value divides every other matrix entry (Why?), and that the situation will then persist when any further elementary operations are applied. We may further arrange, by interchanging rows and/or columns, that an entry with minimal d-value is a_{11} in the upper left-hand corner. Then, since a_{11} divides each a_{1i} in the first row and each a_{i1} in the first column, further elementary operations will reduce the matrix to the form

$$\begin{bmatrix} a_{11} & 0 \\ 0 & C \end{bmatrix},$$

with a_{11} dividing each entry of C. The procedure can now be applied to C and continued until the desired diagonal form is reached. As for uniqueness we may view A as the matrix representing an R-homomorphism f from R^n to R^m. Then B also represents f and by Theorem 5.1 the d_i are the invariant factors of the torsion submodule of $R^m/f(R^n)$, hence are unique by Theorem 3.8.

Theorem 5.2 actually holds true when R is a PID (see Exercise 7.21) but the proof is a bit more complicated and is not in general constructive. Elements with minimal d-value must be replaced by elements with a minimal number of prime factors (counting multiplicities), and elementary row and column operations are not generally sufficient. Matrices in block diagonal form

$$\begin{bmatrix} C & 0 \\ 0 & I \end{bmatrix}, \quad \text{where} \quad C = \begin{bmatrix} a & b \\ c & d \end{bmatrix},$$

with $ad - bc \in U(R)$, can also be necessary; they correspond to changes in bases that result in what are sometimes called *secondary* row and column operations on the representing matrix.

Let us look in detail at a small example. Suppose $f: \mathbb{Z}^2 \to \mathbb{Z}^2$ is given by multiplication by

$$A = \begin{bmatrix} 4 & 6 \\ -6 & 8 \end{bmatrix},$$

so f has matrix A relative to the bases $\{e_1, e_2\}$ in both $M = \mathbb{Z}^2$ and $N = \mathbb{Z}^2$. We wish to perform elementary operations on A to transform it to diagonal form. It will be convenient to compute the matrices P and Q in the process in order to find the new bases for M and N relative to which f is represented by the diagonal matrix. To that end we need only perform successively on identity matrices the same row operations (for P) and column operations (for Q) as are performed on A.

One possible sequence of elementary operations that succeeds for A is given (in order) by $R_{211}, R_{121}, R_{211}, C_{21(-10)}$. Accordingly we have

$$A = \begin{bmatrix} 4 & 6 \\ -6 & 8 \end{bmatrix} \xrightarrow{R_{211}} \begin{bmatrix} 4 & 6 \\ -2 & 14 \end{bmatrix} \xrightarrow{R_{121}} \begin{bmatrix} 2 & 20 \\ -2 & 14 \end{bmatrix} \xrightarrow{R_{211}}$$

$$\begin{bmatrix} 2 & 20 \\ 0 & 34 \end{bmatrix} \xrightarrow{C_{21(-10)}} \begin{bmatrix} 2 & 0 \\ 0 & 34 \end{bmatrix} = D,$$

$$\begin{bmatrix} 1 & 0 \\ 0 & 1 \end{bmatrix} \xrightarrow{R_{211}} \begin{bmatrix} 1 & 0 \\ 1 & 1 \end{bmatrix} \xrightarrow{R_{121}} \begin{bmatrix} 2 & 1 \\ 1 & 1 \end{bmatrix} \xrightarrow{R_{211}} \begin{bmatrix} 2 & 1 \\ 3 & 2 \end{bmatrix} = P,$$

and

$$\begin{bmatrix} 1 & 0 \\ 0 & 1 \end{bmatrix} \xrightarrow{C_{21(-10)}} \begin{bmatrix} 1 & -10 \\ 0 & 1 \end{bmatrix} = Q.$$

Observe that $PAQ = D$, as desired, and hence $AQ = P^{-1}D$. Since D is diagonal the effect of multiplying P^{-1} times D is to multiply the ith column of P^{-1} by the ith diagonal entry of D for each i. Thus in general the bases for M and N relative to which f is represented by D are determined by the columns of Q and of P^{-1}, respectively. For the present example we have

$$\begin{bmatrix} 1 & 0 \\ 0 & 1 \end{bmatrix} \xrightarrow{R_{21(-1)}} \begin{bmatrix} 1 & 0 \\ -1 & 1 \end{bmatrix} \xrightarrow{R_{12(-1)}} \begin{bmatrix} 2 & -1 \\ -1 & 1 \end{bmatrix} \xrightarrow{R_{21(-1)}}$$

$$\begin{bmatrix} 2 & -1 \\ -3 & 2 \end{bmatrix} = P^{-1},$$

so the bases consist of

$$x_1 = \begin{bmatrix} 1 \\ 0 \end{bmatrix}, \quad x_2 = \begin{bmatrix} -10 \\ 1 \end{bmatrix} \quad \text{for } M$$

and

$$y_1 = \begin{bmatrix} 2 \\ -3 \end{bmatrix}, \quad y_2 = \begin{bmatrix} -1 \\ 2 \end{bmatrix} \quad \text{for } N.$$

If we set $E = \text{Im}(f) \subseteq N$ and $G = N/E$, then, by the proof of Theorem 5.1, $G = \langle y_1 + E \rangle \oplus \langle y_2 + E \rangle$, G has invariant factors 2 and 34, and $|G| = 68$.

The matrix $B = PAQ$ that is equivalent to A in Theorem 5.2 is called the *Smith normal form* of A. The thrust of Theorem 5.2 is that any R-homomorphism $f: M \to N$, where R is Euclidean and M and N are free of finite rank, can be represented by a matrix in Smith normal form. Exercise 7.21 indicates the same result for any PID.

An important application is the solution of a system of linear equations with coefficients in $\mathbb{Z}$ (or in any Euclidean ring). The system

$$\begin{aligned} a_{11}x_1 + a_{12}x_2 + \cdots + a_{1n}x_n &= c_1, \\ a_{21}x_1 + a_{22}x_2 + \cdots + a_{2n}x_n &= c_2, \\ &\vdots \\ a_{m1}x_1 + a_{m2}x_2 + \cdots + a_{mn}x_n &= c_m \end{aligned}$$

can be written as $AX = C$, where $A = (a_{ij})$ is the coefficient matrix, $X = (x_1, x_2, \ldots, x_n)^t$ is the column vector of unknowns, and $C = (c_1, c_2, \ldots, c_m)^t$. We may change variables via $X = QY$, with Q invertible over $\mathbb{Z}$, so that the system becomes $AQY = C$, which is equivalent with $PAQY = PC$ if P is also invertible. For appropriate P and Q we have $PAQ = B$ in Smith normal form, so the system is $BY = PC$, which is easily solved if solutions exist, and the solutions X to the original system are obtained as $X = QY$.

For example, let us find all integral solutions of the system

$$\begin{aligned} -33x_1 + 42x_2 - 20x_3 &= -26, \\ 21x_1 - 27x_2 + 13x_3 &= 16. \end{aligned}$$

This can be written as $AX = C$ with

$$A = \begin{bmatrix} -33 & 42 & -20 \\ 21 & -27 & 13 \end{bmatrix},$$

$$X = (x_1, x_2, x_3)^t, \quad \text{and} \quad C = (-26, 16)^t.$$

Change-of-basis matrices

$$P = \begin{bmatrix} 2 & 3 \\ 1 & 2 \end{bmatrix} \quad \text{and} \quad Q = \begin{bmatrix} 0 & 1 & 2 \\ 1 & 2 & 3 \\ 2 & 3 & 3 \end{bmatrix}$$

transform A to Smith normal form

$$B = \begin{bmatrix} 1 & 0 & 0 \\ 0 & 3 & 0 \end{bmatrix}$$

(verify). If we change variables $X = QY$ we obtain $PAQY = BY = PC = \left[\begin{smallmatrix} -4 \\ 6 \end{smallmatrix}\right]$. Thus the system becomes $1 \cdot y_1 = -4, 3 \cdot y_2 = 6$, so that $y_1 = -4$, $y_2 = 2$, and y_3 is unrestricted. We may set $y_3 = a$, an arbitrary integer, and we find that

$$\begin{bmatrix} x_1 \\ x_2 \\ x_3 \end{bmatrix} = \begin{bmatrix} 0 & 1 & 2 \\ 1 & 2 & 3 \\ 2 & 3 & 3 \end{bmatrix} \begin{bmatrix} -4 \\ 2 \\ a \end{bmatrix} = \begin{bmatrix} 2+2a \\ 3a \\ -2+3a \end{bmatrix} = \begin{bmatrix} 2 \\ 0 \\ -2 \end{bmatrix} + a \begin{bmatrix} 2 \\ 3 \\ 3 \end{bmatrix},$$

with $a \in \mathbb{Z}$ arbitrary, is the full set of solutions to the system $AX = C$.

Consider now the ring $R = F[x]$, F a field, and an R-module V_T. Choose a basis $\{v_1, v_2, \ldots, v_n\}$ for V so that T is represented by a matrix A. Let $N = R^n$, the free R-module consisting of all column vectors with n entries from R. In this setting we have the following theorem.

Theorem 5.3. Let E be the submodule of N generated by the columns of the matrix $A - xI$. Then V_T is R-isomorphic with N/E.

Proof. If we set $e_1 = (1, 0, \ldots, 0)^t$, $e_2 = (0, 1, 0, \ldots, 0)^t, \ldots$, as usual, then $\{e_1, e_2, \ldots, e_n\}$ is an R-basis for N. The mapping ϕ: $N \to V_T$ determined by $\phi(e_i) = v_i$ is an R-homomorphism from N onto V_T, so it will suffice, by the FHTM, to show that $\ker \phi = E$. For $1 \le i \le n$ let z_i be the ith column of $A - xI$, i.e.,

$$z_i = \sum_{j=1}^{n} a_{ji} e_j - x e_i \in N.$$

Then

$$\phi(z_i) = \sum_{j=1}^{n} a_{ji} v_j - x v_i = \sum_{j=1}^{n} a_{ji} v_j - T(v_i) = 0$$

for all i, since A represents T relative to $\{v_1, \ldots, v_n\}$. Thus $E \subseteq \ker \phi$.

On the other hand let W be the F-subspace of N/E spanned by $\{e_i + E: 1 \le i \le n\}$. Observe that

$$\begin{aligned} x(e_i + E) = x e_i + E &= \sum_j a_{ji} e_j - \left(\sum_j a_{ji} e_j - x e_i \right) + E \\ &= \sum_j a_{ji} e_j - z_i + E = \sum_j a_{ji}(e_j + E) \in W \end{aligned}$$

since $z_i \in E$, so W is in fact a submodule of N/E. But $N = R\langle e_1, \ldots, e_n \rangle$, so $W = N/E$, and if $u \in N$ we may write $u = \sum_i c_i e_i + z$, where $c_i \in F$ and $z \in E$. Consequently

$$\phi(u) = \sum_i c_i \phi(e_i) + \phi(z) = \sum_i c_i v_i$$

since $E \subseteq \ker \phi$, and hence $u \in \ker \phi$ if and only if all $c_i = 0$, i.e., if and only if $u = z \in E$.

Theorems 5.1, 5.2, and 5.3 make possible the computation of invariant factors for any linear transformation T on a finite-dimensional vector space V. If T is represented by a matrix A relative to a basis $\{v_1,\dots,v_n\}$ for V, set $M = N = R^n$, where $R = F[x]$, and let $z_i \in N$ be the ith column of $A - xI$. Define an R-homomorphism $f: M \to N$ by setting $f(e_i) = z_i$, $1 \le i \le n$; set $E = Im(f) = R\langle z_1,\dots,z_n\rangle$, and we have $V_T \cong N/E$ by Theorem 5.3. The matrix representing f relative to the basis $\{e_1,\dots,e_n\}$ for both M and N is just $A - xI$. By Theorem 5.2 we may choose new bases so that f is represented by a diagonal matrix B with each diagonal entry dividing the next, and then see by Theorem 5.1 that the nonzero nonunit diagonal entries are the invariant factors of N/E, and hence of V_T (the diagonal entries will in fact all be nonzero since V_T is a torsion module).

If we write $P(A - xI)Q = B$, diagonal, then $(A - xI)Q = P^{-1}B$, and since B is diagonal the ith column of $P^{-1}B$ is just the ith column of P^{-1} multiplied by the ith diagonal entry of B. Thus the bases for M and N relative to which f is represented by the diagonal matrix B are, respectively, the column vectors of Q and of P^{-1}.

If we write $y_1, y_2,\dots, y_n$ for the columns of P^{-1}, then, by Theorem 5.1,

$$N/E = \oplus\{R\langle y_i + E\rangle : s + 1 \le i \le n\},$$

where the first s diagonal entries of B are units. The isomorphism, call it θ, given in Theorem 5.3 (and its proof) from N/E to V_T is determined by $\theta(e_i + E) = v_i$, so if the column y_i is written as $y_i = \sum_j y_{ji}e_j$, then $\theta(y_i + E) = \sum_j y_{ji}v_j = u_i$, say. Note here that each $y_{ji} \in R = F[x]$ and $v_j \in V$, so $y_{ji}v_j$ means the module action of polynomials on vectors.

We conclude that $V_T = \oplus\{R\langle u_i\rangle : s + 1 \le i \le n\}$. With this information and the aid of Proposition 4.2 and Theorem 4.3 we may obtain the F-basis for V relative to which T is represented by its rational canonical form. Note that the F-dimension of $R\langle u_i\rangle$ is equal to the polynomial degree of its order as an R-module, i.e., the polynomial degree of the corresponding invariant factor.

Once cyclic vectors are obtained for the rational canonical form it is a straightforward matter to obtain a basis for the Jordan canonical form (provided, of course, that it exists). Each cyclic submodule is first decomposed as a direct sum of primary submodules as indicated in Proposition 3.10. If the order of each is a power of a linear factor, then a basis is chosen as in the discussion preceding Theorem 4.5.

A few examples should make the procedures clear.

First take $V = \mathbb{Q}^3$ and let T be represented by the matrix

$$A = \begin{bmatrix} 5 & -8 & 4 \\ 6 & -11 & 6 \\ 6 & -12 & 7 \end{bmatrix}$$

relative to the basis $\{e_1, e_2, e_3\}$. The sequence $C_{13}, C_{212}, C_{31(x-5)/4}, R_{21(-3/2)}, R_{31(x-7)/4}, C_{32(3/2)}, R_{32(-2)}$ of row and column operations yields

$$P = R_{32(-2)}R_{31(x-7)/4}R_{21(-3/2)}I = \begin{bmatrix} 1 & 0 & 0 \\ -3/2 & 1 & 0 \\ (x+5)/4 & -2 & 1 \end{bmatrix},$$

$$Q = IC_{13}C_{212}C_{31(x-5)/4}C_{32(3/2)} = \begin{bmatrix} 0 & 0 & 1 \\ 0 & 1 & 3/2 \\ 1 & 2 & (x+7)/4 \end{bmatrix}$$

for which

$$P(A - xI)Q = \begin{bmatrix} 4 & 0 & 0 \\ 0 & 1-x & 0 \\ 0 & 0 & (1-x^2)/4 \end{bmatrix}$$

(verify). Thus the invariant factors of T are $f_1(x) = m_T(x) = x^2 - 1$ and $f_2(x) = x - 1$.

The rational canonical form for T is thus

$$C = \begin{bmatrix} 0 & 1 & 0 \\ 1 & 0 & 0 \\ 0 & 0 & 1 \end{bmatrix}.$$

Generators for the cyclic submodules are obtained by means of the columns of P^{-1} as discussed above. Since

$$P^{-1} = R_{21(3/2)}R_{31(7-x)/4}R_{322}I = \begin{bmatrix} 1 & 0 & 0 \\ 3/2 & 1 & 0 \\ (7-x)/4 & 2 & 1 \end{bmatrix}$$

we have

$$u_1 = e_1 + \frac{3}{2}e_2 + \frac{7-x}{4}e_3$$

$$= e_1 + \frac{3}{2}e_2 + \frac{1}{4}(7I - A)e_3 = 0,$$

$$u_2 = e_2 + 2e_3,$$

and

$$u_3 = e_3.$$

Applying Proposition 4.2 to the cyclic vectors u_2 and u_3 we obtain a basis $w_1 = u_3$, $w_2 = Tw_1 = (4, 6, 7)^t$, and $w_3 = u_2$. The change-of-basis matrix (from $\{e_1, e_2, e_3\}$ to $\{w_1, w_2, w_3\}$) is then

$$L = \begin{bmatrix} 0 & 4 & 0 \\ 0 & 6 & 1 \\ 1 & 7 & 2 \end{bmatrix},$$

and $L^{-1}AL = C$, the rational canonical form.

Since $m_T(x) = (x - 1)(x + 1)$ has distinct linear factors the Jordan canonical form for T is diagonal, by Theorem 4.6. The primary submodules of $R\langle u_3 \rangle$ have generators

$$v_1 = (x - 1)u_3 = (A - I)u_3 = (4, 6, 6)^t$$

and

$$v_2 = (x + 1)u_3 = (A + I)u_3 = (4, 6, 8)^t,$$

eigenvectors for the eigenvalues $-1, 1$, respectively. Since $v_3 = w_3$ is also an eigenvector for eigenvalue 1 the Jordan form for T, relative to $\{v_1, v_2, v_3\}$, is

$$J = \begin{bmatrix} -1 & 0 & 0 \\ 0 & 1 & 0 \\ 0 & 0 & 1 \end{bmatrix}.$$

The matrix M with columns v_1, v_2, v_3 serves as a change-of-basis matrix. That end result would probably have been reached more quickly and more directly by means of the standard techniques of linear algebra. The basis vectors v_1, v_2, v_3 are simply linearly independent solutions to the homogeneous systems of equations $(A - \lambda I)X = 0$, $\lambda = -1, 1$.

Next take $V = \mathbb{Q}^4$, and suppose that $T\colon V \to V$ is represented by

$$A = \begin{bmatrix} 3 & -1 & 1 & -1 \\ 0 & 3 & -1 & 1 \\ 2 & -1 & 3 & -4 \\ 3 & -3 & 3 & -4 \end{bmatrix}$$

relative to $\{e_1, e_2, e_3, e_4\}$. Then

$$P(A - xI)Q = \begin{bmatrix} 1 & 0 & 0 & 0 \\ 0 & 1 & 0 & 0 \\ 0 & 0 & 1 & 0 \\ 0 & 0 & 0 & (x + 1)(x - 2)^3/27 \end{bmatrix},$$

where

$$P = R_{43(x-2)^3/27} R_{341} R_{43(-1)} R_{42(6-3x)} R_{32(x^2-7x+9)} R_{41(-3)} R_{31(x-3)} R_{211} I,$$

$$Q = C_{43(x+1)/27} C_{34} C_{43(3x-15)} C_{34} C_{32(x-2)} C_{23} C_{32(-1)} C_{411} C_{31(x-3)} C_{211} C_{13} I.$$

Thus

$$m_T(x) = f_1(x) = (x+1)(x-2)^3 = -8 + 4x + 6x^2 - 5x^3 + x^4$$

is the only invariant factor of T, and the rational canonical form of T is

$$C = \begin{bmatrix} 0 & 0 & 0 & 8 \\ 1 & 0 & 0 & -4 \\ 0 & 1 & 0 & -6 \\ 0 & 0 & 1 & 5 \end{bmatrix}.$$

In this case

$$P^{-1}$$
$$= R_{21(-1)}R_{31(3-x)}R_{413}R_{32(-x^2+7x-9)}R_{42(3x-6)}R_{431}R_{341}R_{43[(x-2)^3/-27]}I$$
$$= \begin{bmatrix} 1 & 0 & 0 & 0 \\ -1 & 1 & 0 & 0 \\ 3-x & -x^2+7x-9 & 1-(x-2)^3/27 & 1 \\ 3 & 3x-6 & 1-2(x-2)^3/27 & 2 \end{bmatrix}.$$

Thus $u_1 = p^{-1}e_4 = (0,0,1,2)^t$ is a cyclic vector for the rational canonical form of T, and u_1, together with $u_2 = Tu_1 = (-1,1,-5,-5)^t$, $u_3 = Tu_2 = (-4,3,2,-1)^t$, and $u_4 = Tu_3 = (-12,6,-1,-11)^t$ provides a basis relative to which T is represented by the rational canonical form C.

The minimal polynomial $m_T(x)$ factors as $p_1(x)p_2(x)$, with $p_1(x) = x+1$, $p_2(x) = (x-2)^3$. The corresponding primary submodules have cyclic vectors $(x-2)^3u_1 = (0,0,-81,-81)^t$ and $(x+1)u_1 = (-1,1,-4,-3)^t$. For arithmetical convenience we may replace $(0,0,-81,-81)^t$ by $v_1 = (0,0,1,1)^t$. Then v_1, together with

$$v_2 = (x+1)u_1 = (-1,1,-4,-3)^t,$$
$$v_3 = (x-2)v_2 = (-3,2,5,0)^t,$$

and

$$v_4 = (x-2)v_3 = (0,-3,-3,0)^t$$

provides a basis for the Jordan canonical form

$$J = \begin{bmatrix} -1 & 0 & 0 & 0 \\ 0 & 2 & 0 & 0 \\ 0 & 1 & 2 & 0 \\ 0 & 0 & 1 & 2 \end{bmatrix}.$$

For a final example take $V = F^4$, $\mathbb{Q} \subseteq F \subseteq \mathbb{C}$, and suppose that $T: V \to V$ is represented by the matrix

$$A = \begin{bmatrix} 3 & -3 & 3 & -5 \\ 2 & -1 & 1 & -3 \\ 2 & -1 & 1 & -4 \\ 2 & -2 & 3 & -4 \end{bmatrix}$$

relative to $\{e_1, e_2, e_3, e_4\}$. Then

$$P(A - xI)Q = \begin{bmatrix} 1 & 0 & 0 & 0 \\ 0 & -1 & 0 & 0 \\ 0 & 0 & 1 & 0 \\ 0 & 0 & 0 & (x^2+1)^2 \end{bmatrix},$$

where

$$\begin{aligned} P &= R_{34} R_{34x} R_{43(-1)} R_{42(1+3x)} R_{32(1+x-x^2)} R_{24(-1)} \\ &\quad \times R_{41(-3)} R_{31(x-2)} R_{21(-3)} R_{12} I, \\ Q &= C_{43(x^3+x^2+3)} C_{34(-1)} C_{42(1-x)} C_{32(x-1)} C_{41(-3)} \\ &\quad \times C_{312} C_{21(-1-x)} C_{13} I. \end{aligned}$$

Thus $m_T(x) = (x^2+1)^2 = x^4 + 2x^2 + 1$.

The vector $u_1 = P^{-1}e_4 = (1,0,1,1)^t$ is a cyclic vector for V_T, and u_1, together with

$$\begin{aligned} u_2 &= Au_1 = (1,0,0,1)^t, \\ u_3 &= Au_2 = (-2,-1,-2,-2)^t \end{aligned}$$

and

$$u_4 = Au_3 = (1,1,1,0)^t$$

provides a basis relative to which T is represented by its rational canonical form

$$C = \begin{bmatrix} 0 & 0 & 0 & -1 \\ 1 & 0 & 0 & 0 \\ 0 & 1 & 0 & -2 \\ 0 & 0 & 1 & 0 \end{bmatrix}.$$

If $F \subseteq \mathbb{R}$, then $x^2 + 1$ is irreducible in $F[x]$ and T has no Jordan canonical form. If, on the other hand, $x^2 + 1$ splits in $F[x]$, then $m_T(x) = (x-i)^2 (x+i)^2$, where $i^2 + 1 = 0$. As cyclic vectors for the primary submodules $V_T[(x+i)^2]$ and $V_T[(x-i)^2]$ we may take

$$v_1 = (x - i)^2 u_1 = (-3-2i, -1, -3, -3-2i)^t$$

and

$$w_1 = (x + i)^2 u_1 = (-3 + 2i, -1, -3, -3 + 2i)^t,$$

respectively. Then

$$\begin{aligned} v_1, v_2 &= (x + i)v_1 = (2 + i, 1 + i, 1 + i, 1 + i)^t, \\ v_3 &= w_1, \end{aligned}$$

and

$$v_4 = (x - i)w_1 = (2 - i, 1 - i, 1 - i, 1 - i)^t$$

provide a basis for the Jordan canonical form

$$J = \begin{bmatrix} -i & 0 & 0 & 0 \\ 1 & -i & 0 & 0 \\ 0 & 0 & i & 0 \\ 0 & 0 & 1 & i \end{bmatrix} \quad \text{for} \quad T.$$

The Smith normal form can also be used to compute invariants of finitely generated abelian groups given by presentations. Much as in Section 10 of Chapter I we may write an additive presentation

$$G = \left\langle a_1, \ldots, a_m \,\middle|\, \sum_{j=1}^{m} r_{ji} a_j = 0, 1 \le i \le n \right\rangle,$$

whereby we view G as the quotient of the free abelian group N with basis $\{a_1, \ldots, a_m\}$ by the subgroup

$$E = \left\langle \sum_j r_{ji} a_j : 1 \le i \le n \right\rangle.$$

If we set $M = \mathbb{Z}^n$ and define $f \in \operatorname{Hom}_{\mathbb{Z}}(M, N)$ by setting

$$f(e_i) = \sum_j r_{ji} a_j, \qquad 1 \le i \le n,$$

then $\operatorname{Im}(f) = E$ and f is represented by the matrix $A = [r_{ij}]$. Theorems 5.1 and 5.2 apply directly, so we may determine the invariant factors and the torsion-free rank of G.

For example, take G to have generators a_1, a_2, a_3, a_4, subject to the relations

$$\begin{aligned} 4a_1 - 4a_2 - 6a_3 + a_4 &= 0, \\ 6a_1 - 8a_2 - 12a_3 + a_4 &= 0, \\ 4a_1 - 10a_2 - 12a_3 + a_4 &= 0. \end{aligned}$$

Then the representing matrix

$$A = \begin{bmatrix} 4 & 6 & 4 \\ -4 & -8 & -10 \\ -6 & -12 & -12 \\ 1 & 1 & 1 \end{bmatrix}$$

is equivalent with

$$PAQ = \begin{bmatrix} 1 & 0 & 0 \\ 0 & 2 & 0 \\ 0 & 0 & 6 \\ 0 & 0 & 0 \end{bmatrix}$$

by means of the sequence $R_{14}, R_{41(-4)}, R_{316}, R_{214}, C_{21(-1)}, C_{31(-1)}, C_{23(-1)}, R_{24(-1)}, R_{32(-2)}, R_{231}$, and R_{24}. Thus the invariant factors of G are 2 and 6, and the torsion-free rank of G is 1, by Theorem 5.1. The torsion subgroup is isomorphic with $\mathbb{Z}_2 \oplus \mathbb{Z}_6$, of order 12, and $G \cong \mathbb{Z}_2 \oplus \mathbb{Z}_6 \oplus \mathbb{Z}$.

6. TENSOR PRODUCTS

Suppose V is a finite-dimensional vector space over the real field $\mathbb{R}$ and that $T: V \to V$ is a linear transformation whose minimal polynomial is $m_T(x) = x^2 + 1$. Then T has no eigenvalues in $\mathbb{R}$, but the equation $m_T(x) = 0$ can of course be solved in the extension field $\mathbb{C}$. It is in fact customary to say that T has $\pm i$ as complex eigenvalues. Corresponding eigenvectors would necessarily allow scalar multiplication by nonreal complex numbers, so it would seem desirable to extend V to a vector space over $\mathbb{C}$. In particular it should then be possible to multiply elements of the original vector space V by elements of the extension field $\mathbb{C}$.

Phrased somewhat more abstractly we wish to be able to "multiply" together elements of two different $\mathbb{R}$-modules $M = \mathbb{C}$ and $N = V$. The tensor product of modules is a useful device that makes such multiplication possible in great generality.

Suppose R is a ring, M is a right R-module, and N is a left R-module. If A is an abelian group, then a function $f: M \times N \to A$ is called a *balanced map* if

(i) $f(x + u, y) = f(x, y) + f(u, y)$,
(ii) $f(x, y + v) = f(x, y) + f(x, v)$, and
(iii) $f(xr, y) = f(x, ry)$

for all $x, u \in M$, all $y, v \in N$, and all $r \in R$.

A *tensor product* of a right R-module M and a left R-module N is an abelian group T together with a balanced map $t: M \times N \to T$ such that given any abelian group A and any balanced map $f\colon M \times N \to A$ there is a unique homomorphism $g\colon T \to A$ such that $f = gt$, i.e., the diagram

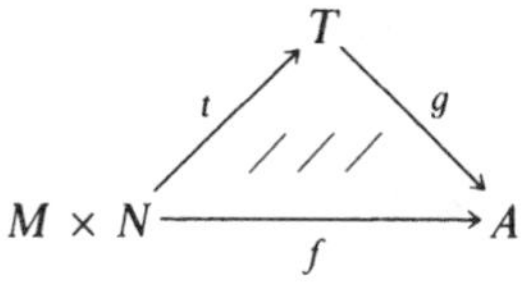

is commutative.

Proposition 6.1. If a tensor product exists it is unique up to isomorphism.

Exercise 6.1. Prove Proposition 6.1.

Exercise 6.2. If $M = \mathbb{Q}$ and $N = \mathbb{Z}_2$, both viewed as abelian groups (i.e., $\mathbb{Z}$-modules), show that $T = 0$ is a tensor product for M and N.

Theorem 6.2. If R is a ring, M is a right R-module, and N is a left R-module, then a tensor product of M and N over R exists.

Proof. Let F be the free abelian group based on the Cartesian product $M \times N$; a typical element of F is thus a finite sum $\sum_i n_i(x_i, y_i)$, $n_i \in \mathbb{Z}$, $x_i \in M$, $y_i \in N$. Let H be the subgroup of F generated by all elements of the forms

(i) $(x + u, y) - (x, y) - (u, y)$,
(ii) $(x, y + v) - (x, y) - (x, v)$, and
(iii) $(xr, y) - (x, ry)$

for all $x, u \in M$, all $y, v \in N$, and all $r \in R$. Set $T = F/H$ and define $t\colon M \times N \to T$ by setting $t(x, y) = (x, y) + H$. It is clear from the definition of H that t is a balanced map. Suppose A is an abelian group and $f\colon M \times N \to A$ is a balanced map. If we define $h\colon F \to A$ by setting

$$h\left(\sum n_i(x_i, y_i)\right) = \sum n_i f(x_i, y_i),$$

then $h \in \mathrm{Hom}(F, A)$ and $H \le \ker h$ since f is balanced. Thus we may define g from $F/H = T$ to A by setting $g(a + H) = h(a)$ for each $a \in F$. Then $g \in \mathrm{Hom}(T, A)$ and

$$gt(x, y) = g\big((x, y) + H\big) = h(x, y) = f(x, y),$$

so $gt = f$. For the uniqueness of g note that $t(M \times N)$ generates T, so if $g' \in \mathrm{Hom}(T, A)$ and $g't = f$, then $g't = gt$, and hence $g' = g$. Consequently T is a tensor product for M and N.

Because of Proposition 6.1 we may take the group T constructed in the proof of Theorem 6.2 and call it *the* tensor product of M and N over R. For a more convenient notation let us write $x \otimes y$ for $(x, y) + H \in T$, and consequently write $\sum_i x_i \otimes y_i$, a finite sum, for a typical element of T. We write $M \otimes_R N$ for T itself (as constructed).

The notation $x \otimes y$ suggests that we have indeed "multiplied" together the elements $x \in M$ and $y \in N$. Observe that

$$(x + u) \otimes y = x \otimes y + u \otimes y,$$
$$x \otimes (y + v) = x \otimes y + x \otimes v,$$

and

$$xr \otimes y = x \otimes ry$$

for all $x, u \in M$, all $y, v \in N$, and all $r \in R$, giving the basic properties of the multiplication.

Exercise 6.3. Show that $x \otimes 0 = 0 \otimes y = 0$ for all $x \in M$, all $y \in N$.

For example, take $R = \mathbb{Z}$, $M = \mathbb{Q}$, and $N = \mathbb{Z}_n$ for some integer $n > 0$. Then for any $x \in M$, $y \in N$ we may write

$$x \otimes y = (x/n) \cdot n \otimes y = (x/n) \otimes ny = (x/n) \otimes 0 = 0.$$

Thus $M \otimes_R N = 0$ in this case.

Exercise 6.4. Suppose R is a ring, M_1 and M_2 are right R-modules, N_1 and N_2 are left R-modules, $f \in \mathrm{Hom}_R(M_1, M_2)$, and $g \in \mathrm{Hom}_R(N_1, N_2)$.

(1) Show that there exists a unique $h \in \mathrm{Hom}_{\mathbb{Z}}(M_1 \otimes_R N_1, M_2 \otimes_R N_2)$ such that $h(x \otimes y) = f(x) \otimes g(y)$ for all $x \in M_1$, $y \in N_1$. (*Hint*: define a (balanced) map from $M_1 \times N_1$ to $M_2 \otimes_R N_2$ via $(x, y) \mapsto f(x) \otimes g(y)$ and see the definition of a tensor product.) The unique homomorphism h is denoted by $f \otimes g$.

(2) Suppose further that $f' \in \mathrm{Hom}_R(M_2, M_3)$ and $g' \in \mathrm{Hom}_R(N_2, N_3)$. Show that $(f' \otimes g')(f \otimes g) = f'f \otimes g'g$.

Suppose S and R are rings. An abelian group M is called an *S-R-bimodule* if it is simultaneously a left S-module and a right R-module, satisfying the further requirement that $s(xr) = (sx)r$ for all $s \in S$, $x \in M$, and $r \in R$.

Suppose now that M is an S-R-bimodule and N is a left R-module. We may make a left S-module of the tensor product $M \otimes_R N$, as follows. For each $s \in S$ define $f_s\colon M \times N \to M \otimes_R N$ via $f_s(x, y) = sx \otimes y$. Then f_s is a balanced map (verify), and there is a unique homomorphism $g_s\colon M \otimes_R N \to M \otimes_R N$ such that $g_s(x \otimes y) = sx \otimes y$ for all $x \in M$, $y \in N$. As a consequence we may define a module action on $M \otimes_R N$ by setting

$$s\left(\sum_i x_i \otimes y_i\right) = \sum_i g_s(x_i \otimes y_i) = \sum_i sx_i \otimes y_i$$

for each $s \in S$.

Exercise 6.5. Verify that $M \otimes_R N$ is a left S-module as indicated above. If M is a unitary left S-module show that $M \otimes_R N$ is also a unitary S-module.

Proposition 6.3. Suppose R is a ring with 1, N is a unitary left R-module, and R is viewed as an R-R-bimodule with ordinary multiplication in R as the module actions. Then $R \otimes_R N$ is R-isomorphic with N.

Proof. The map $f\colon R \times N \to N$ defined by $f(r, y) = ry$ is easily seen to be balanced. Thus there is a homomorphism $g\colon R \otimes_R N \to N$ such that $g(r \otimes y) = ry$ for all $r \in R$ and $y \in N$. It is easy to check that g is an R-homomorphism, and g is onto since N is unitary. If $\sum_i r_i \otimes y_i$ is in $\ker g$, then $\sum_i r_i y_i = 0$. But then

$$\sum_i r_i \otimes y_i = \sum_i 1 \cdot r_i \otimes y_i = \sum_i 1 \otimes r_i y_i = 1 \otimes 0 = 0,$$

so $\ker g = 0$ and the proposition is proved.

Proposition 6.4. Suppose M is a right R-module and N_1 and N_2 are left R-modules. Then

$$M \otimes_R (N_1 \oplus N_2) \cong (M \otimes_R N_1) \oplus (M \otimes_R N_2).$$

Proof. Set $N = N_1 \oplus N_2$ and define homomorphisms p_1 and p_2 from N to N via $p_1(y_1, y_2) = (y_1, 0)$ and $p_2(y_1, y_2) = (0, y_2)$. Clearly $p_i(N) \cong N_i$. Set $f_i = 1 \otimes p_i\colon M \otimes_R N \to M \otimes_R N$ [see Exercise (6.4)]. It is easy to check that $f_1 + f_2 = 1$, the identity map, that $f_1 f_2 = f_2 f_1 = 0$, and that $f_i^2 = f_i$ for each i. As a consequence, if we set $K_i = \operatorname{Im}(f_i)$, then $M \otimes_R N = K_1 \oplus K_2$ (verify). Thus it will suffice to show that $K_i \cong M \otimes_R N_i$, $i = 1, 2$. Define $t_1\colon M \times N_1 \to K_1$ by setting

$$t_1(x, y_1) = x \otimes (y_1, 0) = f_1(x \otimes (y_1, 0)).$$

Then t_1 is balanced. Suppose that $f\colon M \times N_1 \to A$ is a balanced map for some abelian group A. Define $h\colon M \times N \to A$ by setting $h(x, (y_1, y_2)) = f(x, y_1)$, then h is also balanced. Thus there exists a unique homomorphism k from $M \otimes_R N$ to A such that $h = kt$. Let $g = k \mid K_1$, so $g \in \operatorname{Hom}(K_1, A)$, and observe that

$$\begin{aligned} gt_1(x, y_1) &= g(x \otimes (y_1, 0)) = k(x \otimes (y_1, 0)) \\ &= kt(x, (y_1, 0)) = h(x, (y_1, 0)) = f(x, y_1), \end{aligned}$$

so $gt_1 = f$. That g is unique follows from the fact that $\operatorname{Im}(t_1)$ generates K_1. Thus $K_1 \cong M \otimes_R N_1$. A completely analogous argument shows that $K_2 \cong M \otimes_R N_2$, and the proposition is proved.

The diagram below may help in following the proof just completed.

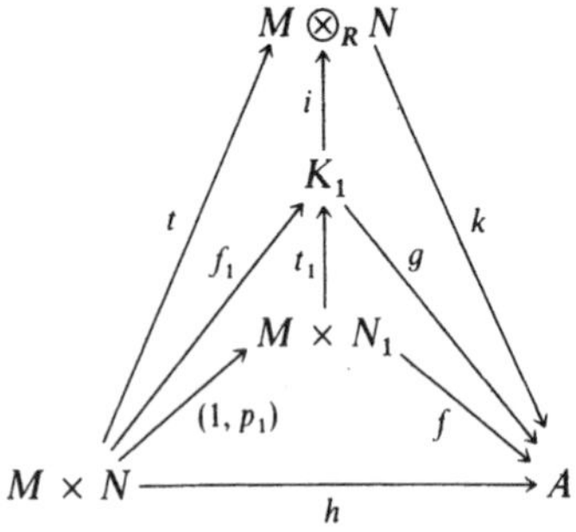

Remark: Clearly a result analogous to Proposition 6.4 holds if $M = M_1 \oplus M_2$, and if the M_i are S-R-bimodules, then the isomorphism is an S-isomorphism.

We may now solve the problem alluded to in the first paragraph of this section, viz., extending the ground field of a vector space.

Suppose V is a vector space over a field F and K is an extension field of F. We shall assume that V is finite dimensional, although the assumption is not at all necessary. Let $\{x_1, \ldots, x_n\}$ be an F-basis for V, so $V = Fx_1 \oplus \cdots \oplus Fx_n \cong F^n$. We may view K as a unitary K-F-bimodule. Then $V^K = K \otimes_F V$ is a unitary K-module, i.e., a K-vector space. Note that

$$V^K = K \otimes_F V \cong K \otimes_F (F^n) \cong (K \otimes_F F)^n \cong K^n$$

by (analogs to) Propositions 6.3 and 6.4. Thus $\dim_K(V^K) = \dim_F V$.

Given $x = \sum_{i=1}^m a_i \otimes v_i \in V^K$ we may write $v_i = \sum_{j=1}^n b_{ji} x_j$, with $b_{ji} \in F$, and hence

$$x = \sum_i a_i \otimes \left(\sum_j b_{ji} x_j\right) = \sum_{j=1}^n \left(\sum_i b_{ji} a_i\right)(1 \otimes x_j).$$

Consequently $\{1 \otimes x_1, \ldots, 1 \otimes x_n\}$ is a K-basis for V^K. The function $v \mapsto 1 \otimes v$ is a 1–1 map from V into V^K, so we may identify V with the image and view V as a subset (*not* a K-subspace) of V^K.

Exercise 6.6. If K is an extension field of F, V is an F-vector space, and $V^K = K \otimes_F V$, then V^K is also an F-vector space. Show that $\dim_F(V^K) = [K:F] \dim_F V$.

If F is a field and V, W are F-vector spaces, then arguments analogous to those above show that

$$\dim_F(V \otimes_F W) = (\dim_F V)(\dim_F W).$$

If $\{x_1,\dots,x_n\}$ and $\{y_1,\dots,y_m\}$ are bases for V and W, respectively, then

$$\{x_i \otimes y_j : 1 \le i \le n, 1 \le j \le m\}$$

is a basis for $V \otimes_F W$. Suppose that $S\colon V \to V$ and $T\colon W \to W$ are linear transformations, and that they are represented by the matrices $A = (a_{ij})$ and $B = (b_{ij})$ relative to the bases $\{x_i\}$ are $\{y_j\}$. Recall from Exercise 6.4 that $S \otimes T\colon V \otimes_F W \to V \otimes_F W$ is characterized by the fact that $(S \otimes T)(v \otimes w) = S(v) \otimes T(w)$ for all $v \in V$, $w \in W$. If the basis $\{x_i \otimes y_j\}$ is ordered

$$x_1 \otimes y_1, x_1 \otimes y_2, \dots, x_1 \otimes y_m, \quad x_2 \otimes y_1, \dots, x_2 \otimes y_m, \dots,$$
$$x_n \otimes y_1, \dots, x_n \otimes y_m,$$

then the matrix $A \otimes B$ that represents $S \otimes T$ has the block form $(a_{ij}B)$, which is commonly called the *Kronecker product* of A and B. This construction is of considerable importance in the theory of group representations, and thereby in the study of quantum mechanics.

Exercise 6.7. Verify that $S \otimes T$, as above, is represented by the Kronecker product of A and B.

If R is a commutative ring then a ring A is called an *R-algebra* if A is a left R-module such that $r(ab) = (ra)b = a(rb)$ for all $r \in R$ and all $a, b \in A$.

Examples abound. A $\mathbb{Z}$-algebra is nothing but an arbitrary ring A. *If* R is a commutative ring with 1, then any polynomial ring $R[x_1,\dots,x_n]$ is an R-algebra. If F is a field and K is an extension of F, then K is an F-algebra. If V is an F-vector space, then the ring $A = \mathrm{Hom}_F(V, V)$ is an F-algebra.

If R is a commutative ring, A and B are R-algebras, and A is an R-R-bimodule, then $A \otimes_R B$ is an R-module. Let us show that it can be made into an R-algebra as well. Fix $a \in A$, $b \in B$, and define $f_{ab}\colon A \times B \to A \otimes_R B$ by setting $f_{ab}(c, d) = ac \otimes bd$. Then f_{ab} is a balanced map and hence there is a unique homomorphism $g_{ab}\colon A \otimes_R B \to A \otimes_R B$ such that $g_{ab}(c \otimes d) = ac \otimes bd$ for all $c \in A$, $d \in B$. We define

$$\left(\sum_i a_i \otimes b_i\right) \cdot \left(\sum_j c_j \otimes d_j\right) = \sum_i g_{a_i b_i}\left(\sum_j c_j \otimes d_j\right) = \sum_{i,j} a_i c_j \otimes b_i d_j.$$

Since $\sum_i g_{a_i b_i}$ is an R-homomorphism it is a routine matter to check that the multiplication, so defined on $A \otimes_R B$, makes $A \otimes_R B$ an R-algebra. It is unitary if R is a ring with 1 and A is unitary as a left R-module.

Theorem 6.5. Suppose R is a commutative ring with 1, and I and J are ideals in R. Then $(R/I) \otimes_R (R/J)$ and $R/(I + J)$ are isomorphic R-algebras.

Proof. Define $t\colon R/I \times R/J \to R/(I + J)$ by setting $t(r + I, s + J) = rs + I + J$. Then t is a balanced map. If A is an abelian group and $f\colon R/I \times R/J \to A$ is a balanced map define $g\colon R/(I + J) \to A$ by setting

$$g(r + I + J) = f(r + I, 1 + J).$$

It is perhaps not obvious that g is well defined. If $r + I + J = r' + I + J$, then $r' = r + a + b$, with $a \in I$ and $b \in J$. Thus

$$r' + I = r + b + I,$$

$$f(r' + I, 1 + J) = f(r + b + I, 1 + J) = f(r + I, 1 + J) + f(b + I, 1 + J),$$

and

$$f(b + \mathrm{I}, 1 + J) = f(1 + I, b + J) = f(1 + I, 0) = 0,$$

so g is well defined. It is clear that g is a homomorphism (of abelian groups), and

$$\begin{aligned} gt(r + I, s + J) &= g(rs + I + J) = f(rs + I, 1 + J) \\ &= f\big((r + I)s, 1 + J\big) = f\big(r + I, s(1 + J)\big) = f(r + I, s + J) \end{aligned}$$

since f is balanced. Thus $gt = f$, and g is uniquely determined since t maps $R/I \times R/J$ onto $R/(I + J)$. Thus $R/(I + J)$ and $(R/I) \otimes_R (R/J)$ are isomorphic abelian groups by Proposition 6.1. We remark that $r + I + J \mapsto (r + I) \otimes (1 + J)$ is an explicit isomorphism between them, and that it is an R-algebra isomorphism.

Corollary. $\mathbb{Z}_m \otimes_{\mathbb{Z}} \mathbb{Z}_n \cong \mathbb{Z}_{(m,n)}$.

Proof. $\mathbb{Z}_m = \mathbb{Z}/(m)$, $\mathbb{Z}_n = \mathbb{Z}/(n)$, and $(\mathrm{m}) + (n) = (m, n)$.

7. FURTHER EXERCISES

1. If R is a ring with 1 and M is an R-module that is not unitary show that $Rm = 0$ for some nonzero $m \in M$.

2. Give an example of an R-module M having R-isomorphic submodules N_1 and N_2 such that M/N_1 and M/N_2 are not isomorphic.

3. Suppose V is a finite-dimensional vector space over a field F, viewed as an F-module. Describe a composition series for V and determine its factors.

4. A sequence $K \xrightarrow{f} M \xrightarrow{g} N$ of R-homomorphisms of R-modules is *exact* at M if $\mathrm{Im}(f) = \ker g$. A *short exact sequence* $0 \to K \to M \to N \to 0$ is exact at K, M, and N. If $0 \to K \to M \to N \to 0$ is short exact show that M is Noetherian if and only if K and N are both Noetherian.

5. Suppose M_1, M_2 and N are submodules of an R-module M, with $M_1 \subseteq M_2$. Show that there is an exact sequence

$$0 \to (M_2 \cap N)/(M_1 \cap N) \to M_2/M_1 \to (M_2 + N)/(M_1 + N) \to 0.$$

6. Let F be a free abelian group with countably infinite basis $\{a_1, a_2, a_3, \ldots\}$, and let $R = \mathrm{End}(F)$. Show that R, as a free R-module, has one basis $B_1 = \{1_R\}$, but also another basis $B_2 = \{\phi_1, \phi_2\}$, where $\phi_1(a_{2n}) = a_n$, $\phi_1(a_{2n-1}) = 0$, $\phi_2(a_{2n}) = 0$, and $\phi_2(a_{2n-1}) = a_n$, $n = 1, 2, \ldots$.

7. Give an example of an R-module M over a commutative ring R where the set $T(M)$ of torsion elements of M is not a submodule.

8. If F is a field set $R = F[x_1, x_2, x_3, \ldots]$, the ring of polynomials in a countably infinite set of distinct indeterminates. Let I be the ideal $(x_1, x_2, \ldots)$ in R. If $M = R$ and $N = I$ show that M is a finitely generated R-module but N is a submodule that is not finitely generated. Is N free?

9. Suppose R is a PID and $M = R\langle a \rangle$ is a cyclic R-module of order r, $0 \neq r \in R$. Show that if N is a submodule of M, then N is cyclic of order s for some divisor s of r. Conversely, M has a cyclic submodule N of order s for each divisor s of r in R.

10. Suppose R is a commutative ring and M is an R-module. A submodule N is called *pure* if $rN = rM \cap N$ for all $r \in R$.

(i) Show that any direct summand of M is pure.

(ii) If M is torsion free and N is a pure submodule show that M/N is torsion free.

(iii) If M/N is torsion free show that N is pure.

11. Suppose L, M, and N are R-modules and $f\colon M \to N$ is an R-homomorphism. Define $f^*\colon \mathrm{Hom}_R(N, L) \to \mathrm{Hom}_R(M, L)$ via $f^*(\phi)\colon m \mapsto \phi(f(m))$ for all $\phi \in \mathrm{Hom}_R(N, L)$, $m \in M$.

(i) Show that f^* is a $\mathbb{Z}$-homomorphism.

(ii) If R is commutative show that f^* is an R-homomorphism.

(iii) If K, M, and N are R-modules show that $K \to M \to N \to 0$ is an exact sequence of R-homomorphisms if and only if

$$0 \to \mathrm{Hom}_{\mathbb{Z}}(N, L) \to \mathrm{Hom}_{\mathbb{Z}}(M, L) \to \mathrm{Hom}_{\mathbb{Z}}(K, L)$$

is an exact sequence of $\mathbb{Z}$-homomorphisms for all R-modules L.

12. (i) If R is an integral domain show that free R-modules are torsion free.

(ii) If R is an integral domain with 1 that is not a field exhibit a torsion free R-module that is not free.

13. Suppose R is a commutative ring with 1 and M is an R-module. Then the R-module $M^* = \mathrm{Hom}_R(M, R)$ is called the *dual module* of M. The elements of M^* are commonly called *R-linear functionals* on M. If M is free of finite rank, with basis $\{x_1, x_2, \ldots, x_n\}$, show that M^* is also free, with basis $\{f_1, f_2, \ldots, f_n\}$, where

$$f_i(x_j) = \begin{cases} 1 & \text{if } \; i = j \\ 0 & \text{if } \; i \neq j \end{cases},$$

(the *dual basis*). Conclude that M and M^* are R-isomorphic in that case.

14. If R is a commutative ring with 1 and M is an R-module define a function $\phi\colon x \to \hat{x}$ from M to its *double dual* $M^{**} = (M^*)^*$ by setting $\hat{x}(f) = f(x)$, all

$f \in M^*$ (see Exercise 13). Show that ϕ is an R-homomorphism. Under what circumstances is ϕ a monomorphism?

15. Use invariant factors to describe all abelian groups of orders 144 and 168.

16. Use elementary divisors to describe all abelian groups of orders 144 and 168.

17. If p and q are distinct primes use invariant factors to describe all abelian groups of order

$$\text{(i)}\quad p^2q^2, \qquad \text{(ii)}\quad p^4q, \qquad \text{and} \qquad \text{(iii)}\quad p^n, \quad 1 \le n \le 6.$$

18. If p and q are distinct primes use elementary divisors to describe all abelian groups of order p^3q^2.

19. Suppose R is a PID and M is a finitely generated torsion R-module. Show that the elementary divisors of M are just the prime power factors (with multiplicities) that result when the invariant factors of M are factored into prime powers. Show on the other hand that the invariant factors can be recovered from the list of elementary divisors as follows: the last invariant factor is the LCM of all the elementary divisors. If the elementary divisors whose product is the last invariant factor are removed from the list, then the LCM of those that remain is the next invariant factor, etc.

20. If R is a PID and M is a finitely generated torsion R-module show that the decomposition of M by means of its invariant factors expresses M as a direct sum of the smallest possible number of cyclic submodules, and that its decomposition by means of its elementary divisors expresses M as the direct sum of the largest possible number of (nonzero) cyclic submodules.

21. Generalize Theorem 5.2 to modules over any PID, using secondary row and column operations when necessary.

22. Find all solutions $X \in \mathbb{Z}^3$ to the system of equations $AX = 0$ if A is

$$\text{(i)}\quad \begin{bmatrix} 1 & -1 & 1 \\ 1 & 0 & 2 \end{bmatrix}, \qquad \text{(ii)}\quad \begin{bmatrix} 0 & 2 & -1 \\ 1 & -1 & 0 \\ 2 & 0 & -1 \end{bmatrix}, \qquad \text{(iii)}\quad \begin{bmatrix} -3 & -4 & 1 \\ 0 & 0 & 0 \\ 3 & 4 & -1 \end{bmatrix}.$$

23. Find all integral solutions to the following systems $AX = B$ of equations:

$$\text{(i)}\quad A = \begin{bmatrix} 1 & -1 & 1 \\ 1 & 0 & 2 \end{bmatrix}, \qquad B = \begin{bmatrix} 4 \\ 5 \end{bmatrix},$$

$$\text{(ii)}\quad A = \begin{bmatrix} 0 & 2 & -1 \\ 1 & -1 & 0 \\ 2 & 0 & -1 \end{bmatrix}, \qquad B = \begin{bmatrix} 5 \\ 1 \\ 7 \end{bmatrix},$$

(iii) $A = \begin{bmatrix} 0 & 2 & 3 \\ 1 & 4 & 6 \\ 2 & 6 & 9 \end{bmatrix}, \qquad B = \begin{bmatrix} 2 \\ 3 \\ 3 \end{bmatrix},$

(iv) $A = \begin{bmatrix} 8 & 19 & 30 \\ 6 & 14 & 22 \end{bmatrix}, \qquad B = \begin{bmatrix} 5 \\ 7 \end{bmatrix}.$

24. What are the invariant factors and the elementary divisors of a diagonal matrix over a field F?

25. If a matrix A over a field F has minimal polynomial $m(x)$ and characteristic polynomial $f(x)$ show that $f(x)$ is a divisor of $m(x)^k$ in $F[x]$ for some positive integer k.

26. Suppose $V = \{(a, b)^t : a, b \in \mathbb{Q}\}$, $T: V \to V$ is multiplication by

$$A = \begin{bmatrix} -10 & 18 \\ -6 & 11 \end{bmatrix},$$

and M is the $\mathbb{Q}[x]$-module V_T. Find the elementary divisors of M and write M as a direct sum of cyclic submodules.

27. Determine whether or not

$$A = \begin{bmatrix} 3 & 0 & 2 \\ 0 & 1 & -1 \\ -4 & 0 & 3 \end{bmatrix} \quad \text{and} \quad B = \begin{bmatrix} 5 & -8 & 4 \\ 6 & -11 & 6 \\ 6 & -12 & 7 \end{bmatrix}$$

are similar over $\mathbb{Q}$.

28. Find an explicit similarity transform relating

$$A = \begin{bmatrix} 0 & -2 & -4 \\ 0 & 2 & 4 \\ 0 & -1 & -2 \end{bmatrix} \quad \text{and} \quad A^t = \begin{bmatrix} 0 & 0 & 0 \\ -2 & 2 & -1 \\ -4 & 4 & -2 \end{bmatrix}.$$

(*Hint*: They have the same rational canonical form.)

29. If A is an $n \times n$ matrix over a field F show that A is similar (over F) to its transpose A^t (see Exercise 28).

30. Find the characteristic polynomial, invariant factors, elementary divisors, rational canonical form, and Jordan canonical form (when possible) over $\mathbb{Q}$ for each of the following matrices:

(i) $\begin{bmatrix} 0 & -4 \\ 1 & -4 \end{bmatrix}$, (ii) $\begin{bmatrix} c+6 & -9 \\ 4 & c-6 \end{bmatrix}$ $(c \in \mathbb{Q})$,

(iii) $\begin{bmatrix} 3 & -2 & -4 \\ 0 & 2 & 4 \\ 0 & -1 & -2 \end{bmatrix}$, (iv) $\begin{bmatrix} -2 & 3 & -2 \\ -1 & 2 & -1 \\ 0 & 1 & 0 \end{bmatrix}$,

(v) $\begin{bmatrix} 1 & 0 & -2 \\ 2 & 6 & 8 \\ -1 & -3 & -4 \end{bmatrix}$, (vi) $\begin{bmatrix} 1 & 0 & 0 \\ -4 & 7 & -4 \\ -8 & 12 & -7 \end{bmatrix}$,

(vii) $\begin{bmatrix} 1 & 2 & 2 \\ 2 & 1 & 2 \\ 2 & 2 & 1 \end{bmatrix}$, (viii) $\begin{bmatrix} 3 & 2 & -1 \\ -2 & -2 & 2 \\ -1 & -5 & 5 \end{bmatrix}$,

(ix) $\begin{bmatrix} 0 & -1 & 2 \\ 3 & 8 & -14 \\ 3 & 6 & -10 \end{bmatrix}$, (x) $\begin{bmatrix} 4 & -2 & -1 \\ 5 & -2 & -1 \\ -2 & 1 & 1 \end{bmatrix}$,

(xi) $\begin{bmatrix} -2 & 0 & 0 & 1 \\ 1 & 1 & 0 & 1 \\ 2 & 0 & 1 & -2 \\ -1 & 0 & 0 & 0 \end{bmatrix}$, (xii) $\begin{bmatrix} 1 & 0 & 1 & 0 \\ 4 & 3 & -2 & 0 \\ -2 & 1 & 5 & 0 \\ 2 & 0 & -1 & 3 \end{bmatrix}$.

31. An $n \times n$ matrix A over a field F is called *idempotent* if $A^2 = A$.

(i) What are the possible minimal polynomials for an idempotent matrix?

(ii) Show that an idempotent matrix is similar over F to a diagonal matrix.

(iii) Show that idempotent $n \times n$ matrices A and B are similar over F if and only if they have the same rank.

32. An $n \times n$ matrix A over a field F is called *nilpotent* if $A^k = 0$ for some positive integer k.

(i) If A is nilpotent and $A \neq 0$ show that A is not similar to a diagonal matrix.

(ii) Show that a nilpotent matrix A has a Jordan canonical form over F, and list the possible Jordan forms for A.

33. Suppose A and B are $n \times n$ matrices over a field F and that $\left[\begin{smallmatrix} A & 0 \\ 0 & A \end{smallmatrix}\right]$ and $\left[\begin{smallmatrix} B & 0 \\ 0 & B \end{smallmatrix}\right]$ are similar over F. Show that A and B are similar over F.

34. Suppose A and B are finitely generated abelian groups and that $A \oplus A \cong B \oplus B$. Show that $A \cong B$.

35. Compute the invariants of the abelian groups with generators $a_1, \ldots, a_n$, subject to the following relations:

(i)
$$n = 2,$$
$$a_1 - 5a_2 = 0,$$
$$3a_1 + 7a_2 = 0,$$

(ii)
$$n = 3,$$
$$a_1 \qquad - 3a_3 = 0,$$
$$a_1 + 2a_2 + 5a_3 = 0,$$

(iii) $n = 3$,

$$\begin{aligned} 2a_1 - a_2 &= 0, \\ a_1 - 3a_2 &= 0, \\ a_1 + a_2 + a_3 &= 0. \end{aligned}$$

(iv) $n = 4$,

$$\begin{aligned} 2a_1 - a_2 + 5a_3 &= 0, \\ a_2 - 3a_3 &= 0, \\ 3a_1 - 7a_3 &= 0. \end{aligned}$$

36. If R is a commutative ring and M, N are R-R-bimodules show that $M \otimes_R N$ and $N \otimes_R M$ are isomorphic as R-modules.

37. If R is a commutative ring with 1 and x_1, x_2 are distinct indeterminates show that $R[x_1, x_2]$ and $R[x_1] \otimes_R R[x_2]$ are isomorphic as R-algebras.

38. If A and B are finitely generated abelian groups show that $A \otimes_{\mathbb{Z}} B$ is a finitely generated abelian group. Find its rank and elementary divisors in terms of those of A and B.

39. Suppose A is a finitely generated abelian group.
   (i) Compute $A \otimes_{\mathbb{Z}} \mathbb{Q}$.
   (ii) Define $f: A \to A \otimes_{\mathbb{Z}} \mathbb{Q}$ by setting $f(a) = a \otimes 1$ for all $a \in A$. Show that f is a homomorphism. Under what circumstances is f a monomorphism?

40. If A is an abelian group show that $\mathbb{Z}_n \otimes_{\mathbb{Z}} A \cong A/nA$.

41. If F is a field and K is an extension field show that $M_n(K) \cong K \otimes_F M_n(F)$ (as F-algebras).

42. Suppose that F is a field, K is a finite Galois extension, and that $G = G(K:F)$ is a direct product of two subgroups G_1 and G_2. Set $L_1 = \mathscr{F}G_1$ and $L_2 = \mathscr{F}G_2$. Show that the ring $L_1 \otimes_F L_2$ is a field, with a subfield isomorphic with F, and that if $L_1 \otimes_F L_2$ is viewed as an extension of F, then K and $L_1 \otimes_F L_2$ are F-isomorphic.

43. Suppose R is a commutative ring with 1 and E, F are free R-modules, with respective bases $\{x_\alpha : \alpha \in A\}$ and $\{y_\beta : \beta \in B\}$. Show that $E \otimes_R F$ is free with basis $\{x_\alpha \otimes y_\beta : \alpha \in A, \beta \in B\}$.

44. Suppose R is a commutative ring and L, M, N are R-modules. Show that $\operatorname{Hom}_R(L, \operatorname{Hom}_R(M, N))$ and $\operatorname{Hom}_R(L \otimes_R M, N)$ are isomorphic R-modules. Conclude in particular that $(L \otimes_R M)^* \cong \operatorname{Hom}_R(L, M^*)$ (see Exercise 13). (*Hint*: If $f \in \operatorname{Hom}_R(L, \operatorname{Hom}_R(M, N))$ and $a \in L$ write $f(a)$ as f_a. Try mapping $f \to \hat{f}$, where $\hat{f}(a \otimes b) = f_a(b)$ for $a \in L$, $b \in M$.)

45. If $K \to M \to N \to 0$ is an exact sequence of (left) R-modules and L is a right R-module show that $L \otimes_R K \to L \otimes_R M \to L \otimes_R N \to 0$ is an exact sequence of abelian groups (see Exercise 6.4 for the homomorphisms).

46. Suppose R is a ring with 1. A unitary R-module P is called *projective* if given an exact sequence $M \xrightarrow{g} N \to 0$ of R-modules and an R-homomorphism $f: P \to N$, then there is an R-homomorphism $h: P \to M$ such that $f = gh$, i.e.,

the diagram

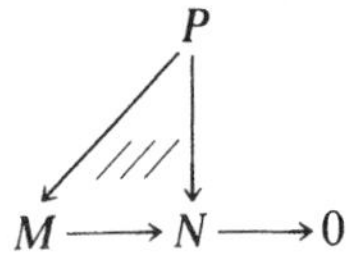

is commutative.

(i) Show that free modules are projective.

(ii) If $P = P_1 \oplus P_2$ show that P is projective if and only if both P_1 and P_2 are projective. Generalize to arbitrary direct sums.

47. Show that an R-module P is projective if and only if P is a direct summand of some free module F (see Theorem 2.8).

48. Show that an R-module P is projective if and only if given any short exact sequence $0 \to K \to M \to P \to 0$ we may conclude that $M \cong K \oplus P$ (we say that the sequence *splits*).

49. (i) If R is a PID or a division ring show that every projective R-module is free.

(ii) Give an example of a projective module that is not free.

50. If P is a projective R-module show that there is a free R-module F such that $F \cong F \oplus P$ (see Exercise 47).

51. Let R be a ring with 1. A unitary R-module Q is called *injective* if given any exact sequence $0 \to K \xrightarrow{g} M$ of R-modules and a homomorphism $f: K \to Q$, then there is a homomorphism h: $M \to Q$ such that $f = hg$, i.e., the diagram

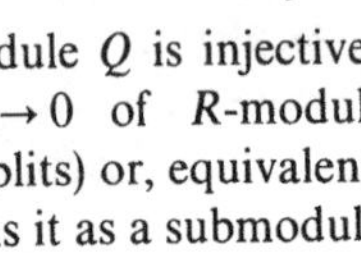

is commutative. If $Q = Q_1 \times Q_2$ show that Q is injective if and only if both Q_1 and Q_2 are injective. Generalize to arbitrary direct products.

52. Show that if an R-module Q is injective, then given any short exact sequence $O \to Q \to M \to N \to 0$ of R-modules we may conclude that $M \cong Q \oplus N$ (the sequence splits) or, equivalently, Q is a direct summand of every R-module that contains it as a submodule.

53. An (additive) abelian group A is called *divisible* if $nA = A$ for all nonzero $n \in \mathbb{Z}$.

(i) Show that $A = \mathbb{Q}$ is divisible.

(ii) Show that any homomorphic image of a divisible group is divisible.

Thus, for example, $\mathbb{Q}/\mathbb{Z}$ is divisible.

(iii) Show that no finitely generated abelian group $A(\neq 0)$ can be divisible.

54. If A is an abelian group show that A has a unique largest divisible subgroup D, and that $A = D \oplus B$, where B has no nonzero divisible subgroups. (*Hint*: Take D to be the subgroup generated by the union of all divisible subgroups of A.)

55. If A is any abelian group there is a divisible group D with a subgroup isomorphic with A. (*Hint*: There is a free abelian group F with a subgroup K such that $A \cong F/K$; F is a direct sum of copies of $\mathbb{Z}$, and $\mathbb{Z} \leq \mathbb{Q}$.)

56. If an abelian group A is injective as a $\mathbb{Z}$-module show that A is divisible. (Use Exercise 55.)

Chapter V Structure of Rings and Algebras

1. PRELIMINARIES

We begin by describing an important class of examples. Suppose G is a finite group and R is a commutative ring with 1. Denote by RG the set of all functions $a\colon G \to R$. With the usual addition of functions and multiplication of functions by elements of R it is easy to see that RG is a unitary R-module. By a mild abuse of notation we may view each $x \in G$ as an element of RG, viz., we may take x to be the function from G to R such that $x(x) = 1$ and $x(y) = 0$ if $x \neq y \in G$. Then clearly each $a \in RG$ can be written uniquely as $a = \sum\{a_x \cdot x\colon x \in G\}$, with each $a_x \in R$, from which it follows that the elements of G constitute a basis for RG, and RG is a free R-module of rank $|G|$.

The basis G is closed under (group) multiplication, and that multiplication extends naturally to a multiplication on all of RG by means of

$$(\sum a_x x)(\sum b_y y) = \sum\{a_x b_y \cdot xy\colon x, y \in G\} = \sum\{(\sum\{a_x b_y\colon xy = u\}) \cdot u\colon u \in G\}.$$

Exercise 1.1. Verify that RG is a ring, and in fact an R-algebra, and that the multiplication in RG is given by

$$ab(x) = \sum\{a(xy^{-1})b(y)\colon y \in G\}$$

for all $a, b \in RG$, $x \in G$. (This multiplication is sometimes called the *convolution* product.)

The R-algebra RG is called the *group algebra* of G over R.

Exercise 1.2. Let $G = \{1, x, x^2\}$ be a cyclic group of order 3 and let $R = \mathbb{Z}_2$. Write out addition and multiplication tables for the group algebra RG.

If R is any ring and M is a left R-module we define the *annihilator* of M in R to be

$$A_R(M) = A(M) = \{r \in R : rx = 0, \text{ all } x \in M\}.$$

Then $A(M)$ is an ideal (two sided) in R. We say that M is a *faithful* R-module if $A(M) = 0$. Since $A(M) \cdot M = 0$ there is an obvious $R/A(M)$-module action on M, viz., $[r + A(M)] \cdot x = rx$ for all $r \in R$, $x \in M$. The annihilator of M in $R/A(M)$ is just the zero element $A(M)$, so M is always a faithful $R/A(M)$-module.

Recall (Exercise IV.1.1) that an abelian group M is an R-module if and only if there is a ring homomorphism $f: R \to \text{End}(M)$. The annihilator $A(M)$ is simply $\ker f$, and M is faithful if and only if f is a monomorphism. The fact that M is a faithful $R/A(M)$-module is thus a consequence of the FHT for rings.

If M is a left R-module we define the *centralizer* of M to be

$$C(M) = \text{Hom}_R(M, M) = \{g \in \text{End}(M) : g(rx) = rg(x), \text{ all } r \in R, x \in M\}.$$

Note that $C(M)$ is a subring of $\text{End}(M)$.

If we recall again that there is a homomorphism $f: R \to \text{End}(M)$, say with $f: r \mapsto f_r$ and $f_r(x) = rx$, then $g \in C(M)$ if and only if $gf_r(x) = g(rx) = rg(x) = f_r g(x)$ for all r and x, i.e., if and only if g commutes with all f_r. Hence the name "centralizer."

An R-module M is called *simple*, or *irreducible*, if $A(M) \neq R$ and the only submodules of M are 0 and M. Note that the condition $A(M) \neq R$ serves only to rule out the trivial module action $rx = 0$, all $r \in R$.

Recall that a *division ring* is a ring R with 1 in which every nonzero element is invertible, i.e., $U(R) = R^*$. A division ring clearly can have no zero divisors.

Proposition 1.1 (Schur's Lemma). If R is a ring and M is a simple R-module, then $C(M)$ is a division ring. If N is another simple R-module, then either M and N are R-isomorphic or else $\text{Hom}_R(M, N) = 0$.

Proof. If $0 \neq f \in C(M)$, then $\ker f$ and $\text{Im}(f)$ are submodules of M, and $\ker f \neq M$ so $\ker f = 0$. Also $\text{Im}(f) \neq 0$ so $\text{Im}(f) = M$, and hence f is invertible in $\text{End}(M)$. It is easy to check that in fact $f^{-1} \in C(M)$, so $C(M)$ is a division ring. The proof of the second statement is similar; if $0 \neq f \in \text{Hom}_R(M, N)$ then $\ker f = 0$ and $\text{Im}(f) = N$, so f is an R-isomorphism.

Suppose that F is a field, that the ring R is a unitary F-algebra with 1, and that M is a unitary R-module. Then M is automatically also an F-vector space, with scalar multiplication given by $a \cdot x = (a \cdot 1) \cdot x$ for all $a \in F$ and all $x \in M$.

Proposition 1.2. Suppose F is an algebraically closed field, R is a unitary F-algebra with 1, and M is a simple unitary R-module that is finite dimensional as an F-vector space. Then $C(M) \cong F$. In fact, $C(M) = \{a \cdot I: a \in F\}$, I being the identity map on M.

Proof. If $g \in C(M)$, then g is an F-linear transformation from M to M. Let $f(x) = m_g(x)$ be its minimal polynomial over F. Since F is algebraically closed we may write

$$f(x) = \prod\{(x - a_i): 1 \leq i \leq k\}, \qquad \text{with } a_i \in F.$$

By definition of the minimal polynomial

$$f(g) = \prod\{(g - a_i I): 1 \leq i \leq k\} = 0.$$

But each $g - a_i I \in C(M)$, and $C(M)$ is a division ring by Schur's Lemma. Thus in fact $f(x)$ must be linear, and $g = aI$ for some $a \in F$.

An R-module M is called *semisimple* (or *completely reducible*) if it is a direct sum of simple submodules, or if $M = 0$.

Exercise 1.3. If M is a semisimple R-module and $0 \neq x \in M$ show that $Rx \neq 0$. (*Hint*: First assume that M is simple.)

Proposition 1.3. Suppose L and N are submodules of M with $L \subseteq N$, and that there are submodules K and P of M such that $M = L \oplus K = N \oplus P$. If we set $Q = N \cap K$, then $N = L \oplus Q$.

Proof. It is clear that $L + Q \subseteq N$ and that $L \cap Q = 0$, so $L + Q = L \oplus Q \subseteq N$. If $x \in N$ write $x = y + z$, with $y \in L$ and $z \in K$. Then $z = x - y \in K \cap N = Q$, so $N \subseteq L \oplus Q$.

Corollary. If M is a module in which every submodule is a direct summand, then every submodule has the same property.

Proposition 1.4. Suppose $M \neq 0$ is an R-module such that if $0 \neq x \in M$, then $Rx \neq 0$. Then the following are equivalent.

(i) M is semisimple,
(ii) M is a sum (not necessarily direct) of simple submodules, and
(iii) every submodule of M is a direct summand.

Proof. (i) $\Rightarrow$ (ii) is obvious.
(ii) $\Rightarrow$ (iii): Say $M = \sum\{M_a: a \in A\}$, with each M_a simple, and let N be a submodule of M. By Zorn's Lemma there is a maximal collection $\{M_b: b \in B \subseteq A\}$ such that $N \cap \sum\{M_b: b \in B\} = 0$. Set $K = \sum\{M_b: b \in B\}$; then $N + K = N \oplus K$. If $N \oplus K \neq M$, then some M_a, $a \in A \setminus B$, is not contained in $N \oplus K$, and consequently $(N \oplus K) \cap M_a = 0$ since M_a is simple. But then $N \cap (M_a + K) = 0$ (Why?), contradicting the maximality of $\{M_b: b \in B\}$. Thus $M = N \oplus K$ and N is a direct summand of M.

(iii) $\Rightarrow$ (i) By Zorn's Lemma there is a submodule P of M that is maximal with respect to being a direct sum of simple submodules of M (conceivably $P = 0$). If $P \neq M$ write $M = P \oplus N$ and choose $x \in N$, $x \neq 0$. By another application of Zorn's Lemma there is a submodule L of N that is maximal with respect to $x \notin L$. If we write $M = L \oplus K$ and set $Q = N \cap K$ then $N = L \oplus Q$ by Proposition 1.3. Let us show that Q is simple. Clearly $Q \neq 0$, or else $N = L$, contradicting $x \in N \setminus L$. If Q is not simple it has a proper nonzero submodule Q_1, and we may write $Q = Q_1 \oplus Q_2$ by the corollary to Proposition 1.3. But now the element x cannot be in both $L \oplus Q_1$ and $L \oplus Q_2$, since $(L \oplus Q_1) \cap (L \oplus Q_2) = L$ and $x \notin L$. Thus one of $L \oplus Q_1$ and $L \oplus Q_2$ would violate the maximality of L, and so Q is simple. (We have incidentally shown that M has simple submodules, and hence $P \neq 0$.) But now the submodule $P \oplus Q$ violates the maximality of P, so $P = M$ and the proof is complete.

Question: Where was the hypothesis used that $Rx \neq 0$ if $0 \neq x \in M$?

Corollary. If M is a semisimple module then submodules and homomorphic images of M are also semisimple.

Proof. For submodules apply the corollary to Proposition 1.3. If K is a homomorphic image of M, say $K = f(M)$, set $N = \ker f$. We may write $M = N \oplus L$, and then $K \cong L$, which is semisimple, so K is semisimple.

Proposition 1.5. Suppose M is a simple R-module and I is a left ideal in R that is not contained in $A(M)$. Then $Ix = M$ for some $x \in M$. If I is a two-sided ideal not contained in $A(M)$, then $Iy = M$ for all nonzero $y \in M$.

Proof. Choose $x \in M$ such that $Ix \neq 0$. Then $Ix = M$ since Ix is a submodule. If I is two-sided and $0 \neq y \in M$ then Iy is a submodule so $Iy = 0$ or $Iy = M$. If $Iy = 0$ set $N = \{z \in M : Iz = 0\}$. Then N is a submodule and $N \neq 0$ since $y \in N$, so $N = M$ and hence $I \subseteq A(M)$, a contradiction. Thus $Iy = M$.

Corollary. If M is simple, then $Rx = M$ for all nonzero $x \in M$.

A left ideal I in a ring R is called *modular* if there is an element $e \in R$ such that $r - re \in I$ for all $r \in R$. Equivalently, $r + I = (r + I)(e + I)$ for all $r \in R$. We call e a *right relative unit* for the modular left ideal R. Similarly a right ideal J is called modular if there is a corresponding left relative unit. Note that if R is a ring with 1 then every proper ideal is modular, with $e = 1$. Note also that if I is a modular (left or right) ideal with relative unit e and $I \neq R$, then $e \notin I$.

Proposition 1.6. An R-module M is simple if and only if $M \cong R/I$ for some maximal modular left ideal I of R.

Proof. $\Rightarrow$: Choose $x \in M$, $x \neq 0$, and define $f: R \to M$ by setting $f(r) = rx$. Then $\text{Im}(f) = M$ by the corollary to Proposition 1.5, so $M \cong R/\ker f$ by the FHT for modules. Set $I = \ker f$ and note that I, as a submodule of R, is a left ideal. Note also that I is maximal since M is simple (see Proposition IV.1.3). Choose $e \in R$ such that $f(e) = x$, or $ex = x$. If $r \in R$, then $f(r - re) = rx - rex = 0$, so $r - re \in I$, and I is modular.

$\Leftarrow$: We must show that $M = R/I$ is a simple R-module if I is a maximal modular left ideal. If e is a relative right unit corresponding to I then $e \notin I$, and

$$e(e + I) = e^2 + I = e + (e^2 - e) + I = e + I,$$

so $e \notin A(M)$ and $A(M) \neq R$. The only submodules of $M = R/I$ are 0 and M since I is maximal (again see Proposition IV.1.3), and so M is simple.

Exercise 1.4. Show that every proper modular (left or right) ideal in a ring R is contained in a maximal modular (left or right) ideal.

Theorem 1.7 (Maschke). Suppose G is a finite group, F is a field, and either char $F = 0$ or char $F = p$ with $p \nmid |G|$. If R is the group algebra FG and M is a unitary R-module with finite F-dimension then M is semisimple.

Proof. Suppose $N \neq 0$ is a submodule of M. Choose a vector subspace V of M such that $M = N \oplus V$ (as a vector space over F). Let $f: M \to N$ be the projection of M onto N along V, i.e., if $x \in M$ and $x = y + z$, with $y \in N$ and $z \in V$ then $f(x) = y$. Define $g: M \to M$ by means of

$$g(m) = |G|^{-1} \sum\{xf(x^{-1}m) : x \in G\}$$

for all $m \in M$, and observe that g is an F-linear transformation from M to M. If $y \in G$, then

$$yg(y^{-1}m) = |G|^{-1} \sum\{yxf((yx)^{-1}m) : x \in G\} = g(m),$$

all $m \in M$, since yx ranges over all of G as x ranges over G with y fixed. We may conclude that $g \in C_R(M)$, since G is a basis for $R = FG$.

If $m \in M$ and $x \in G$ then $xf(x^{-1}m) \in xN \subseteq N$, so $g(N) \subseteq N$. If $n \in N$ and $x \in G$, then $x^{-1}n \in N$, so $f(x^{-1}n) = x^{-1}n$, and therefore $xf(x^{-1}n) = xx^{-1}n = n$, so $g(n) = n$, and therefore $g^2 = g$. If we set $K = (I - g)M$, then K is an R-module since $I - g \in C_R(M)$. If $m \in N \cap K$, then $m = g(m)$, and also $m = (I - g)(m')$ for some $m' \in M$, and hence $m = g(I - g)(m') = (g - g^2)(m') = 0$. Thus $M = N \oplus K$, and M is semisimple by Proposition 1.4(iii).

Maschke's theorem is of considerable importance in the representation theory of finite groups.

2. THE JACOBSON RADICAL

If R is a ring, then the *Jacobson radical* of R is the ideal

$$J(R) = \bigcap\{I : I = A(M) \text{ for some simple } R\text{-module } M\}.$$

If there are no simple R-modules we define $J(R) = R$ and say that R is a *radical ring*. If $J(R) = 0$ we say that R is a *semisimple ring*.

A rather basic point of view in the structure theory of rings is that semisimple rings are "manageable," and hence that the Jacobson radical $J(R)$ is a measure of deviation from manageability. It will be convenient to have several alternate descriptions of $J(R)$.

If I is a left ideal in a ring R set $(I:R) = \{r \in R : rR \subseteq I\}$. This is sometimes called the *quotient* of I by R. It should not, of course, be confused with the notion of a quotient ring.

Proposition 2.1. If I is a left ideal in a ring R and M is the R-module R/I, then $A_R(M) = (I:R)$, and in particular $(I:R)$ is an ideal in R.

Proof. If $r \in R$ we have $r \in A(M)$ if and only if $r(R/I) = I$ if and only if $rR \subseteq I$ if and only if $r \in (I:R)$.

Exercise 2.1. Show that if I is a modular left ideal in a ring R then $(I:\mathrm{R}) \subseteq I$, and if J is a two-sided ideal in R with $J \subseteq I$ then $J \subseteq (I:R)$.

Theorem 2.2. If R is not a radical ring, then $J(R)$ is the intersection of all quotients $(I:R)$ of maximal modular left ideals I in R.

Proof. Propositions 1.6 and 2.1.

If R is a ring define a new binary operation $*$ on R by setting $r * s = r + s + rs$ for all $r, s \in R$. An element $s \in R$ is called *left quasi-regular* if there exists $r \in R$ such that $r * s = 0$, in which case r is called a *left adverse* for s; r is then *right quasi-regular* and s is a *right adverse* for r. An element of R that is both left and right quasi-regular is said to be *quasi-regular*. A left (right) ideal in R is called *left (right)-quasi-regular* if each of its elements is left (right) quasi-regular.

If R is a ring with 1 note that $(1 + r)(1 + s) = 1$ if and only if $r + s + rs = 0$, and hence $s \in R$ is left (right) quasi-regular if and only if $1 + s$ has a left (right) inverse. Thus s is quasi-regular if and only if $1 + s \in U(R)$.

Exercise 2.2. (1) Show that $(R, *)$ is a semigroup with identity 0.

(2) If R_{qr} denotes the set of quasi-regular elements in the ring R, show that $(R_{\mathrm{qr}}, *)$ is a group.

Proposition 2.3. Let J denote the intersection of all the maximal modular left ideals in a ring R, or $J = R$ if R has no maximal modular left ideals. Then J is left quasi-regular.

Proof. If $s \in J$ set $I' = \{r + rs : r \in R\}$. Then I' is a modular left ideal with right relative unit $e = -s$. If $I' \neq R$, then I' is contained in a maximal modular left ideal I (see Exercise 1.4), and $s \in J \subseteq I$, so $-rs \in I$, for all $r \in R$. But then $r = -rs + (r + rs) \in I$ for all $r \in R$, a contradiction. Thus $I' = R$, and in particular there is an element $r \in R$ such that $-s = r + rs$, i.e., $r * s = 0$.

Theorem 2.4. If R is not a radical ring, then $J(R)$ is the intersection of all the maximal modular left ideals I in R.

Proof. As in the proof of Proposition 2.3 let J be the intersection of all the maximal modular left ideals I of R. Since $(I:R) \subseteq I$ for each I we have $J(R) \subseteq J$ by Theorem 2.2. If $r \in R \backslash J(R)$, then $rM \neq 0$ for some simple R-module M, and hence $rx \neq 0$ for some $x \in M$. But then $I = \{s \in R : sx = 0\}$ is a maximal modular left ideal in R (see the proof of Proposition 1.6), and $r \notin I$ so $r \notin J$. Thus $J(R) = J$.

Corollary. $J(R)$ is left quasi-regular.

Theorem 2.5. If I is any left quasi-regular left ideal in R then $I \subseteq J(R)$. Consequently $J(R)$ is the unique maximal element (with respect to set inclusion) in the set of all left quasi-regular ideals in R.

Proof. If not, then $IM \neq 0$ for some simple R-module M, and hence $Ix = M$ for some $x \in M$ by Proposition 1.5. Choose $s \in I$ such that $sx = -x$, and let $r \in R$ be a left adverse for s. Then

$$0 = (r * s)x = (r + s + rs)x = rx + sx + rsx = rx - x - rx = -x,$$

and hence $x = 0$, a contradiction since $M \neq 0$.

Note that if $s \in J(R)$ then $r * s = 0$ for some $r \in R$, and then $r = -s - rs \in J(R)$. Thus r is also left quasi-regular, so in fact r is quasi-regular with (unique) adverse s [see Exercise 2.2(2)], and the same applies to s. Thus in fact $J(R)$ is a quasi-regular ideal in R, and $(J(R), *)$ is a group. Furthermore, $J(R)$ is the unique largest ideal of R with that property, by Theorem 2.5.

If I and J are left ideals in R denote by IJ the left ideal in R generated by all elements ab, $a \in I$, $b \in J$. Thus IJ is the set of all finite sums $\sum_{i=1}^{k} a_i b_i$, where $1 \leq i \in \mathbb{Z}$, $a_i \in I$, and $b_i \in J$. Thus $IJ \subseteq J$. Completely analogous definitions apply if I and J are right ideals (then $IJ \subseteq I$), or I and J are two-sided ideals (then $IJ \subseteq I \cap J$). In particular we may define $I^2 \subseteq I$, and by induction $I^n \subseteq I$ for any ideal I (left, right, or two sided).

An element r in a ring R is called *nilpotent* if $r^k = 0$ for some integer $k \geq 1$. For example

$$r = \begin{bmatrix} 2 & -1 \\ 4 & -2 \end{bmatrix}$$

is a nilpotent element in the ring of 2×2 matrices over $\mathbb{Z}$, since $r^2 = 0$.

Exercise 2.3. If r and s are nilpotent elements of a ring R and $rs = sr$ show that $r + s$ is nilpotent. Show by example that the result may fail if $rs \neq sr$.

If every element of a left (or right) ideal I is nilpotent, then I is called a *nil* left (or right) ideal. For example, the principal ideal (2) is a nil ideal in $\mathbb{Z}_8$.

A left (or right) ideal I in R is called a *nilpotent* left (or right) ideal if $I^n = 0$ for some integer $n \geq 1$.

Proposition 2.6. A nilpotent ideal I (left, right, or two-sided) is nil.

Proof. If $I^n = 0$, then $a_1 a_2 \cdots a_n = 0$ for all choices of $a_i \in I$, and in particular $a^n = 0$ for all $a \in I$.

Proposition 2.7. If r is a nilpotent element in a ring R, then r is quasi-regular. Consequently every left, right, and two-sided nil ideal in R is contained in $J(R)$.

Proof. If $r^n = 0$ set $s = -r + r^2 - r^3 + \cdots \pm r^{n-1}$ and verify that s is both a left and right adverse for r. Apply Theorem 2.5.

Theorem 2.8. If R is a ring, then $J(R)$ consists of the set of all elements $s \in R$ such that rs is left quasi-regular for all $r \in R$.

Proof. Set

$$J = \{s \in R : rs \text{ is left quasi-regular for all } r \in R\}.$$

Then $J(R) \subseteq J$ is clear since $J(R)$ is a left quasi-regular ideal. If $J \neq J(R)$ choose $r \in J \backslash J(R)$ such that $rM \neq 0$ for some simple R-module M. Choose $x \in M$ such that $rx \neq 0$. Then $Rrx = M$ by Proposition 1.5, and we may choose $a \in R$ for which $arx = -x$. Then ar has a left adverse b and we have

$$0 = 0 \cdot x = (b * ar)x = (b + ar + bar)x = bx - x - bx = -x,$$

a contradiction.

Theorem 2.9. If R is any ring, then $J(J(R)) = J(R)$, i.e., $J(R)$ is a radical ring.

Proof. $J(J(R)) \subseteq J(R)$ by definition. If $s \in J(R)$, then rs is left quasi-regular for all $r \in J(R)$ by Theorem 2.8, and the left adverse of rs is in $J(R)$ [recall that $(J(R), *)$ is a group]. Thus $s \in J(J(R))$, again by Theorem 2.8.

Theorem 2.10. If R is a ring, then $J(R/J(R)) = 0$, so $R/J(R)$ is a semisimple ring.

Proof. We may assume that $J(R) \neq R$. If M is any simple R-module, then $J(R) \subseteq A(M)$, so M is also a simple $R/J(R)$-module. Conversely if M is a simple $R/J(R)$-module, then M is also an R-module with the action defined by $rx = [r + J(R)]x$, and as such M is a simple R-module. The annihilator in $R/J(R)$ of M is just $A_R(M)/J(R)$, and

$$\begin{aligned} J(R/J(R)) &= \bigcap\{A_R(M)/J(R) : M \text{ simple}\} \\ &= \left(\bigcap\{A_R(M) : M \text{ simple}\}\right)/J(R) = J(R)/J(R) = 0 \end{aligned}$$

in $R/J(R)$.

Theorem 2.11. If R is a ring and I is an ideal in R, then $J(I) = I \cap J(R)$.

Proof. If $s \in I \cap J(R)$ and $r \in I$, then rs is left quasi-regular (in R). If $a \in R$ is a left adverse, then $a = -rs - ars \in I$, so rs is in fact left quasi-regular in I. Thus $I \cap J(R) \subseteq J(I)$ by Theorem 2.8. Suppose then that $r \in J(I)$, and consider the left ideal Rr. Note that $(Rr)^2 = RrRr \subseteq RIr \subseteq Ir$, and that Ir is a left quasi-regular left ideal in I, since $r \in J(I)$. Thus $(Rr)^2$ is a left quasi-regular left ideal in R, and hence $(Rr)^2 \subseteq J(R)$. Let us show that in fact $Rr \subseteq J(R)$. If not consider $[Rr + J(R)]/J(R)$; it is a nonzero left ideal in $R/J(R)$ whose square is 0. Since $[Rr + J(R)]/J(R)$ is nilpotent it is nil, by Proposition 2.6; hence it is contained in $J(R/J(R))$ by Proposition 2.7. That is a contradiction, since $R/J(R)$ is semisimple by Theorem 2.10. Thus $Rr \subseteq J(R)$, so sr is left quasi-regular for every $s \in R$, and hence $r \in J(R)$ by Theorem 2.8. The proof is complete.

Corollary 1. If R is a radical ring and I is an ideal in R, then I is a radical ring.

Corollary 2. If R is a semisimple ring and I is an ideal in R, then I is a semisimple ring.

For example, let $R = \mathbb{Z}_n$, and suppose $n = p_1^{e_1} p_2^{e_2} \cdots p_k^{e_k}$ is the prime factorization of n. Set $m = p_1 p_2 \cdots p_k$. Since $\mathbb{Z}_n = \mathbb{Z}/(n)$ the ideals of $\mathbb{Z}_n$ all have the form $(a)/(n)$, where (a) is a principal ideal in $\mathbb{Z}$ with $(a) \supseteq (n)$, i.e., $a \mid n$ in $\mathbb{Z}$. Clearly then the maximal ideals in $\mathbb{Z}_n$ are $(p_1)/(n), \ldots, (p_k)/(n)$, and hence

$$J(\mathbb{Z}_n) = \bigcap\{(p_i)/(n) : 1 \le i \le k\} = \left(\bigcap\{(p_i) : 1 \le i \le k\}\right)/(n) = (m)/(n).$$

Consequently

$$\mathbb{Z}_n/J(\mathbb{Z}_n) = \left(\mathbb{Z}/(n)\right)/\left((m)/(n)\right) \cong \mathbb{Z}_m$$

by the Freshman Theorem for rings. We conclude that $\mathbb{Z}_n$ is semisimple if and only if n is square-free.

3. THE DENSITY THEOREM

A ring R is called *simple* if $R^2 \neq 0$ and the only (two-sided) ideals in R are 0 and R. Note that $R^2 = 0$ for a ring R if and only if $rs = 0$ for all r and s in R, i.e., the multiplication is trivial. Every additive subgroup of R is an ideal when the multiplication is trivial. Thus the restriction $R^2 \neq 0$ simply rules out the additive groups $\mathbb{Z}_p$, p prime, with trivial multiplication, as simple rings.

Note that $R^2 = R$ for a simple ring, since R^2 is an ideal.

Clearly every simple ring must be either semisimple or radical, since $J(R)$ is an ideal. An example of a radical simple ring has been given by Sasiada and Cohn [33].

Suppose D is a division ring and let $R = M_n(D)$, the set of all $n \times n$ matrices with entries from D. With the usual matrix operations R is a ring and a unitary left D-module, even when D is not commutative. If we denote by $E_{ij} \in R$ the matrix having $1 \in D$ as its ijth entry and 0s elsewhere it is easy to see that R has $\{E_{ij}\}$ as a D-basis, so R is a free D-module of rank n^2, i.e., a (left) D-vector space of dimension n^2. If D is a field, then R is a D-algebra.

Let us see that $R = M_n(D)$ is a simple ring. Clearly $R^2 \neq 0$, since R is a ring with 1. Suppose $J \neq 0$ is an ideal in R. If $0 \neq A \in J$ write $A = (a_{ij})$ and suppose in particular that a_{ij} is a nonzero entry. Then $E_{ii}AE_{jj} = a_{ij}E_{ij} \in J$, and hence $a_{ij}^{-1}(a_{ij}E_{ij}) = E_{ij} \in J$. But then for any k and m we have $E_{ki}E_{ij}E_{jm} = E_{km} \in J$, and so $J = R$.

A ring R is called *primitive* if there exists a faithful simple R-module M. An ideal P in a ring R is called a *primitive ideal* if R/P is a primitive ring. Observe that a primitive ring R is semisimple, for if M is a faithful simple R-module, $A(M) = 0$ and hence $J(R) = 0$ since $J(R) \subseteq A(M)$.

Proposition 3.1. If a ring R is both simple and semisimple, then R is primitive.

Proof. Since R is semisimple there is a simple R-module M and the ideal $A(M)$ is not all of R. Thus $A(M) = 0$ since R is simple, and so M is faithful.

Proposition 3.2. An ideal P in a ring R is primitive if and only if $P = (I:R)$ for some maximal modular left ideal I of R, hence if and only if $P = A(M)$ for some simple R-module M.

Proof. The two conclusions are equivalent by Propositions 1.6 and 2.1. If P is a primitive ideal let M be a faithful simple R/P-module. Then M is a simple R-module, with $P \subseteq A_R(M)$. In fact $P = A_R(M)$, for otherwise M would not be faithful as an R/P-module. Conversely, if $P = A(M)$ for some simple R-module M then M is a faithful simple R/P-module, so P is a primitive ideal.

Corollary. If R is not a radical ring then $J(R)$ is the intersection of all the primitive ideals in R.

Recall that if M is a simple R-module, then by Schur's Lemma $D = C(M)$ is a division ring. We may endow M with the structure of a unitary left D-module (i.e., a left vector space over D) by defining $f \cdot m = f(m)$ for all $m \in M, f \in D$. If $r \in R$ define $T_r: M \to M$ by setting $T_r(m) = rm$, all $m \in M$. Note then that if $f \in D = C(M)$ we have

$$T_r(fm) = r \cdot f(m) = f(rm) = f \cdot T_r(m),$$

and so each T_r is a D-linear transformation on M. The map $r \mapsto T_r$ is a ring homomorphism from R into $\text{Hom}_D(M, M)$, with kernel $A(M)$. If R is primitive and M is a faithful simple R-module, then $A(M) = 0$, so $r \mapsto T_r$ is a monomorphism and we may (and shall) view R as a subring of $\operatorname{Hom}_D(M, M)$ by identifying r with T_r.

Suppose M is a simple R-module and $D = C(M)$. We say that R acts *densely*, or that R is *dense*, on M if given any finite D-linearly independent set $\{m_1, \ldots, m_k\} \subseteq M$, and any other elements $n_1, \ldots, n_k$ (not necessarily distinct) in M, there is some $r \in R$ such that $T_r(m_i) = rm_i = n_i$, $1 \leq i \leq k$.

Exercise 3.1. Suppose M is a simple R-module and $D = C(M)$. If $\dim_D(M)$ is finite and R is dense on M show that $r \mapsto T_r$ is an epimorphism. If R is primitive, with $A(M) = 0$, conclude that $R = \operatorname{Hom}_D(M, M)$.

Proposition 3.3. Suppose M is a simple R-module, $D = C(M)$, V is a finite-dimensional D-subspace of M, and $m \in M \backslash V$. Then there is an element $r \in A_R(V)$ such that $rm \neq 0$.

Proof. Induction on $n = \dim(V)$. The result is true by the definition of simplicity if $n = 0$, so suppose $n > 0$ and assume the result for subspaces of dimension $n - 1$. Choose a D-subspace W of V, with $\dim_D(W) = n - 1$, and choose $v \in V \backslash W$, so that $V = W \oplus Dv$. Suppose by way of contradiction that $A(V)m = 0$. By induction there is some $r \in R$ such that $rW = 0$ but $rv \neq 0$, and hence $A(W)v \neq 0$. But $A(W)v$ is a submodule of M, so $A(W)v = M$. Define $f: M \to M$ by setting $f(sv) = sm$ for all $s \in A(W)$. (Why is f well defined?) Note that if $r \in R$, then

$$f(r(sv)) = f((rs)v) = (rs)m = r(sm) = rf(sv),$$

so $f \in C(M) = D$. If $r \in A(W)$, then $f(rv) = rm = rf(v)$, so $r(m - fv) = 0$ for all $r \in A(W)$. It follows from the induction hypothesis that $m - fv \in W$. But then $m \in W + fv \subseteq W + Dv = V$, a contradiction, and the proposition is proved.

Theorem 3.4 (The Jacobson–Chevalley Density Theorem). If M is a simple R-module, then R acts densely on M. In particular, if R is a primitive ring, M is a faithful simple R-module, and $D = C(M)$, then R is (isomorphic with) a dense ring of D-linear transformations on M.

Proof. Suppose $\{m_1, \ldots, m_k\}$ is a D-linearly independent subset of M and $n_1, \ldots, n_k \in M$. For each i, $1 \leq i \leq k$, set

$$V_i = \oplus\{Dm_j : i \neq j, 1 \leq j \leq k\}.$$

Then $m_i \notin V_i$, so by Proposition 3.3 we may choose $r_i \in A(V_i)$ such that $r_i m_i \neq 0$. Consequently, by the corollary to Proposition 1.5, $Rr_i m_i = M$, and so $s_i r_i m_i = n_i$ for some $s_i \in R$ (note, though, that $s_i r_i V_i = 0$, and hence $s_i r_i m_j = 0$ if $i \neq j$). If we set $r = \sum_{j=1}^{k} s_j r_j$ it follows that $rm_i = n_i$, $1 \leq i \leq k$, as desired.

Theorem 3.5. Suppose R is a primitive ring. Then there is a division ring D and a (left) vector space M over D such that either

(i) $\dim_D(M)$ is finite and R is isomorphic with the ring $\mathrm{Hom}_D(M,M)$ of all D-linear transformations of M, or

(ii) there are ascending chains $R_1 \subseteq R_2 \subseteq \cdots$ of subrings of R and D-subspaces $M_1 \subseteq M_2 \subseteq \cdots$ of M, with $\dim_D(M_k) = k$, such that for each k there is an epimorphism $f_k : R_k \to \mathrm{Hom}_D(M_k, M_k)$.

Proof. Since R is primitive there is a faithful simple R-module M. Set $D = C(M)$, a division ring by Schur's Lemma. If $\dim_D(M) = n$, finite, then by the Density Theorem and Exercise 3.1 we see that $R \cong \mathrm{Hom}_D(M,M)$.

If $\dim_D(M)$ is not finite let $\{m_1, m_2, \ldots\}$ be an infinite D-linearly independent subset of M, and for each $k \geq 1$ set $M_k = \oplus\{Dm_i : 1 \leq i \leq k\}$. Set $R_k = \{r \in R : rM_k \subseteq M_k\}$ for each k. It follows from the Density Theorem that every D-linear transformation $T: M_k \to M_k$ has the form $T = T_r | M_k$ for some $r \in R$ with $T_r m = rm$ for all $m \in M_k$, and so $r \mapsto T_r | M_k$ maps R_k onto $\mathrm{Hom}_D(M_k, M_k)$.

If D is a division ring (or *any* ring, for that matter) we may define an *opposite* ring D^{op}, whose underlying set and addition are the same as for D, but with multiplication defined by $a \circ b = ba$. Clearly all ring-theoretic properties of D are inherited by D^{op}, with left and right interchanged. In particular D^{op} is a division ring if D is.

If M is a (left) vector space over a division ring D, then M is naturally also a *right* vector space over $\Delta = D^{op}$, where md, for $m \in M$ and $d \in \Delta$, is defined to be dm. The key point to check is that $m(d_1 \circ d_2) = (d_2 d_1)\, m = d_2(d_1 m) = d_2(md_1) = (md_1)d_2$. It is immediate from the definitions that $\mathrm{Hom}_D(M,M) = \mathrm{Hom}_\Delta(M,M)$, where, of course, $T \in \mathrm{Hom}_\Delta(M,M)$ if and only if $T \in \mathrm{End}(M)$ and $T(md) = T(m)d$ for all $m \in M$, $d \in \Delta$.

If $\dim_\Delta(M)$ is finite choose an ordered Δ-basis $\{m_1 \ldots, m_n\}$ for M. Then each $T \in \mathrm{Hom}_D(M,M)$ is associated with an $n \times n$ matrix (t_{ij}) over Δ as follows: express Tm_j as $\sum_i m_i t_{ij}$, $1 \leq i \leq n$ (compare with p. 145). Then the usual arguments from linear algebra show that $T \mapsto (t_{ij})$ is an isomorphism from the ring $\mathrm{Hom}_D(M, M)$ to the ring $\Delta_n = M_n(\Delta)$ of all $n \times n$ matrices over Δ.

The replacement of D by $\Delta = D^{op}$ and of M as left D-space by M as right Δ-space and the apparent fussiness of the above procedure were required in order that the map $T \mapsto (t_{ij})$ respect the products in the two rings.

Corollary to Theorem 3.5. Suppose R is a primitive ring. Then there is a division ring Δ such that either

(i) R is isomorphic with the ring $\Delta_n = M_n(\Delta)$ of all $n \times n$ matrices over Δ for some n, or

(ii) there is an ascending chain $R_1 \subseteq R_2 \subseteq \cdots$ of subrings of R and for each k an epimorphism from R_k to Δ_k.

4. ARTINIAN RINGS

A descending chain $I_1 \supseteq I_2 \supseteq I_3 \supseteq \cdots$ of left ideals in a ring R is said to *terminate* if it is finite or if there is some index k such that $I_n = I_k$ for all $n \geq k$. If every descending chain of left ideals of R terminates we say that R satisfies the *descending chain condition*, or the DCC. A ring that satisfies the DCC is also called an *Artinian* ring.

For example, $R = \mathbb{Z}$ is *not* an Artinian ring since the chain $(2) \supseteq (4) \supseteq (8) \supseteq (16) \supseteq \cdots$ does not terminate. On the other hand, if $R = M_n(D)$, the ring of $n \times n$ matrices over a division ring D, then any left ideal I of R is a D-subspace. If $I_1 \supseteq I_2 \supseteq \cdots$ is a descending chain then $\dim_D(I_{k+1}) \leq \dim_D(I_k) \leq n^2$ for all k, so the chain must terminate. Thus $R = M_n(D)$ is Artinian.

The argument just given shows that any ring that is a finite-dimensional unitary F-algebra for some field F must be Artinian. In particular, every group algebra FG, for a finite group G, is Artinian.

Strictly speaking, a ring satisfying the DCC for left ideals should be called *left* Artinian, with a corresponding definition for right Artinian rings. The theory of right Artinian rings is completely analogous to the theory of left Artinian rings. It is not the case, however, that the two concepts are equivalent (see Exercise 5.32 for an example).

A ring R is said to satisfy the *minimal condition* (for left ideals) if every nonempty set of left ideals of R contains a left ideal that is minimal with respect to set inclusion.

Proposition 4.1. A ring R is Artinian if and only if it satisfies the minimal condition.

Exercise 4.1. Prove Proposition 4.1.

Theorem 4.2. If R is Artinian, then its Jacobson radical $J(R)$ is nilpotent, and hence nil. Consequently $J(R)$ is the unique largest nilpotent ideal in R.

Proof. Write $J(R) = J$. The chain $J \supseteq J^2 \supseteq J^3 \supseteq \cdots$ must terminate, so $J^{n+k} = J^n$ for some positive integer n and all integers $k \geq 0$. If $J^n \neq 0$ let $\mathcal{S}$ be the set of all left ideals I in R such that $I \subseteq J^n$ and $J^n I \neq 0$. Then $\mathcal{S} \neq \varnothing$ since $J^n \in \mathcal{S}$, so there is a minimal element I_0 in $\mathcal{S}$ by Proposition 4.1. Choose $a \in I_0$ such that $J^n a \neq 0$. Then $J^n a$ is a left ideal, $J^n a \subseteq J^n$, and $J^n \cdot J^n a = J^{2n} a \neq 0$, so $J^n a \in \mathcal{S}$. Since $J^n a \subseteq I_0$, which is minimal in $\mathcal{S}$, we have $J^n a = I_0$. Choose $b \in J^n$ for which $ba = -a$. If c is the adverse of b, then

$$\begin{aligned} 0 &= a + ba = a + ba + c(a + ba) \\ &= a + (b + c + cb)a = a + (c * b)a = a, \end{aligned}$$

a contradiction. Thus $J^n = 0$ and J is nilpotent. For the second statement see Propositions 2.6 and 2.7.

Corollary. If R is Artinian, then every nil ideal in R is nilpotent.

Proposition 4.3. If R is a simple Artinian ring, then R is semisimple.

Proof. Since $J(R)$ is nilpotent by Theorem 4.2 and $R^2 = R$ we have $J(R) \neq R$. But $J(R)$ is an ideal, so $J(R) = 0$.

Theorem 4.4 (Maschke). If G is a finite group and F is a field, with $\text{char}(F) \nmid |G|$, then the group algebra FG is a semisimple ring.

Proof. Since FG has F-dimension $|G|$ and each left ideal is an F-subspace it is clear that FG is Artinian. Set $J = J(FG)$, then J is nil by Theorem 4.2. If we view FG as an FG-module then it is a semisimple module by Theorem 1.7, and J is a submodule, so we may write $FG = J \oplus I$ for some left ideal I, by Proposition 1.4(iii). Write $1 = a + b$, with $a \in J$, $b \in I$. Then $a = a \cdot 1 = a(a + b) = a^2 + ab$, and $a - a^2 = ab \in J \cap I = 0$, so $a = a^2$. But then a can not be nilpotent unless $a = 0$, which says $1 = b \in I$, and hence $I = FG$, $J = 0$.

Theorem 4.5 (Wedderburn–Artin). If R is an Artinian ring, then the following are equivalent:

(i) R is simple.
(ii) R is primitive.
(iii) R is isomorphic with the ring $\Delta_n = M_n(\Delta)$ of all $n \times n$ matrices with entries from some division ring Δ for some positive integer n.

Proof. We saw that (iii) $\Rightarrow$ (i) on page 181, and we know that (i) $\Rightarrow$ (ii) by Propositions 4.3 and 3.1. In order to show that (ii) $\Rightarrow$ (iii) we may view R as a dense subring of $\text{Hom}_D(M, M)$ by the Density Theorem, where M is a faithful simple R-module, $D = C(M)$, and $\Delta = D^{op}$. By Theorem 3.5 it will be sufficient to show that $\dim_D(M)$ is finite. If it is not we may choose an infinite D-linearly independent set $\{m_1, m_2, \ldots\} \subseteq M$. For each positive integer k set $M_k = \oplus\{Dm_j : 1 \leq j \leq k\}$. Since R is dense on M we may choose $r_k \in R$, for each k, such that $r_k m_i = 0$ for $1 \leq i \leq k$ but $r_k m_{k+1} \neq 0$. Thus $r_k \in A(M_k)\backslash A(M_{k+1})$, so $A(M_1) \supseteq A(M_2) \supseteq \cdots$ is a descending chain of ideals in R that does not terminate. That is a contradiction since R is Artinian, and the theorem is proved.

Corollary. If R is Artinian, then every primitive ideal P in R is a maximal modular ideal and every maximal modular ideal is primitive. Thus $J(R)$ is the intersection of all the maximal modular ideals of R $(J(R) = R$ if there are no maximal modular ideals).

Proof. The ring R/P is primitive and Artinian, so it is simple, and hence P is a maximal ideal, modular since R/P is a ring with 1. Apply the corollary to Proposition 3.2.

Note that the word "modular" in the corollary to Theorem 4.5 is necessary. For example, if R is the subring $\{0, 2, 4, 6\}$ of $\mathbb{Z}_8$, then $I = \{0, 4\}$ is the unique maximal ideal in R (it is not modular) but $J(R) = R$.

A nonzero ideal I in a ring R is a *minimal* ideal if the only ideals J with $0 \subseteq J \subseteq I$ are $J = 0$ and $J = I$. Note that if I is a minimal ideal in R, then either $I^2 = 0$ or $I^2 = I$, and that in the latter case I is a simple ring. (Why?) For example if $R = \mathbb{Z}_4$, then $I = (2)$ is a minimal ideal with $I^2 = 0$; if $R = \mathbb{Z}_6$, then $I = (2)$ is a minimal ideal with $I^2 = I$.

Theorem 4.6 (Wedderburn–Artin). If R is a semisimple Artinian ring, then R has only finitely many minimal ideals $I_1, I_2, \ldots, I_k$, and $R = I_1 \oplus I_2 \oplus \cdots \oplus I_k$. Each minimal ideal I_j is a simple Artinian ring, so $I_j \cong D_{n_j}^{(j)}$, the ring of $n_j \times n_j$ matrices over a division ring $D^{(j)}$.

Proof. Since R is semisimple the intersection of all the primitive ideals in R is 0, by the corollary to Proposition 3.2. Let P_1 be a primitive ideal and choose (if possible) a primitive ideal P_2 different from P_1. Then choose (again if possible) a primitive ideal P_3 that does not contain $P_1 \cap P_2$. Inductively choose (as long as possible) a primitive ideal P_{n+1} that does not contain $\bigcap\{P_i : 1 \leq i \leq n\}$. The process must terminate in a finite number of steps since otherwise $\{\bigcap\{P_i : 1 \leq i \leq n\}\}_{n=1}^{\infty}$ would be a descending chain of ideals that does not terminate. Thus we obtain primitive ideals $P_1, \ldots, P_m$ such that $\bigcap\{P_i : 1 \leq i \leq m\} = 0$. Relabel if necessary and choose a set $\{P_1, \ldots, P_k\}$ of primitive ideals such that $\bigcap\{P_i : 1 \leq i \leq k\} = 0$ but no proper subset intersects in 0. Set $I_j = \bigcap\{P_i : 1 \leq i \leq k, i \neq j\}$ for $1 \leq j \leq k$. Since P_j is a maximal ideal (Corollary, Theorem 4.5) and I_j is not contained in P_j we have $I_j + P_j = R$ for each j, and also $I_j \cap P_j = \bigcap\{P_i : 1 \leq i \leq k\} = 0$. Thus $R = I_j \oplus P_j$ for each j, and $I_j \cong R/P_j$. Since P_j is maximal I_j can have only 0 and I_j as ideals, and $I_j^2 \neq 0$ since R is semisimple (see Theorem 4.2). It follows that each I_j is a minimal ideal in R and is a simple Artinian ring.

Set $Q_i = \bigcap\{P_j : 1 \leq j \leq i\}$, for $1 \leq i \leq k$, and note that $R = I_1 \oplus Q_1$. By the Isomorphism Theorem (II.1.6) we have

$$Q_1/(Q_1 \cap P_j) \cong (Q_1 + P_j)/P_j = R/P_j \qquad \text{if} \quad 2 \leq j \leq k,$$

so each $Q_1 \cap P_j$ is a maximal ideal in Q_1, $2 \leq j \leq k$. Furthermore,

$$\bigcap\{Q_1 \cap P_j : 2 \leq j \leq k\} = \bigcap\{P_j : 1 \leq j \leq k\} = 0,$$

and no proper subset of $\{Q_1 \cap P_j : 2 \leq j \leq k\}$ intersects in 0. As above we see that

$$Q_1 = (\bigcap\{Q_1 \cap P_j : 3 \leq j \leq k\}) \oplus (Q_1 \cap P_2) = I_2 \oplus Q_2,$$

and hence $R = I_1 \oplus I_2 \oplus Q_2$. Inductively $R = I_1 \oplus I_2 \oplus \cdots \oplus I_j \oplus Q_j$, and hence $R = I_1 \oplus I_2 \oplus \cdots \oplus I_k$ since $Q_k = 0$.

Since I_j is simple it is isomorphic with the ring $D_{n_j}^{(j)}$ of $n_j \times n_j$ matrices over a division ring $D^{(j)}$ by Theorem 4.5. We note in passing that as a result each I_j is a ring with 1 and hence R is a ring with 1.

If I is any minimal ideal in R, then $RI = I$ and so

$$I = (\oplus\{I_j\}) \cdot I = \oplus\{I_j \cdot I : 1 \leq j \leq k\}.$$

Each direct summand is an ideal in I so exactly one summand is nonzero and $I = I_jI$ for some j. But then $I = I_j$ since I_jI is a nonzero ideal contained in both of the minimal ideals I and I_j. Thus $I_1, \ldots, I_k$ is the full set of all minimal ideals in R.

Corollary. A semisimple Artinian ring is a ring with 1.

Theorem 4.7. Suppose F is an algebraically closed field and A is a finite-dimensional F-algebra that is semisimple as a ring. Then A is isomorphic with $F_{n_1} \oplus F_{n_2} \oplus \cdots \oplus F_{n_k}$, where F_{n_j} is the ring of $n_j \times n_j$ matrices with entries from F.

Proof. By Theorem 4.6 it will suffice to prove the theorem under the further assumption that A is simple, and in fact that $A = \Delta_n$, where $D = C(M)$ and $\Delta = D^{op}$ for some simple A-module M. By Proposition 1.2 we have $D = F \cdot 1 \cong F$ and so $A \cong F_n$.

The final step in describing the structure of semisimple Artinian rings would be a description of all division rings. Unfortunately no description is known. The next theorem does serve as a complete description of all *finite* division rings, in view of Theorem III.3.11.

Theorem 4.8 (Wedderburn). A finite division ring D is a field.

Proof. (Witt [40]). Let $F = \{n \cdot 1 : n \in \mathbb{Z}, 1 \in D\}$. Then, as on page 80, F is a prime field, so $F \cong \mathbb{Z}_p$ for some prime p. If $\dim_F(D) = r$, then $|D| = p^r$. Let $Z = \{x \in D : xy = yx \text{ for all } y \in D\}$, the *center* of D. Then Z is a field, $F \subseteq Z$, and Z^* is the center $Z(D^*)$ of the group D^*. Say $|Z| = p^s = q$. If $\dim_Z(D) = t$, then $st = r$. For each $x \in D$ define the *centralizer* of x in D to be $C_D(x) = \{y \in D : xy = yx\}$. Then $C_D(X)$ is a subdivision ring of D, $Z \subseteq C_D(x)$, and if $x \neq 0$, then $C_D(x)^* = C_{D^*}(x)$ the group-theoretic centralizer. Set $d_x = \dim_Z C_D(x)$. Then $d_x \mid t$ (Proposition III.1.1 applies even if D is not commutative) and if $x \neq 0$, then the number of elements in the conjugacy class of x in D^* is

$$[D^* : C_{D^*}(x)] = (q^t - 1)/(q^{d_x} - 1)$$

by Proposition I.2.3. Suppose $\{x_1, \ldots, x_k\}$ is a set of representatives for the conjugacy classes of noncentral elements in D^*. Then by the class equation (page 14) we have

$$q^t - 1 = q - 1 + \sum\{(q^t - 1)/(q^{dx_i} - 1): 1 \leq i \leq k\}.$$

Set $f(x) = x^t - 1$ and let $g(x) = \Phi_t(x)$ be the tth cyclotomic polynomial. Then $g(x) \in \mathbb{Z}[x]$, $g(x) \mid f(x)$, and if $\omega \in \mathbb{C}$ is a primitive tth root of unity, then

$$g(x) = \prod\{(x - \omega^j): 1 \leq j \leq t \text{ and } (j, t) = 1\}$$

(see page 97). If $d | t$ and $d < t$, then $x^d - 1 \mid x^t - 1$ (by high school algebra) and $(x^d - 1, g(x)) = 1$ since the roots of $x^d - 1$ are not primitive tth roots of unity, so we may write $x^t - 1 = (x^d - 1)g(x)h(x)$, and $h(x) \in \mathbb{Z}[x]$ by the division algorithm. Substituting $x = q$ we see that $(q^t - 1)/(q^d - 1)$ is in $\mathbb{Z}$. Observe also that

$$g(q) = \prod\{(q - \omega^j): 1 \leq j \leq t \text{ and } (j, t) = 1\},$$

and that $|q - \omega^j| > q - 1$ (see Fig. 1), and hence $|g(q)| > q - 1$.

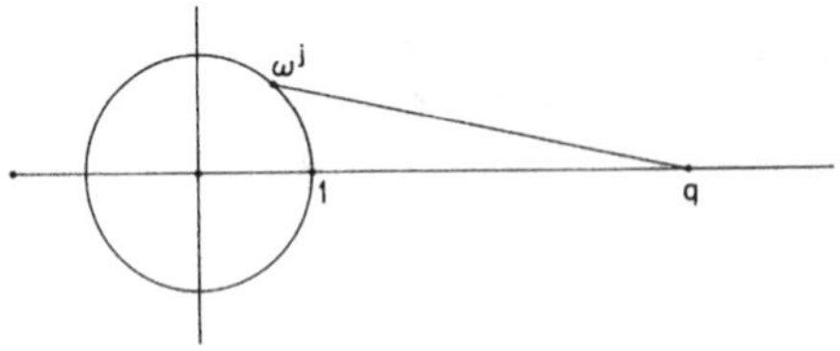

FIGURE 1

Suppose now that $t > 1$. Then $g(q) \mid (q^t - 1)/(q^{dx_i} - 1)$, $1 \leq i \leq k$, and $g(q) \mid q^t - 1$, so by the class equation $g(q) \mid q - 1$. That is in conflict with the fact that $|g(q)| > q - 1$, so $t = 1$ and $D = Z$ is a field.

If F is an algebraic field extension of the real field $\mathbb{R}$, then either $F = \mathbb{R}$ or $F \cong \mathbb{C}$ by the Fundamental Theorem of Algebra (Theorem III.3.10). We are familiar with the division ring $\mathbb{H}$ of quaternions which has $\mathbb{R}$ as a subfield (in fact $\mathbb{R}$ is the center of $\mathbb{H}$), with $\dim_{\mathbb{R}}(\mathbb{H}) = 4$. If $a \in \mathbb{H}$ then left multiplication by a determines an $\mathbb{R}$-linear transformation T of $\mathbb{H}$. If $f(x) \in \mathbb{R}[x]$, then $f(T)$ is left multiplication by $f(a)$, so a is a root in $\mathbb{H}$ of the minimal polynomial of T over $\mathbb{R}$, and hence a is algebraic over $\mathbb{R}$. Consequently $\mathbb{H}$ is algebraic as a ring over $\mathbb{R}$. The next theorem tells us that there are no more such examples.

Theorem 4.9 (Frobenius). Suppose D is a division ring containing $\mathbb{R}$ in its center, with D algebraic over $\mathbb{R}$. Then D is isomorphic either with $\mathbb{R}$, $\mathbb{C}$, or $\mathbb{H}$.

Proof (Palais [29]). As noted above it is sufficient, by the Fundamental Theorem of Algebra, to assume that D is not commutative. Choose $d \in D \backslash \mathbb{R}$ and set $K = \mathbb{R} + \mathbb{R}d$, a two-dimensional $\mathbb{R}$-subspace of D. Note that elements of K commute with one another when multiplied, since $\mathbb{R}$ is central. Choose a subspace F of D that contains K and is maximal with respect to the property that its elements commute when multiplied. By the maximality of F any $a \in D$ that commutes with all elements of F must lie in F, for otherwise $F + \mathbb{R}a$ would be a larger commuting subspace. In particular, if $a,b \in F$, then $ab \in F$, and if $a \neq 0$, then $a^{-1} \in F$, so F is a field. But then F is an algebraic extension of $\mathbb{R}$, and $F \neq \mathbb{R}$, so $F \cong \mathbb{C}$, and in fact $F = K = \mathbb{R} + \mathbb{R}d$.

Since $F \cong \mathbb{C}$ we may choose $i \in F$ such that $i^2 = -1$. Then $F = \mathbb{R} + \mathbb{R}i$, and we may identify F with $\mathbb{C}$. Thus we may also view D as a (left) vector space over $\mathbb{C}$. Define a $\mathbb{C}$-linear transformation $T: D \to D$ by setting $T(a) = ai$ for all $a \in D$. Then $T^2 = -I$, but $T \neq \pm I$, so T has minimal polynomial $x^2 + 1$ and eigenvalues i and $-i$. Write D^+ and D^- for the corresponding eigenspaces, so $D^+ = \{a \in D : ai = ia\}$ and $D^- = \{a \in D : ai = -ia\}$. Clearly $\mathbb{C} \subseteq D^+$ and, by the maximality of $F = \mathbb{C}$, $D^+ \subseteq \mathbb{C}$, so $D^+ = \mathbb{C}$. If $a \in D$, then

$$a = (a - iai)/2 + (a + iai)/2 \in D^+ + D^-$$

since

$$(a - iai)i = ai + ia = i(-iai + a)$$

and

$$(a + iai)i = ai - ia = -i(iai + a).$$

But also $D^+ \cap D^- = 0$, so $D = D^+ \oplus D^-$. Note that, by the definition of D^-, $ab \in D^+$ if $a, b \in D^-$. Choose a nonzero element $b \in D^-$ and define another $\mathbb{C}$-linear transformation $S: D \to D$ by setting $S(a) = ab$ for all $a \in D$. Then S is invertible (its inverse is right multiplication by b^{-1}), and $S(D^-) = D^+$. Thus $\dim_{\mathbb{C}}(D^-) = \dim_{\mathbb{C}}(D^+) = 1$, hence $\dim_{\mathbb{C}}(D) = 2$ and $\dim_{\mathbb{R}}(D) = 4$.

Since $0 \neq b \in D^-$ we see by the observations made at the beginning of the proof that $\mathbb{R} + \mathbb{R}b$ is a field isomorphic with $\mathbb{C}$. Note that

$$b^2 \in D^+ \cap (\mathbb{R} + \mathbb{R}b) = \mathbb{C} \cap (\mathbb{R} + \mathbb{R}b) = \mathbb{R}.$$

If b^2 were positive it would have two square roots in $\mathbb{R}$, but also the two square roots $\pm b$ in $\mathbb{R}b$, hence $x^2 - b^2$ would have four roots in the field $\mathbb{R} + \mathbb{R}b$, contradicting Proposition III.1.7. Thus $b^2 < 0$ in $\mathbb{R}$. Replace b by $j = b/|\sqrt{b^2}|$ to obtain $j \in D^-$ with $j^2 = -1$ and $ji = -ij$, and set $k = ij$. Then $k \in D^-$ and k is not an $\mathbb{R}$-multiple of j, so $\{j, k\}$ is an $\mathbb{R}$-basis for D^-. Thus $\{1, i, j, k\}$ is an $\mathbb{R}$-basis for D. The usual further facts about multiplication of the basis elements i, j, and k in the ring $\mathbb{H}$ of quaternions follow easily, so $D \cong \mathbb{H}$.

5. FURTHER EXERCISES

1. Suppose F is a field and let $R = M_2(F)$, the ring of 2×2 matrices over F. Exhibit a left ideal in R that is not a semisimple ring.

2. If R and S are rings and $f: R \to S$ is a homomorphism show that $f(J(R)) \subseteq J(f(R))$.

3. If $\{R_\alpha : \alpha \in A\}$ is any nonempty family of rings show that

(i) $J(\oplus R_\alpha) = \oplus J(R_\alpha)$, and

(ii) $J(\prod R_\alpha) = \prod J(R_\alpha)$.

Conclude that $\oplus R_\alpha$ is semisimple if and only if each R_α is semisimple, and likewise for $\prod R_\alpha$.

4. Suppose F is a field and $f(x) \in F[x]$, and set $R = F[x]/(f(x))$. Determine $J(R)$ (see the example on p. 180). Under what conditions is R semisimple?

5. If F is a field let R be the ring of $n \times n$ upper triangular matrices with entries from F, i.e., if $A = (a_{ij}) \in R$, then $a_{ij} = 0$ if $i > j$. Find $J(R)$ and show that $R/J(R)$ is commutative (in fact $R/J(R) \cong F^n$).

6. If R is any ring show that $J(M_n(R)) = M_n(J(R))$.

7. An element $a = \sum\{a_x \cdot x : x \in G\}$ in a group algebra FG is called a *class function* (cf) if it is constant on each conjugacy class in G, i.e., $a(y^{-1}xy) = a(x)$ for all $x, y \in G$. Write cf(G) for the set of all class functions on G.

(i) Show that cf(G) is the center of FG.

(ii) For each conjugacy class K in G define $\phi_K \in FG$ by setting $\phi_K(x) = 1$ if $x \in K$, $\phi_K(x) = 0$ otherwise, i.e., $\phi_K = \sum\{x : x \in K\}$. Show that the elements ϕ_K constitute a basis for cf(G). Conclude that the dimension of the center of FG is the class number of G.

8. Suppose R is a commutative ring and I is an ideal in R. Define $\sqrt{I}$ to be $\{a \in R : a^k \in I$ for some integer $k \geq 0\}$ (this is sometimes called the *radical* of the ideal I).

(i) Show that $\sqrt{I}$ is an ideal in R.

(ii) If I and J are ideals in R show that $\sqrt{IJ} = \sqrt{I \cap J} = \sqrt{I} \cap \sqrt{J}$.

(iii) Show that $\sqrt{I + J} = \sqrt{\sqrt{I} + \sqrt{J}}$.

9. Suppose G is a finite group, F is a field, and K an extension field. Show that the K-algebras KG and $K \otimes_F FG$ are K-isomorphic.

10 (Rieffel [32]). Suppose R is a simple ring with 1 and I is a nonzero left ideal in R, viewed as a left R-module. Write R' for $C_R(I)$. Then R' is a ring and I is also a left R'-module. Set $R'' = C_{R'}(I)$ and define $\theta: R \to \operatorname{End}(I)$ by means of $\theta_r(x) = rx$, all $r \in R$, $x \in I$. Also define $\psi: I \to \operatorname{End}(I)$ by means of $\psi_x(y) = yx$, all $x, y \in I$.

(i) Show that $\theta_r \in R''$, all $r \in R$, and $\psi_x \in R'$, all $x \in I$.

(ii) Show that θ is a (ring) monomorphism.

(iii) If $\phi \in R''$ and $x \in I$ show that $\phi\theta_x = \theta_{\phi(x)}$, and conclude that $\theta(I) = \{\theta_x : x \in I\}$ is a left ideal in R''.

(iv) Use the fact that $IR = R$ to show that $Im(\theta)$ is a left ideal in R'', and note that $1_{R''} \in Im(\theta)$.

(v) Conclude that R'' and R are isomorphic rings (this is a version of the "Double Centralizer" property for simple rings).

11. Suppose $p \in \mathbb{Z}$ is a prime dividing the order of a finite group G and F is a field of characteristic p. Show that FG is not semisimple. (*Hint*: Show that $a = \sum\{x : x \in G\}$ is a nonzero nilpotent element in the center of FG and conclude that $a \in J(FG)$.)

12. Suppose R is a ring with 1 and that $I = R \backslash U(R)$ is an ideal in R. Show that $J(R) = I$.

13. Suppose that R is a radical ring that is nilpotent (i.e., $R^k = 0$ for some k, $0 < k \in \mathbb{Z}$). Show inductively that R^i contains the ith term in the descending central series of the group $(R, *)$, and conclude that $(R, *)$ is a nilpotent group.

14. Show that a primitive commutative ring is a field.

15. Suppose $V \neq 0$ is a (left) vector space over a division ring D, and R is a ring of D-linear transformations of V. Suppose further that R is transitive on V, meaning that if $u, v \in V$, with $u \neq 0$, then $ru = v$ for some $r \in R$. Show that R is a primitive ring, with $M = V$ as a faithful simple R-module.

16. Let R be the ring of all matrices of the form

$$\begin{bmatrix} a & b \\ b & a+b \end{bmatrix},$$

$a, b \in \mathbb{Q}$, and let M be the R-module $\mathbb{Q}^2$ of column vectors.

(i) Show that R is a division ring (in fact a field).
(ii) Show that M is a transitive R-module (see Exercise 15).
(iii) Show that M is a simple R-module.
(iv) Show that $C(M) \cong R$.

17. Suppose V is a (left) vector space over a division ring D, and R is a *doubly transitive* ring of D-linear transformations of V, meaning that if $\{u_1, u_2\}$ is D-linearly independent in V and $\{v_1, v_2\} \subseteq V$, then $ru_1 = v_1$ and $ru_2 = v_2$ for some $r \in R$. Show that R is a primitive ring and that $C(V) \cong D$. Conclude that R acts densely on V.

18. Suppose M is a simple R-module and $D = C(M)$. Endow M with the discrete topology, the space M^M of all functions from M to M with the product topology, and $\mathrm{Hom}_D(M, M)$ with the topology it inherits as a subspace of M^M. Show that R acts densely on M if and only if R is topologically dense when viewed as a subring of $\mathrm{Hom}_D(M, M)$.

19. If R is an Artinian ring and $f: R \to S$ is a ring epimorphism show that S is also Artinian.

20. If R is an integral domain with 1 and R is Artinian show that R is a field.

21. If R is a commutative ring with 1 and R is Artinian show that every prime ideal of R is maximal. (You may wish to apply Exercises 19 and 20.)

22. Suppose R is an Artinian ring with 1 and M is a unitary left R-module. Show that M is semisimple if and only if $J(R)M = 0$.

23. Describe all semisimple rings having 1296 elements.

24. If p and q are distinct primes describe all semisimple rings having the following numbers of elements: (a) p^4, (b) p^4q^4, (c) p^{10}, (d) p^8q^5.

25. If V is a finite-dimensional vector space over a field F and $T: V \to V$ is a linear transformation show that the ring $F[T]$ of all polynomials (over F) in T is isomorphic with $F[x]/(m_T(x))$. What is $J(F[T])$? Determine when $F[T]$ is simple and when it is semisimple. (See Exercise 4, above.)

26 (Wedderburn). Suppose that A is a finite-dimensional algebra over a field F and that A has an F-basis consisting of nilpotent elements. Show that A is nilpotent. (Show first that F can be assumed algebraically closed by considering $A^K = K \otimes_F A$, K an algebraic closure of F. Then suppose the result false and use induction on $\dim_F(A)$. Argue that A is semisimple, then that A is simple, hence a matrix algebra. Then use the basis of nilpotent elements to see that *all* elements have trace 0 and arrive at a contradiction.)

27. (Jennings [18]). Suppose G is a finite p-group and F is a field of characteristic p. Show that $J = J(FG)$ has $\{x - 1 : 1 \neq x \in G\}$ as an F-basis, so $\dim_F(J) = |G| - 1$. (*Hint*: Let W be the F-subspace spanned by all $x - 1$. Show that W is an ideal and each $x - 1$ is nilpotent. Apply Exercise 26.)

28. If G is a finite p-group and $R = J(\mathbb{Z}_p G)$ show that the map $x \mapsto 1 - x$ is an isomorphism between G and a subgroup of $(R, *)$ (see Exercise 27).

29. Show that a finite nilpotent group is isomorphic with a subgroup of $(R, *)$ for some finite nilpotent radical ring R (see Exercises 13 and 28).

30. Let $F = \mathbb{Z}_2$, let G_1 be cyclic of order 4, and let G_2 be Klein's 4-group. Show that the group algebras FG_1 and FG_2 are not isomorphic. (*Hint*: Investigate the "degree of nilpotence" of their Jacobson radicals. See Exercise 27.)

31. If G_1 is cyclic of order 4 and G_2 is Klein's 4-group show that $\mathbb{C}G_1$ and $\mathbb{C}G_2$ are isomorphic $\mathbb{Q}$-algebras.

32. Let

$$m = \begin{bmatrix} m_1 \\ m_2 \end{bmatrix}$$

be a fixed nonzero column vector with entries from $\mathbb{Q}$ and let R denote the additive group of all row vectors $a = [a_1, a_2]$ with entries from $\mathbb{Q}$, endowed with a multiplication as follows. If $a = [a_1, a_2]$, $b = [b_1, b_2] \in R$, then $a \circ b = amb$ (ordinary matrix multiplication). Note that if $\lambda(a)$ is defined to be $a_1 m_1 + a_2 m_2$ then $a \circ b = \lambda(a)b$ (ordinary scalar multiplication). Set $n = [-m_2, m_1] \in R$ and let $N = (n)$, the principal ideal generated by n.

(i) Show that R is a ring having left (multiplicative) identities but no right identity.

(ii) Show that every left ideal of R is a $\mathbb{Q}$-vector subspace, and conclude that R is (left) Artinian.

(iii) Show that R does not satisfy the DCC for right ideals (consider additive subgroups of N).

(iv) If $e = [e_1, e_2] \in R$ is chosen so that $\lambda(e) = 1$ show that e is a right relative unit for the modular left ideal N.

(v) Show that 0, N and R are the only (two-sided) ideals in R. Conclude that $N = J(R)$.

33. Suppose F is an algebraically closed field and D is a division ring that is algebraic over F, with F contained in the center of D. Show that $D = F$.

34. Suppose F is an algebraically closed field and A is a finite-dimensional semisimple F-algebra. Show that the dimension of the center of A is equal to the number of simple direct summands of A.

35. Suppose R is an Artinian ring with 1 whose set of left ideals is linearly ordered, i.e., if I_1 and I_2 are left ideals then either $I_1 \subseteq I_2$ or $I_2 \subseteq I_1$. (For an example see Exercise II.8.34.) Write J for $J(R)$.

(i) Show that J is a maximal ideal and a maximal left ideal in R.

(ii) Show that R/J is a division ring.

(iii) Show that every proper left ideal in R is of the form J^n for some $n > 0$ in $\mathbb{Z}$, and conclude that every left ideal is an ideal.

36. If I is a minimal ideal in a ring R and $I^2 \neq 0$ show that I is a simple ring.

Chapter VI | Further Topics

1. INFINITE ABELIAN GROUPS

All groups in this section will be abelian; they will be written additively unless it is explicitly noted otherwise.

Recall from Section 3 of Chapter IV that an abelian group A has a *torsion subgroup* $T = T(A)$ consisting of all of its elements of finite order, and that A/T is torsion free.

If T is a torsion abelian group and $p \in \mathbb{Z}$ is prime let T_p be the set of all elements in T whose orders are powers of p,

$$T_p = \{x \in T : |x| = p^k, \text{ some } k\}.$$

Clearly $T_p \leq T$. If $0 \neq x \in T$, then $\langle x \rangle$ is the direct sum of its Sylow subgroups, and in particular x can be written uniquely as $x = x_1 + x_2 + \cdots + x_k$, with $0 \neq x_i \in T_{p_i}$ for various distinct primes $p_1, \ldots, p_k$. An application of Exercise IV.2.2 yields the following theorem.

Theorem 1.1 (Primary Decomposition). If T is an abelian torsion group and

$$T_p = \{x \in T : |x| = p^k, \text{ some } k\}$$

for each prime $p \in \mathbb{Z}$, then $T = \oplus_p T_p$.

For an important example take $T = \mathbb{Q}/\mathbb{Z}$. Then

$$T_p = \{(a/p^b) + \mathbb{Z} : a, b \in \mathbb{Z}, b \geq 0\}.$$

It is easy to see that we may choose representative elements of the cosets $(a/p^b) + \mathbb{Z}$ of the form c/p^b, where $0 \leq c < p^b$, $b \geq 0$, and $(p, c) = 1$. If we let

$\mathbb{Z}_{p^\infty}$ denote that set of representative elements and define addition in $\mathbb{Z}_{p^\infty}$ to be ordinary addition modulo 1, then clearly $T_p \cong \mathbb{Z}_{p^\infty}$.

Exercise 1.1. (a) If p is a prime let E_p denote the multiplicative group of all p^kth roots of unity in $\mathbb{C}$, for various k, i.e.,

$$E_p = \{e^{2\pi i j/p^k} \in \mathbb{C} : 0 \leq k \in \mathbb{Z}, j \in \mathbb{Z}\}.$$

Show that $E_p \cong \mathbb{Z}_{p^\infty}$.

(b) Show that every proper subgroup of $\mathbb{Z}_{p^\infty}$ is cyclic and that its subgroups are linearly ordered by inclusion. In fact the proper subgroups may be listed as $0 = A_0 \leq A_1 \leq A_2 \leq \cdots$, with $|A_k| = p^k$, and $\mathbb{Z}_{p^\infty} = \bigcup A_k$.

Recall from Chapter IV, Section 2, that an abelian group F is *free* if it is isomorphic with a direct sum of copies of the additive group $\mathbb{Z}$, or equivalently if F has a basis (as a $\mathbb{Z}$-module). The cardinality of any basis is called the *rank* of F. Any function from a basis for F to an abelian group A extends uniquely to a homomorphism from F to A. As a consequence every abelian group A is a homomorphic image of a free abelian group.

An abelian group G is called *projective* if for every exact sequence $A \xrightarrow{f} B \to 0$ of abelian groups and every $g \in \mathrm{Hom}(G, B)$ there is some $h \in \mathrm{Hom}(G, A)$ for which the diagram

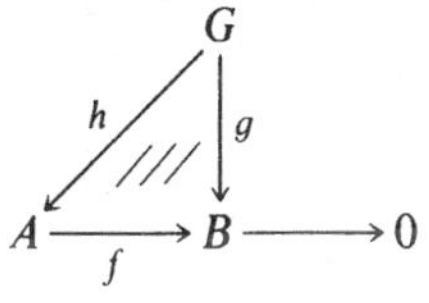

is commutative, i.e., $g = fh$.

Theorem 1.2. An abelian group G is projective if and only if it is free.

Proof. $\Leftarrow$: Suppose G is free with S as a basis. If $A \xrightarrow{f} B \to 0$ is exact and $g \in \mathrm{Hom}(G, B)$ choose $a_s \in A$ for each $s \in S$ such that $f(a_s) = g(s)$. Then the function $h: S \to A$ defined by $h(s) = a_s$ extends to $h \in \mathrm{Hom}(G, A)$, and $fh = g$, so G is projective.

$\Rightarrow$: There is a free abelian group F and an epimorphism $\theta: F \to G$. If $K = \ker\theta$ there is an isomorphism $g: G \to F/K$. If $\eta: F \to F/K$ is the quotient map, then, since G is projective, there is some $h \in \mathrm{Hom}(G, F)$ such that the diagram

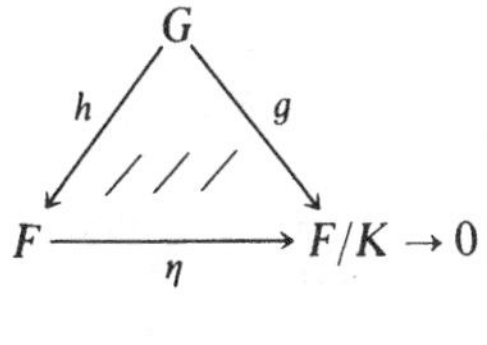

is commutative, i.e., $g = \eta h$. But then h is 1–1 since g *is* 1–1, so G is isomorphic with a subgroup of F, and hence G is free by Theorem IV.2.9.

An abelian group A is called *divisible* if $nA = A$ for every nonzero integer n. It is of course sufficient to require that $pA = A$ for every prime p. If A is divisible, $x \in A$, and $0 \neq n \in \mathbb{Z}$, then $ny = x$ for some $y \in A$, and we may think of y as the result of "dividing x by n." The "quotient" y is not necessarily uniquely determined, however.

Perhaps the simplest example of a divisible group is the additive group $\mathbb{Q}$ of rational numbers.

Proposition 1.3. Suppose A is a divisible group and $f\colon A \to B$ is an epimorphism. Then B is divisible.

Proof. If $0 \neq n \in \mathbb{Z}$, then $nB = nf(A) = f(nA) = f(A) = B$.

As an immediate consequence of Proposition 1.3 we see that $\mathbb{Q}/\mathbb{Z}$ is divisible, and then that $\mathbb{Z}_{p^\infty}$ is divisible.

Proposition 1.4. If A is torsion free and divisible, then $A = \bigoplus_\beta A_\beta$, where each A_β is isomorphic with $\mathbb{Q}$.

Proof. If $0 \neq n \in \mathbb{Z}$ and $x \in A$, then $x = ny$ for some $y \in A$. In this case y is unique, for if also $nz = x$, $z \in A$, then $n(z - y) = 0$ and hence $z = y$ since A is torsion-free. Thus we may write $y = (1/n)x$. But then for any $m/n \in \mathbb{Q}$ we may write $(m/n)x = m \cdot (1/n)x$, and it is easy to verify that A is consequently a $\mathbb{Q}$-vector space. As such it has a basis and the proposition is proved.

An abelian group G is called *injective* if for any abelian groups A, B in an exact sequence $0 \to B \xrightarrow{f} A$, and for any $g \in \mathrm{Hom}(B, G)$, there is some $h \in \mathrm{Hom}(A, G)$ such that $g = hf$, i.e., the diagram

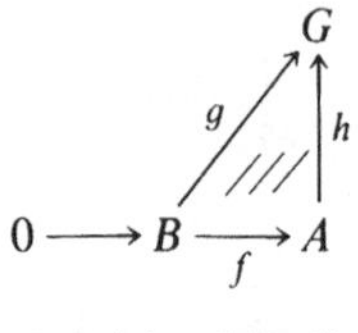

is commutative. Note that this definition "dualizes" (i.e., reverses the arrows in) the definition of a projective abelian group.

In particular, B might be a subgroup of A and f the inclusion mapping, in which case the requirement is that any homomorphism from B to G can be extended to the larger group A.

Theorem 1.5. An abelian group G is injective if and only if it is divisible.

Proof. $\Rightarrow$: If $0 \neq n \in \mathbb{Z}$ and $x \in G$ define a homomorphism g from $B = \langle n \rangle \leq \mathbb{Z} = A$ to G by setting $g(kn) = kx$. Extend g to $h \in \mathrm{Hom}(\mathbb{Z}, G)$ and set

$y = h(1)$. Then

$$ny = nh(1) = h(n) = g(n) = x$$

and G is divisible.

$\Leftarrow$: Assume G is divisible and suppose we wish to complete the diagram

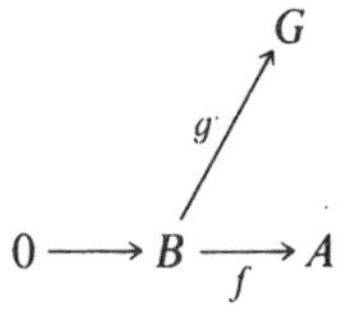

Let

$$\mathscr{S} = \{(C, j) : j \in \text{Hom}(C, G), f(B) \leq C \leq A, jf = g\},$$

the collection of all "partial completions" of the diagram. Then $\mathscr{S} \neq \varnothing$ since $(fB, gf^{-1}) \in \mathscr{S}$, and $\mathscr{S}$ is partially ordered if we agree that $(C_1, j_1) \geq (C_2, j_2)$ provided $C_1 \geq C_2$ and $j_1 | C_2 = j_2$. By Zorn's Lemma there is a maximal element $(D, h) \in \mathscr{S}$, and it will suffice to show that $D = A$. If not, choose $x \in A \backslash D$ and set $E = \langle D, x \rangle = D + \langle x \rangle$. If $D \cap \langle x \rangle = 0$ we may define $h' \in \text{Hom}(E, G)$ by setting $h'(d + mx) = h(d)$, $d \in D$, and check that $(E, h') > (D, h)$, violating maximality. If $D \cap \langle x \rangle \neq 0$, choose k minimal and positive in $\mathbb{Z}$ for which $kx \in D$. Observe then that elements of E have unique representations as $d + mx$, where $d \in D$ and $0 \leq m < k$. Set $a = h(kx) \in G$. Since G is divisible we may write $a = kb$ for some $b \in G$. Define $h'' \in \text{Hom}(E, G)$ by setting $h''(d + mx) = g(d) + mb$, $d \in D$ and $0 \leq m < k$ *in* $\mathbb{Z}$. Again it is easy to verify that $(E, h'') > (D, h)$, violating maximality, so $D = A$.

Corollary. If A is an abelian group and D is a divisible subgroup of A, then D is a direct summand of A.

Proof. If $i: D \to A$ denotes inclusion, then $0 \to D \xrightarrow{i} A$ is exact, and we have a commutative diagram

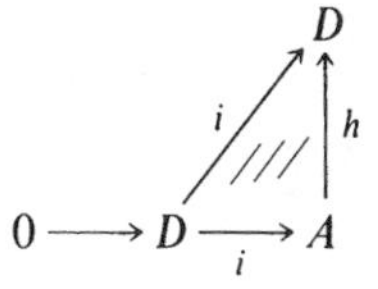

for some $h \in \text{Hom}(A, D)$. Set $K = \ker h$. If $x \in A$, then $x = h(x) + [x - h(x)] \in D + K$ since $h^2 = h$. If $y \in D \cap K$, then $y = h(y) = 0$, so $A = D \oplus K$.

Exercise 1.2. Suppose D is an abelian group that is a direct summand of every abelian group A that contains D as a subgroup. Show that D is divisible.

An abelian group A is called *reduced* if it has no nonzero divisible subgroups. Clearly $\mathbb{Z}$ is reduced, for example.

Proposition 1.6. An abelian group A has a unique maximal divisible subgroup D. If $B \leq A$ and B is divisible, then $B \leq D$, so in fact

$$D = \bigcup \{B \leq A : B \text{ is divisible}\}.$$

Proof. Set

$$D = \langle \bigcup \{B \leq A : B \text{ is divisible}\} \rangle.$$

Then clearly D is divisible and the rest follows.

Corollary. If A is abelian and D is its maximal divisible subgroup, then $A = D \oplus K$, with K reduced.

Proof. See the corollary to Theorem 1.5.

If $p \in \mathbb{Z}$ is a prime and A is an abelian p-group set $A[p] = \{x \in A : px = 0\}$. Clearly $A[p] \leq A$, and it is easy to see that $A[p]$ is quite naturally a vector space over $\mathbb{Z}_p$. Consequently $A[p]$ is a direct sum, with each summand isomorphic with $\mathbb{Z}_p$.

Proposition 1.7. Suppose A and B are divisible abelian p-groups for some prime p. If $A[p] \cong B[p]$, then $A \cong B$.

Proof. Let g: $A[p] \to B[p]$ be an isomorphism, but view g as an element of $\text{Hom}(A[p], B)$, and let i: $A[p] \to A$ be inclusion. By Theorem 1.5 there exists $h \in \text{Hom}(A, B)$ such that the diagram

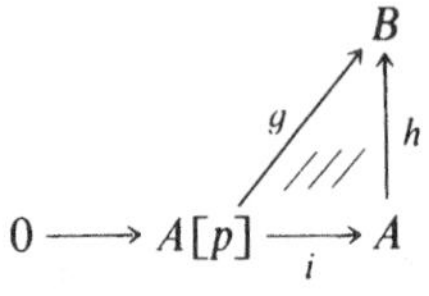

commutes, i.e., $h \mid A[p] = g$. We show by induction that $\ker h = 0$. Take $x \in$ *ker h*. If $|x| \leq p$, then $x \in A[p]$, so $h(x) = g(x) = 0$ and hence $x = 0$. Suppose then that $|x| = p^n$, $n > 1$, and assume that the only $y \in \ker h$ with $|y| \leq p^{n-1}$ is $y = 0$. Set $y = px$. Then $y \in \ker h$ and $|y| = p^{n-1}$, so $n = 1$, a contradiction. Thus $\ker h = 0$.

Next take any $y \in B$, say $|y| = p^n$. Again we shall use induction to see that h is onto. If $n = 1$, then $y \in B[p] = \text{Im}(g) \leq \text{Im}(h)$. Suppose $n > 1$ and that all elements of B having order p^{n-1} or less are in $\text{Im}(h)$. Since $p^{n-1}y \in B[p]$ we have $p^{n-1}y = g(x) = h(x)$ for some $x \in A[p]$. Choose $z \in A$ such that $p^{n-1}z = x$. Then

$$p^{n-1}[y - h(z)] = p^{n-1}y - h(p^{n-1}z) = p^{n-1}y - h(x) = 0,$$

so by induction there is some $u \in A$ with $h(u) = y - h(z)$. Thus $h(u + z) = y \in \text{Im}(h)$, and the proof is complete.

An abelian group A with torsion subgroup T is said to *split* if T is a direct summand of A, in which case $A = T \oplus B$ for some torsion-free subgroup B of A. Note, for example, that every finitely generated abelian group splits, by Theorem IV.3.3.

Exercise 1.3. Set

$$A = \prod\{\mathbb{Z}_p\text{:all primes } p \in \mathbb{Z}\}.$$

Show that the torsion subgroup of A is $T = \bigoplus_p \mathbb{Z}_p$, that A is reduced, and that A/T is divisible. Conclude that A does not split.

Proposition 1.8. If A is divisible and has torsion subgroup T, then T is divisible, and consequently A splits.

Proof. If $0 \neq n \in \mathbb{Z}$ and $x \in T$, then $ny = x$ for some $y \in A$. But then $|y| \leq n|x|$ so $y \in T$ and T is divisible. Apply the corollary to Theorem 1.5.

Theorem 1.9. Suppose A is a divisible abelian group. Then there are index sets Γ and Δ_p for each prime p so that

$$A = (\bigoplus\{A_\gamma : \gamma \in \Gamma\}) \oplus (\bigoplus\{T_p : p \text{ prime}\}),$$

where each $A_\gamma \cong \mathbb{Q}$ and $T_p \cong \bigoplus\{\mathbb{Z}_{p^\infty} : \delta(p) \in \Delta_p\}$ for all p. Furthermore, A is determined up to isomorphism by the cardinalities $|\Gamma|$ and $|\Delta_p|$, all p.

Proof. By Proposition 1.8 A splits as $A = G \oplus T$, where T is torsion and divisible, G is torsion-free and divisible. By Proposition 1.4 $G = \bigoplus\{A_\gamma : \gamma \in \Gamma\}$, with each $A_\gamma \cong \mathbb{Q}$, and $|\Gamma|$ is uniquely determined as the dimension of G as a $\mathbb{Q}$-vector space. By Theorem 1.1 we have $T = \bigoplus_p T_p$, and T_p is divisible for each prime p. For each p choose a set Δ_p for which $|\Delta_p|$ is the (uniquely determined) $\mathbb{Z}_p$-dimension of $T_p[p]$, and set

$$H_p = \bigoplus\{\mathbb{Z}_{p^\infty} : \delta(p) \in \Delta_p\}.$$

Then clearly $H_p[p] \cong T_p[p]$ for all p, and so $H_p \cong T_p$ by Proposition 1.7. The theorem follows.

Exercise 1.4. Let $\mathbb{C}^*$ be the (multiplicative) group of nonzero complex numbers and let

$$C = \{z \in \mathbb{C}^* : |z| = 1\},$$

also a multiplicative group. Show that $C \cong \mathbb{C}^*$. (*Hint*: Apply Theorem 1.9 to both groups.)

Exercise 1.5. Show that every abelian group A is a subgroup of a divisible group. (*Hint*: We may assume that $A = F/K$, F free abelian, and that $F = \bigoplus_\alpha \mathbb{Z}$. Use the fact then that $\mathbb{Z} \leq \mathbb{Q}$.)

2. PÓLYA–REDFIELD ENUMERATION

Suppose G is a finite group acting on a finite set S. The *character* θ of the permutation representation is defined by

$$\theta(x) = |\{s \in S : xs = s\}|$$

for all $x \in G$. Thus θ is a function from G to the nonnegative integers; it counts the *fixed points* in S for each $x \in G$.

Proposition 2.1. The character θ of a permutation representation of G on S is a class function, i.e., θ is constant on conjugacy classes in G.

Proof. If $x,y \in G$ and $s \in S$, then $xs = s$ if and only if $x^y(y^{-1}s) = y^{-1}s$, so $s \mapsto y^{-1}s$ determines a 1–1 correspondence between the sets of fixed points for x and its conjugate x^y.

Theorem 2.2 (Burnside's Orbit Formula). If a finite group G acts on a finite set S with character θ, then the number of G-orbits in S is

$$|G|^{-1}\sum\{\theta(x) : x \in G\}.$$

Proof. Set $\mathscr{S} = \{(x,s) \in G \times S : xs = s\}$. For fixed $x \in G$ there are $\theta(x)$ such ordered pairs; for fixed $s \in S$ there are $|\mathrm{Stab}_G(s)|$ of them. Thus

$$|\mathscr{S}| = \sum\{\theta(x) : x \in G\} = \sum\{|\mathrm{Stab}_G(s)| : s \in S\}.$$

Let $S_1, S_2, \ldots, S_k$ be the distinct G-orbits in S and choose $s_i \in S_i$ for each i. Then

$$\sum\{|\mathrm{Stab}_G(s)| : s \in S\} = \sum_{i=1}^{k} |S_i|\,|\mathrm{Stab}_G(s_i)|$$

$$= \sum_{i=1}^{k} [G:\mathrm{Stab}(s_i)]|\mathrm{Stab}(s_i)| = k|G|,$$

so $k = |G|^{-1}\sum\{\theta(x) : x \in G\}$.

Exercise 2.1. If G is transitive on S and $|S| > 1$ show that there is some $x \in G$ having no fixed points, i.e., $\theta(x) = 0$.

As an example of how Burnside's Orbit Formula can be used, imagine coloring a regular tetrahedron with two colors, say red (r) and blue (b), each face being a solid color. There are $2^4 = 16$ actually different colorings, since there are 2 choices for each of the 4 faces, but they are not all distinguishable. For example, the 4 different colorings with 1 red face and 3 blue faces are indistinguishable if we allow the tetrahedron to move about. Clearly two colorings are indistinguishable precisely when one can be rotated to the other. Note that we are viewing the rotation group G of the tetrahedron as acting not on the set of faces but on the set S of 16 distinct colorings of the faces. Thus the number of distinguishable colorings is the number of G-orbits in S.

Recall (Exercise I.12.27) that G is isomorphic with the alternating group A_4. It is easy to verify that $(12)(34) \in A_4$ corresponds to a rotation of 180° about an axis joining the midpoints of opposite edges, and that the 3 elements of order 2 are all conjugate to (12)(34). There are 2 conjugacy classes of elements of order 3, each having 4 elements. One contains (123), which corresponds to a rotation through 120° about an axis through a vertex and the center of the opposite face; the other class contains its inverse, (132). In the action on the set of 16 colorings it is easy to see that (12)(34) fixes 4 colorings, since faces adjacent at the axis edges must be the same for a fixed coloring. Similarly (123) and (132) each fix 4 colorings, and of course the identity 1 fixes all 16 colorings. Let us tabulate the information obtained thus far.

σ	1	(12)(34)	(123)	(132)
$\lvert \mathrm{cl}(\sigma) \rvert$	1	3	4	4
$\theta(\sigma)$	16	4	4	4

Because of Proposition 2.1 the information in the table is sufficient for computation of the number k of orbits by means of Burnside's Orbit Formula. In fact,

$$k = \tfrac{1}{12}(16 + 3 \cdot 4 + 4 \cdot 4 + 4 \cdot 4) = 5,$$

and there are 5 distinguishable colorings. They are, of course, rrrr, rrrb, rrbb, rbbb, and bbbb.

Exercise 2.2. Suppose three colors, red (r), blue (b), and yellow (y), are used to color a tetrahedron. Show that there are 15 distinguishable colorings, and list them.

The main result of this section, the Redfield–Pólya Theorem, is essentially a refinement of Burnside's Orbit Formula, which makes it possible to count orbits in a variety of situations. One further bit of technical apparatus, the *cycle index*, will be needed.

If the finite group G acts on a set S with n elements, each $x \in G$ corresponds to a permutation σ of S, which can be written uniquely as a product of disjoint cycles. If σ has cycle type $(j_1, j_2, \ldots, j_n)$, we will say that $x \in G$ has cycle type $(j_1, j_2, \ldots, j_n)$. Recall from Chapter I that conjugate elements have the same cycle type. Recall also that if $(j_1, j_2, \ldots, j_n)$ is a cycle type, then $\sum_{i=1}^{n} i \cdot j_i = n$.

If G acts on S, $|S| = n$, and $x \in G$ has cycle type $(j_1, j_2, \ldots, j_n)$ we define the *monomial* of x to be

$$\mathrm{mon}(x) = t_1^{j_1} t_2^{j_2} \cdots t_n^{j_n},$$

where $t_1, t_2, \ldots, t_n$ are n distinct (commuting) indeterminates. Then define the *cycle index* of the action of G on S to be the polynomial (say over the rational field $\mathbb{Q}$) in $t_1, t_2, \ldots, t_n$ given by

$$\mathscr{Z} = \mathscr{Z}_{G,S}(t_1, \ldots, t_n) = |G|^{-1} \sum\{\text{mon}(x) : x \in G\}.$$

Note that if G has conjugacy classes $K_1, K_2, \ldots, K_m$, with $x_i \in K_i$ for all i, then

$$\mathscr{Z} = |G|^{-1} \sum_{i=1}^{m} |K_i| \,\text{mon}(x_i).$$

Let us compute several examples of cycle indices. They will prove useful later. It will be convenient to tabulate information regarding conjugacy classes and cycle types.

EXAMPLES

1. Let $G = S_3$ and $S = \{1, 2, 3\}$.

σ	1	(12)	(123)
$\lvert\text{cl}(\sigma)\rvert$	1	3	2
Type	(3, 0, 0)	(1, 1, 0)	(0, 0, 1)

Thus $\mathscr{Z} = \frac{1}{6}(t_1^3 + 3t_1t_2 + 2t_3)$.

2. Let $G = A_4$, $S = \{1, 2, 3, 4\}$.

σ	1	(12)(34)	(123)	(132)
$\lvert\text{cl}(\sigma)\rvert$	1	3	4	4
Type	(4, 0, 0, 0)	(0, 2, 0, 0)	(1, 0, 1, 0)	(1, 0, 1, 0)

Thus $\mathscr{Z} = \frac{1}{12}(t_1^4 + 3t_2^2 + 4t_1t_3 + 4t_1t_3)$. Note that this is the same as the cycle index for the action of the rotation group of the tetrahedron acting on the set of 4 faces. It would be instructive at this point for the reader to compute the cycle index for the action of that group on the set of 6 edges of the tetrahedron.

The next few examples involve geometric symmetry groups. The reader is urged to construct cardboard models of the appropriate geometric objects and verify the tabulated information.

3. Let G be the rotation group of a cube (see Exercise I.1.3). We shall compute its cycle indices for 3 different permutation representations: (i) on the set F of 6 faces, (ii) on the set V of 8 vertices, and (iii) on the set E of 12 edges. Recall (Exercise I.12.28) that $G \cong S_4$, so we may list elements of S_4 as representatives of the conjugacy classes. Observe that (1234) corresponds to a 90° rotation and (13)(24) to its square, whereas (12) corresponds to the other type of 180° rotation, whose axis joins midpoints of opposing edges.

σ	1	(12)	(123)	(1234)	(13)(24)
$\lvert\text{cl}(x)\rvert$	1	6	8	6	3
F-type	(6, 0, ...)	(0, 3, 0, ...)	(0, 0, 2, 0, ...)	(2, 0, 0, 1, ...)	(2, 2, 0, ...)
V-type	(8, 0, ...)	(0, 4, 0, ...)	(2, 0, 2, 0, ...)	(0, 0, 0, 2, ...)	(0, 4, 0, ...)
E-type	(12, 0, ...)	(2, 5, 0, ...)	(0, 0, 4, 0, ...)	(0, 0, 0, 3, ...)	(0, 6, 0, ...)

Thus we have

(i) $\mathscr{Z}_{G,F} = \frac{1}{24}(t_1^6 + 6t_2^3 + 8t_3^2 + 6t_1^2t_4 + 3t_1^2t_2^2)$,
(ii) $\mathscr{Z}_{G,V} = \frac{1}{24}(t_1^8 + 9t_2^4 + 8t_1^2t_3^2 + 6t_4^2)$, and
(iii) $\mathscr{Z}_{G,E} = \frac{1}{24}(t_1^{12} + 6t_1^2t_2^5 + 8t_3^4 + 6t_4^3 + 3t_2^6)$.

4. Let $G = C_n = \langle a \mid a^n = 1\rangle$, a cyclic group of order n, acting as a plane rotation group on the set S of n vertices of a regular n-gon. If we label the vertices consecutively as $1, 2, 3, \ldots, n$ we may assume that $a = (123\cdots n) \in S_n$. We know from Section 1, Chapter 1 that C_n has a unique cyclic subgroup of order d for each divisor d of n, and that a cyclic group of order d has $\phi(d)$ generators, i.e., C_n has $\phi(d)$ elements of order d. Furthermore, each $a^k \in C_n$ has order $n/(n,k)$, and in fact a^k is a product of (n,k) disjoint $n/(n,k)$-cycles. Thus if $1 \le d \le n$ and $d \mid n$ there are $\phi(d)$ elements of order d, each having cycle type $(0, \ldots, n/d, \ldots, 0)$. Consequently

$$\mathscr{Z} = (1/n)\sum\{\phi(d)t_d^{n/d} : 1 \le d \le n, d \mid n\}.$$

Thus, for example, if $n = 12$ then

$$\mathscr{Z} = \tfrac{1}{12}(t_1^{12} + t_2^6 + 2t_3^4 + 2t_4^3 + 2t_6^2 + 4t_{12}).$$

5. Let $G = D_n$, the dihedral group of order $2n$, acting on the set of vertices of a regular n-gon. In the presentation

$$G = \langle a,b \mid a^n = b^2 = 1, b^{-1}ab = a^{-1}\rangle$$

we may assume that a is a generating rotation as in Example 4 above, so the cycle structure of each power a^k has already been determined. Write $A = \langle a\rangle$. Take b to be a reflection whose mirror passes through one of the vertices. Since $a^{-k}ba^k = ba^{2k}$ we have $\mathrm{cl}(b) = \{ba^{2k} : 0 \le k < n\}$. If n is odd, then $\langle a\rangle = \langle a^2\rangle$, so in that case $\mathrm{cl}(b)$ is the entire coset bA (all reflections are conjugate). Each reflection then fixes the vertex on the mirror and interchanges the remaining vertices in pairs, so the cycle type is $(1, (n-1)/2, 0, \ldots)$.

If n is even there are two classes of reflections, viz., $\mathrm{cl}(b)$ and $\mathrm{cl}(ba)$. The reflections in $\mathrm{cl}(b)$ have mirrors through 2 opposite vertices and cycle type $(2, (n-2)/2, 0, \ldots)$. Those in $\mathrm{cl}(ba)$ have mirrors through midpoints of opposite edges and cycle type $(0, n/2, 0, \ldots)$.

Thus if n is odd, then

$$\mathscr{Z} = (1/2n)\left[\sum\{\phi(d)t_d^{n/d} : 1 \le d \le n, d \mid n\} + nt_1t_2^{(n-1)/2}\right];$$

if n is even, then

$$\mathscr{Z} = (1/2n)\left[\sum\{\phi(d)t_d^{n/d} : 1 \le d \le n, d \mid n\} + (n/2)t_1^2t_2^{(n-1)/2} + (n/2)t_2^{n/2}\right].$$

In our discussion above of colorings of a tetrahedron we passed from a permutation action on the set of 4 faces to an action on the set of $2^4 = 16$

distinct colorings. Suppose the faces to be numbered 1, 2, 3, 4 and write $D = \{1, 2, 3, 4\}$ for the set of faces. Let $R = \{\mathrm{r}, \mathrm{b}\}$, the set of colorings. Then the set of colorings can be identified with the set of all functions $f: D \to R$, since a coloring is simply an assignment of a color for each of the faces.

In general suppose that D (the *domain*) and R (the *range*) are two finite sets, and write R^D for the set of all functions from D to R. Note that $|R^D| = |R|^{|D|}$. If G is a finite group that acts on D, *then* G also acts on R^D as follows: if $\sigma \in G$, $f \in R^D$, and $d \in D$, then $(\sigma f)(d) = f(\sigma^{-1} d)$, giving $\sigma f \in R^D$.

Exercise 2.3. Verify that the action described above of G on R^D is a permutation representation.

The G-orbits in the set R^D of functions are commonly called *patterns*.

Suppose S is a commutative ring with 1 containing the rational field $\mathbb{Q}$ as a subring, with $1_S = 1_{\mathbb{Q}}$ (S will often be a polynomial ring, for example). A function $\mathbf{w}: R \to S$ is called a *weight assignment* on R. For each $r \in R$ the ring element $\mathbf{w}(r)$ is called the *weight* of r. A weight assignment $\mathbf{w}$ on R induces a weight assignment W on R^D by means of

$$W(f) = \prod\{\mathbf{w}(f(d)) : d \in D\} \qquad \text{for each} \quad f: D \to R.$$

Suppose $\mathbf{w}$ is a weight assignment on R, and $f_1, f_2 \in R^D$ are in the same pattern, say with $f_1 = \sigma f_2$, $\sigma \in G$. Then

$$\begin{aligned} W(f_1) &= \prod\{\mathbf{w}(f_1(d)) : d \in D\} = \prod\{\mathbf{w}(\sigma f_2(d)) : d \in D\} \\ &= \prod\{\mathbf{w}(f_2(\sigma^{-1} d)) : d \in D\} = W(f_2), \end{aligned}$$

since σ^{-1} acts as a permutation of D. Thus the induced weight assignment W is constant over patterns in R^D. If F is a pattern in R^D define $W(F) = W(f)$ for any $f \in F$.

If U is any subset of R define the *inventory* of U to be $\mathrm{Inv}(U) = \sum\{\mathbf{w}(\mathrm{r}): \mathrm{r} \in U\}$. Similarly if T is a subset of R^D define its *inventory* to be $\mathrm{Inv}(T) = \sum\{W(f) : f \in T\}$. We agree in general that $\mathrm{Inv}(\varnothing) = 0$.

In our earlier example (coloring the tetrahedron) we had $D = \{1, 2, 3, 4\}$ and $R = \{\mathrm{r}, \mathrm{b}\}$. Take S to be the ring $\mathbb{Q}[t_1, t_2]$ of polynomials in two indeterminates, and define a weight assignment $\mathbf{w}: R \to S$ by means of $\mathbf{w}(\mathrm{r}) = t_1$, $\mathbf{w}(\mathrm{b}) = t_2$. As we saw, there are five patterns $F_1, \ldots, F_5$, represented by $f_1 = \mathrm{rrrr} \in F_1$, $f_2 = \mathrm{rrrb} \in F_2$, $f_3 = \mathrm{rrbb} \in F_3$, $f_4 = \mathrm{rbbb} \in F_4$, and $f_5 = \mathrm{bbbb} \in F_5$. It is easy to check that $|F_1| = |F_5| = 1$, $|F_2| = |F_4| = 4$, and $|F_3| = 6$, and it is clear that $W(F_1) = W(f_1) = t_1^4$, $W(F_2) = t_1^3 t_2$, $W(F_3) = t_1^2 t_2^2$, $W(F_4) = t_1 t_2^3$, and $W(F_5) = t_2^4$. Consequently

$$\begin{aligned} \mathrm{Inv}(R^D) &= t_1^4 + 4t_1^3 t_2 + 6t_1^2 t_2^2 + 4t_1 t_2^3 + t_2^4 \\ &= (t_1 + t_2)^4 = [\mathrm{Inv}(R)]^{|D|}. \end{aligned}$$

The example is a special case of the next result.

Proposition 2.3. Suppose R and D are finite sets and $\mathbf{w}: R \to S$ is a weight assignment on R. Suppose that D is partitioned into a union of disjoint sets $D_1, D_2, \ldots, D_m$ and let T be the set of functions $f \in R^D$ that are constant on each of the subsets D_i. Then

$$\operatorname{Inv}(T) = \prod_{i=1}^{m} \sum\{\mathbf{w}(r)^{|D_i|} : r \in R\}.$$

Proof. Define $\psi: D \to \{1,2,\ldots,m\}$ by agreeing that $\psi(d) = j$ if and only if $d \in D_j$. For each $f \in T$ define $\phi = \phi_f: \{1,2,\ldots,m\} \to R$ by means of $\phi(j) = f(d)$ if and only if $d \in D_j$. Then each $f \in T$ clearly factors as $f = \phi_f \psi$. If $R = \{r_1, r_2, \ldots, r_k\}$, then the right hand side of the equation to be proved is the product

$$\prod_{i=1}^{m} \{\mathbf{w}(r_1)^{|D_i|} + \cdots + \mathbf{w}(r_k)^{|D_i|}\}.$$

To multiply we choose one term from each factor, in all possible ways, multiply the terms together, and then add the resulting products. To choose a term from each factor is to choose a function $\phi: \{1,2,\ldots,m\} \to R$, then multiplying yields $\prod_{i=1}^{m}\{\mathbf{w}(\phi(i))^{|D_i|}\}$, after which we add the results for all such functions ϕ. But

$$\mathbf{w}(\phi(i))^{|D_i|} = \prod\{\mathbf{w}(\phi(\psi(d))) : d \in D_i\} = \prod\{\mathbf{w}(f(d)) : d \in D_i\},$$

where $f = \phi\psi \in T$. The choice of all possible ϕ results in all $f \in T$, so the result of adding is

$$\sum_{f \in T} \prod_{i=1}^{m} \prod_{d \in D_i} \mathbf{w}(f(d)) = \sum_{f \in T} W(f) = \operatorname{Inv}(T).$$

Corollary. $\operatorname{Inv}(R^D) = [\operatorname{Inv}(R)]^{|D|}$.

Proof. If $D = \{d_1, \ldots, d_n\}$ take $D_i = \{d_i\}$, $1 \le i \le n$.

Suppose now that D and R are finite sets, $\mathbf{w}: R \to S$ is a weight assignment, and G is a finite group acting on D, and hence on R^D. Recall that if $f \in R^D$, then the weight of f is $W(f) = \prod\{\mathbf{w}(f(d)) : d \in D\}$, and if F is a pattern (i.e., a G-orbit in R^D), then the weight of F is $W(F) = W(f)$ for any $f \in F$. Define the *pattern inventory* of G to be

$$\mathrm{PI} = \sum\{W(F) : \text{all patterns } F \subseteq R^D\}.$$

Suppose, for example, that we define $\mathbf{w}(r) = 1 \in \mathbb{Q}$ for all $r \in R$. Then $W(f) = W(F) = 1$ for all $f \in R^D$ and all patterns F, and consequently the pattern inventory PI is just the number of patterns. Thus PI, if it can be computed, contains at least as much information as is afforded by the Burnside Orbit Formula applied to the action of G on R^D.

Theorem 2.4 (Redfield–Pólya). Suppose G is a finite group acting on a finite set D, $|D| = k$, with cycle index $\mathscr{Z}$, and R is a finite set with a weight assignment $\mathbf{w}\colon R \to S$. Then the pattern inventory for the action of G on R^D is

$$\mathrm{PI} = \mathscr{Z}\left(\sum_{r\in R} \mathbf{w}(r), \sum_{r\in R} \mathbf{w}(r)^2, \ldots, \sum_{r\in R} \mathbf{w}(r)^k\right).$$

Proof. Let $W_1, W_2, \ldots$ be the distinct weights of patterns in R^D, and set $\mathscr{F}_i = \{f \in R^D : W(f) = W_i\}$, $i = 1, 2, \ldots$. Clearly $\mathscr{F}_i$ is a union of patterns; say there are m_i different patterns in $\mathscr{F}_i$. Thus $\mathrm{PI} = \sum_i m_i W_i$. Note that G acts on each $\mathscr{F}_i$, and that the G-orbits in $\mathscr{F}_i$ are just the m_i patterns whose union is $\mathscr{F}_i$. If θ_i is the character of the action of G on $\mathscr{F}_i$ then, by the Burnside Orbit Formula, $m_i = |G|^{-1}\sum\{\theta_i(\sigma):\sigma \in G\}$, and hence

$$\mathrm{PI} = \sum_i m_i W_i = |G|^{-1}\sum\{\sum_i \theta_i(\sigma) W_i : \sigma \in G\}.$$

Fix $\sigma \in G$. Then $\sum_i \theta_i(\sigma) W_i$ is the inventory of $\{f \in R^D : \sigma f = f\}$, since $\theta_i(\sigma)$ is just the number of $f \in \mathscr{F}_i$ fixed by σ, and so

$$\mathrm{PI} = |G|^{-1}\sum_{\sigma\in G}(\sum\{W(f): f \in R^D \text{ and } \sigma f = f\}).$$

If $\sigma f = f$ and $d \in D$, then

$$f(d) = \sigma^{-1}f(d) = f(\sigma d) = \sigma^{-1}f(\sigma d) = f(\sigma^2 d) = \cdots,$$

and f is constant on the σ-cycles in D. Conversely, if $f \in R^D$ is constant on the σ-cycles in D, then $\sigma f = f$. (Why?) Thus if the σ-cycles partition D into disjoint sets $D_1, D_2, \ldots$, then $\{f : \sigma f = f\}$ is just the set of all $f \in R^D$ that are constant on each of the subsets D_i, i.e., the set $T \subseteq R^D$ in Proposition 2.3. By that proposition

$$\sum\{W(f):\sigma f = f\} = \prod_i(\sum\{\mathbf{w}(r)^{|D_i|}: r \in R\})$$

for each fixed $\sigma \in G$. If the cycle type of σ is $(j_1, j_2, \ldots, j_k)$, then among the sets D_i there are j_1 for which $|D_i| = 1$, j_2 for which $|D_i| = 2$, etc. Thus

$$\prod_i(\sum\{\mathbf{w}(r)^{|D_i|}: r \in R\}) = \left[\sum_{r\in R}\mathbf{w}(r)\right]^{j_1}\left[\sum_{r\in R}\mathbf{w}(r)^2\right]^{j_2}\cdots\left[\sum_{r\in R}\mathbf{w}(r)^k\right]^{j_k}$$

for each σ, and consequently

$$\mathrm{PI} = |G|^{-1}\sum_{\sigma\in G}\left[\sum_{r\in R}\mathbf{w}(r)\right]^{j_1(\sigma)}\left[\sum_{r\in R}\mathbf{w}(r)^2\right]^{j_2(\sigma)}\cdots\left[\sum_{r\in R}\mathbf{w}(r)^k\right]^{j_k(\sigma)}$$

$$= \mathscr{Z}\left(\sum_{r\in R}\mathbf{w}(r), \sum_{r\in R}\mathbf{w}(r)^2, \ldots, \sum_{r\in R}\mathbf{w}(r)^k\right).$$

Corollary. The number of different patterns in R^D is $\mathscr{Z}(|R|,|R|,\ldots,|R|)$.

Proof. Set $\mathbf{w}(\mathrm{r}) = 1$ for all $r \in R$.

For an example let $D = F$ be the set of 6 faces of a cube and let $R = \{$red, blue, yellow$\}$. Let $S = \mathbb{Q}(\mathrm{r}, \mathrm{b}, \mathrm{y})$, where r, b, and y are distinct indeterminates, and assign weights by means of $\mathbf{w}(\text{red}) = \mathrm{r}$, $\mathbf{w}(\text{blue}) = \mathrm{b}$, $\mathbf{w}(\text{yellow}) = \mathrm{y}$. If G is the rotation group of the cube, then the cycle index $\mathscr{Z}_{G,F}$ is

$$\mathscr{Z} = \tfrac{1}{24}(t_1^6 + 6t_2^3 + 8t_3^2 + 6t_1^2t_4 + 3t_1^2t_2^2)$$

(see Example 3, above). By the Redfield–Pólya Theorem

$$\begin{aligned}
\mathrm{PI} &= \tfrac{1}{24}[(\mathrm{r}+\mathrm{b}+\mathrm{y})^6 + 6(\mathrm{r}^2+\mathrm{b}^2+\mathrm{y}^2)^3 + 8(\mathrm{r}^3+\mathrm{b}^3+\mathrm{y}^3)^2 \\
&\quad + 6(\mathrm{r}+\mathrm{b}+\mathrm{y})^2(\mathrm{r}^4+\mathrm{b}^4+\mathrm{y}^4) + 3(\mathrm{r}+\mathrm{b}+\mathrm{y})^2(\mathrm{r}^2+\mathrm{b}^2+\mathrm{y}^2)^2] \\
&= \mathrm{r}^6 + \mathrm{b}^6 + \mathrm{y}^6 + \mathrm{r}^5\mathrm{b} + \mathrm{r}^5\mathrm{y} + \mathrm{r}\mathrm{b}^5 + \mathrm{r}\mathrm{y}^5 + \mathrm{b}^5\mathrm{y} + \mathrm{b}\mathrm{y}^5 \\
&\quad + 2(\mathrm{r}^3\mathrm{b}^3 + \mathrm{r}^3\mathrm{y}^3 + \mathrm{b}^3\mathrm{y}^3 + \mathrm{r}^4\mathrm{b}^2 + \mathrm{r}^4\mathrm{y}^2 + \mathrm{b}^4\mathrm{y}^2 + \mathrm{b}^2\mathrm{y}^4 \\
&\quad + \mathrm{r}^2\mathrm{b}^4 + \mathrm{r}^2\mathrm{y}^4 + \mathrm{r}^4\mathrm{b}\mathrm{y} + \mathrm{r}\mathrm{b}^4\mathrm{y} + \mathrm{r}\mathrm{b}\mathrm{y}^4) \\
&\quad + 3(\mathrm{r}^3\mathrm{b}^2\mathrm{y} + \mathrm{r}^3\mathrm{b}\mathrm{y}^2 + \mathrm{r}^2\mathrm{b}^3\mathrm{y} + \mathrm{r}^2\mathrm{b}\mathrm{y}^3 + \mathrm{r}\mathrm{b}^3\mathrm{y}^2 + \mathrm{r}\mathrm{b}^2\mathrm{y}^3) + 6\mathrm{r}^2\mathrm{b}^2\mathrm{y}^2.
\end{aligned}$$

(Verify.) The total number of distinguishable colorings (i.e., patterns) is obtained by setting $\mathrm{r} = \mathrm{b} = \mathrm{y} = 1$; there are 57 of them. Note that the individual terms in the polynomial PI give the number of patterns of various types. For example, the term r^6 represents the unique pattern with all faces red; the term $3\mathrm{r}^3\mathrm{b}^2\mathrm{y}$ represents 3 distinct patterns each having 3 red faces, 2 blue faces, and 1 yellow face, etc.

For a second example let us determine the number of distinguishable necklaces that can be made by using 9 red or blue beads. We assume that the beads of the same color are indistinguishable, and that 2 necklaces are indistinguishable if one can be made to look like the other by rotating it or turning it over. In effect, then, we imagine each necklace laid out so the beads are at the vertices of a regular 9-gon, and that the beads are acted on by the dihedral group D_9. Take D to be the set of 9 beads, $R = \{$red, blue$\}$, $S = \mathbb{Q}(\mathrm{r}, \mathrm{b})$ as above, and $\mathbf{w}(\text{red}) = \mathrm{r}$, $\mathbf{w}(\text{blue}) = \mathrm{b}$. The cycle index for D_9 is

$$\mathscr{Z} = \tfrac{1}{18}(t_1^9 + 2t_3^3 + 6t_9 + 9t_1t_2^4),$$

so the pattern inventory is

$$\begin{aligned}
\mathrm{PI} &= \tfrac{1}{18}[(\mathrm{r}+\mathrm{b})^9 + 2(\mathrm{r}^3+\mathrm{b}^3)^3 + 6(\mathrm{r}^9+\mathrm{b}^9) + 9(\mathrm{r}+\mathrm{b})(\mathrm{r}^2+\mathrm{b}^2)^4] \\
&= \mathrm{r}^9 + \mathrm{r}^8\mathrm{b} + 4\mathrm{r}^7\mathrm{b}^2 + 7\mathrm{r}^6\mathrm{b}^3 + 10\mathrm{r}^5\mathrm{b}^4 \\
&\quad + 10\mathrm{r}^4\mathrm{b}^5 + 7\mathrm{r}^3\mathrm{b}^6 + 4\mathrm{r}^2\mathrm{b}^7 + \mathrm{r}\mathrm{b}^8 + \mathrm{b}^9.
\end{aligned}$$

Set $\mathrm{r} = \mathrm{b} = 1$ to see that the number of distinguishable necklaces is 46, and observe, for example, that there are 7 patterns each using 3 red beads and 6 blue beads.

Exercise 2.4. Let p be an odd prime. Discuss the possible necklaces with p beads (i) if red and blue beads are used, or (ii) if red, blue, and yellow beads are used. In particular determine the number of distinguishable necklaces.

Exercise 2.5. In how many distinguishable ways can the faces of a tetrahedron be colored if n different colors are available?

A different sort of weight assignment is often useful in combinatorics. Suppose D, R, and G are as usual, let x be an indeterminate, and set $S = \mathbb{Q}[x]$. Let $k: R \to \mathbb{Z}$ be a function, with $k(\mathrm{r}) \geq 0$ for all $\mathrm{r} \in R$, and call $k(\mathrm{r})$ the *content of R*. For $f \in R^D$ define the *content* of f to be $K(f) = \sum\{k(f(d)): d \in D\}$. Assign weights on R by means of $\mathbf{w}(\mathrm{r}) = x^{k(\mathrm{r})}$ for all $\mathrm{r} \in R$. Note that then, for $f \in R^D$,

$$W(f) = \prod\{\mathbf{w}(f(d)): d \in D\} = \prod\{x^{k(f(d))}: d \in D\} = x^{\Sigma_d k(f(d))} = x^{K(f)}.$$

For each nonnegative integer m let $a_m = |\{\mathrm{r} \in R: k(\mathrm{r}) = m\}|$, and define the *generating function* for R (and k) to be

$$a(x) = \sum\{a_m x^m: 0 \leq m \in \mathbb{Z}\},$$

a polynomial whose coefficients count the numbers of elements in R having various contents.

If f_1 and f_2 are in the same pattern F in R^D it is easily verified that $K(f_1) = K(f_2)$, so we may define $K(F) = K(f)$ for any $f \in F$. If $0 \leq m \in \mathbb{Z}$ set

$$A_m = |\{F: F \text{ is a pattern and } K(F) = m\}|,$$

and define the *pattern counting function* to be

$$A(x) = \sum_m A_m x^m \in \mathbb{Z}[x].$$

Theorem 2.5. Suppose R and D are finite sets and G is a finite group acting on D, with cycle index $\mathscr{Z}$, generating function $a(x)$, and pattern counting function $A(x)$. Then

$$A(x) = \mathscr{Z}(a(x), a(x^2), \ldots, a(x^k)).$$

Proof. Since $\mathrm{PI} = \sum\{W(F): \text{all patterns } F\}$ and $W(F) = x^{K(F)}$, we have

$$\mathrm{PI} = A_0 + A_1 x + A_2 x^2 + \cdots = A(x).$$

But by the Redfield–Pólya Theorem

$$\begin{aligned}\mathrm{PI} &= \mathscr{Z}\left(\sum_{\mathrm{r}} \mathbf{w}(\mathrm{r}), \sum_{\mathrm{r}} \mathbf{w}(\mathrm{r})^2, \ldots\right)\\ &= \mathscr{Z}(a_0 + a_1 x + a_2 x^2 + \cdots, a_0 + a_1 x^2 + a_2 x^4 + \cdots, \cdots)\\ &= \mathscr{Z}(a(x), a(x^2), \ldots, a(x^k)).\end{aligned}$$

If $|D| = k$, then for each integer m, $0 \le m \le k$, define $D^{(m)}$ to be the set of all m-sets in D, i.e., the set of all subsets of size m in D. Thus $|D^{(m)}| = \binom{k}{m}$. If a group G acts on D, then it acts naturally on each $D^{(m)}$ by means of

$$\sigma\{d_1, d_2, \ldots, d_m\} = \{\sigma d_1, \sigma d_2, \ldots, \sigma d_m\} \qquad \text{for each} \quad \sigma \in G.$$

Theorem 2.6. Suppose G acts on D, $|D| = k$, with cycle index $\mathscr{Z}$. If x is an indeterminate and $0 \le m \le k$, then the number of G-orbits in $D^{(m)}$ is the coefficient of x^m in

$$\mathscr{Z}(1 + x, 1 + x^2, \ldots, 1 + x^k).$$

Proof. Take $R = \{0, 1\}$, define a content $k\colon R \to \mathbb{Z}$ via $k(0) = 0$, $k(1) = 1$, and hence assign weights via $\mathbf{w}(0) = 1$, $\mathbf{w}(1) = x$. If $f \in R^D$, then

$$W(f) = \prod\{\mathbf{w}(f(d)) : d \in D\} = \prod_d x^{k(f(d))} = x^{\Sigma_d k(f(d))} = x^{\Sigma_d f(d)}.$$

Thus $W(f) = x^m$ if and only if precisely m elements of D are mapped by f to 1, i.e., if and only if $f^{-1}(1) \in D^{(m)}$. There is thus a 1–1 correspondence between G-orbits of m-sets in $D^{(m)}$ and patterns of content m in R^D. But the number of patterns of content m is the coefficient A_m in the pattern-counting function $A(x)$. Since the generating function is $a(x) = 1 + x$ the result follows from Theorem 2.5.

Corollary 1. The number of G-orbits in D is the coefficient of x in $\mathscr{Z}(1 + x, 1 + x^2, \ldots, 1 + x^k)$.

Proof. Identify D with $D^{(1)}$.

Corollary 2. The total number of orbits of G acting simultaneously on all $D^{(m)}$, $0 \le m \le k$, is

$$A_0 + A_1 + A_2 \cdots + A_k = \mathscr{Z}(1 + 1, 1 + 1^2, \ldots) = \mathscr{Z}(2, 2, \ldots, 2).$$

For an example, let G be the rotation group of a cube, acting on the set $D = V$ of vertices. The cycle index is $\mathscr{Z} = \frac{1}{24}[t_1^8 + 9t_2^4 + 8t_1^2t_3^2 + 6t_4^2]$. Thus

$$\begin{aligned} A(x) &= \mathscr{Z}(1 + x, 1 + x^2, \ldots, 1 + x^8) \\ &= \tfrac{1}{24}[(1 + x)^8 + 9(1 + x^2)^4 + 8(1 + x)^2(1 + x^3)^2 + 6(1 + x^4)^2] \\ &= 1 + x + 3x^2 + 3x^3 + 7x^4 + 3x^5 + 3x^6 + x^7 + x^8. \end{aligned}$$

(Verify.) Thus, for example, there are 3 orbits of 2-sets: one orbit, of size 12, consists of pairs of vertices that share an edge; another, of size 12, consists of pairs of vertices at opposite corners of faces; and the third orbit, of size 4, consists of pairs of vertices at opposite corners of the cube.

Exercise 2.6. Describe the 3 orbits of 3-sets and the 7 orbits of 4-sets in the example above.

Exercise 2.7. If the rotation group G of the cube acts on the set E of edges compute the numbers of orbits of m-sets, $0 \leq m \leq 12$, and describe the orbits of 2-sets explicitly.

Exercise 2.8. Suppose a hollow regular tetrahedron has thin walls, and both the inside and outside of each face is to be colored red or blue. How many distinguishable colorings are there, and how many of them have 3 or fewer of the faces red?

3. PSL(V)

Let V be a vector space of finite dimension n over a field F. Write $\mathrm{GL}(V)$ for the group of all invertible linear transformations of V. Any choice of basis in V provides an isomorphism between $\mathrm{GL}(V)$ and the group $\mathrm{GL}(n, F)$ of invertible $n \times n$ matrices with entries from F. Both are called the *general linear group.* The *special linear group* is the subgroup $\mathrm{SL}(V)$ of $\mathrm{GL}(V)$ [or $\mathrm{SL}(n, F)$ of $\mathrm{GL}(n, F)$] consisting of those linear transformations (or matrices) having determinant 1.

When F is a finite field with q elements it is customary to denote $\mathrm{GL}(n, F)$ and $\mathrm{SL}(n, F)$ by $\mathrm{GL}(n, q)$ and $\mathrm{SL}(n, q)$, respectively.

Exercise 3.1. Show that the center of $\mathrm{GL}(V)$ is the set of scalar transformations, i.e.,

$$Z(\mathrm{GL}(V)) = \{a1 : 0 \neq a \in F\},$$

and also that

$$Z(\mathrm{SL}(V)) = \{a1 : a^n = 1 \in F\}.$$

If $0 \neq v \in V$ write $[v]$ for $Fv = \{av : a \in F\}$, the line through the origin spanned by v, and call $[v]$ a *projective point.* The set of all distinct projective points $[v]$ is called the *$(n-1)$-dimensional projective space* based on V, denoted by $P_{n-1}(V)$. There is a natural permutation action of $\mathrm{GL}(V)$ on $P_{n-1}(V)$, viz., $T[v] = [Tv]$ for $T \in \mathrm{GL}(V)$ and $[v] \in P_{n-1}(V)$.

It is clear from Exercise 3.1 that $Z(\mathrm{GL}(V))$ acts trivially on $P_{n-1}(V)$, i.e., that it is in the kernel of the permutation action. Suppose conversely that $T \in \mathrm{GL}(V)$ acts trivially on every projective point $[v]$. Thus $Tv = av$ for each nonzero $v \in V$, where $a \in F$ may depend on v. If $v_1, v_2 \in V$ are linearly independent suppose $Tv_1 = a_1 v_1$, $Tv_2 = a_2 v_2$, and $T(v_1 + v_2) = a_{12}(v_1 + v_2)$. But then, since $T(v_1 + v_2) = Tv_1 + Tv_2$, we may conclude that $a_1 = a_{12} = a_2$, and hence conclude that $T = a_1 1 \in Z(\mathrm{GL}(V))$. It also follows that $Z(\mathrm{SL}(V))$ is the kernel of the permutation action of $\mathrm{SL}(V)$ on $P_{n-1}(V)$.

Define the *projective general linear group* $\mathrm{PGL}(V)$ to be the quotient group $\mathrm{GL}(V)/Z[\mathrm{GL}(V)]$, and the *projective special linear* group $\mathrm{PSL}(V)$ to be

$SL(V)/Z[SL(V)]$. It is a consequence of the remarks above that $PGL(V)$ and $PSL(V)$ act faithfully as permutation groups on the projective space $P_{n-1}(V)$.

There are of course isomorphic versions of the projective groups:

$$PGL(n, F) = GL(n, F)/Z[GL(n, F)]$$

and

$$PSL(n, F) = SL(n, F)/Z[SL(n, F)],$$

which are often denoted $PGL(n, q)$ and $PSL(n, q)$ when F has q elements.

When F is finite the groups are finite. Let us compute their orders.

Proposition 3.1. Suppose $|F| = q$. Then

(i) $|GL(n, q)| = q^{n(n-1)/2} \prod\{q^k - 1 : 1 \leq k \leq n\}$,
(ii) $|SL(n, q)| = |PGL(n, q)| = |GL(n, q)|/(q - 1)$,
and
(iii) $|PSL(n, q)| = |SL(n, q)|/(n, q - 1)$.

Proof. (i) The first row of a matrix in $GL(n, q)$ can be any one of $q^n - 1$ nonzero row vectors. There are $q^n - q$ choices for the second row since it must be independent of the first. Similarly there are $q^n - q^2$ choices for the third row in order that the first three are linearly independent, etc. Thus

$$\begin{aligned} |GL(n, q)| &= \prod\{q^n - q^k : 0 \leq k \leq n - 1\} \\ &= q^{1+2+\cdots+(n-1)} \prod\{q^k - 1 : 1 \leq k \leq n\} \\ &= q^{n(n-1)/2} \prod\{q^k - 1 : 1 \leq k \leq n\}. \end{aligned}$$

(ii) Since $|Z[GL(n, q)]| = |F^*| = q - 1$ it is clear that $|PGL(n, q)| = |GL(n, q)|/(q - 1)$. The determinant map is a homomorphism from $GL(n, q)$ onto F^* with kernel $SL(n, q)$, so

$$|GL(n, q)|/|SL(n, q)| = |F^*| = q - 1.$$

(iii) Since F^* is cyclic of order $q - 1$ we see by Exercise 3.1 that $a \cdot 1 \in Z(SL(n, q))$ if and only if $a^n = a^{q-1} = 1$, i.e., $a^{(n, q-1)} = 1$, hence if and only if a is in the unique subgroup of order $(n, q - 1)$ in F^*.

Although SL and PGL have the same order they are usually not isomorphic.

Exercise 3.2. If F is any field denote by $\mathcal{GL}_2(F)$ the group of *linear fractional transformations* of F, i.e., all rational functions $f(x) \in F(x)$ of the form $f(x) = (ax + b)/(cx + d)$, where $a, b, c, d \in F$ and $ad - bc \neq 0$, with composition of functions as the group operation. Show that $\mathcal{GL}_2(F) \cong PGL(2, F)$. Let $\mathcal{L}_2(F)$ be the subgroup of $\mathcal{GL}_2(F)$ consisting of those $f(x)$ for which $ad - bc$ is a square in F. Show that $\mathcal{L}_2(F) \cong PSL(2, F)$.

If $\dim V = n$, then a *hyperplane* in V is any subspace W of dimension $n - 1$. A linear transformation $T \neq 1$ of V is called a *transvection* if there is a hyperplane W such that $T|W = 1_W$, i.e., T fixes each vector in W, and $Tv - v \in W$ for all $v \in V$.

If T is a transvection of V with fixed hyperplane W choose a basis for V by first choosing $v_1 \in V \backslash W$ and then choosing a basis $\{v_2, \ldots, v_n\}$ for W. The resulting matrix representation for T makes it clear that $\det T = 1$, so $T \in SL(V)$.

Exercise 3.3. (1) Show that the inverse of a transvection is a transvection.

(2) Suppose V is a subspace of V_1, $v \in V_1 \backslash V$, and T is a transvection on V with hyperplane W. Show that T extends to a transvection T_1 on V_1 whose fixed hyperplane W_1 contains W and v.

Proposition 3.2. If v and w are linearly independent in V, then there is a transvection T of V for which $Tv = w$.

Proof. Choose a hyperplane W with $v - w \in W$, $v \notin W$ (hence $w \notin W$), and define T by means of $T|W = 1_W$, $Tv = w$. If $x \in V$ write $x = av + u$, with $a \in F$, $u \in W$. Then

$$Tx - x = aw + u - (av + u) = a(w - v) \in W,$$

so T is a transvection.

Proposition 3.3. Suppose W_1 and W_2 are distinct hyperplanes in V and $v \in V \backslash (W_1 \cup W_2)$. Then there is a transvection T of V with $T(W_1) = W_2$ and $Tv = v$.

Proof. Note that $W_1 + W_2 = V$, so $\dim(W_1 \cap W_2) = n - 2$. Thus $W = (W_1 \cap W_2) + Fv$ is another hyperplane. Since $W_1 + W_2 = V$ there are $x \in W_1$ and $y \in W_2$ with $x + y = v$. Note that $x \notin W_2$ (or else $v \in W_2$), so $W_1 = (W_1 \cap W_2) + Fx$, and similarly $W_2 = (W_1 \cap W_2) + Fy$. Thus $V = (W_1 \cap W_2) + Fx + Fy$. It follows that $x \notin W$, for otherwise also $y = v - x \in W$ and hence $V \subseteq W$, a contradiction. Define T by means of $T|W = 1_W$ and $Tx = -y$. As in the proof of Proposition 3.2 we see that T is a transvection. Also

$$T(W_1) = T(W_1 \cap W_2 + Fx) = W_1 \cap W_2 + F(-y) = W_2,$$

and $Tv = v$ since $v \in W$.

Theorem 3.4. The set of transvections of V generates $SL(V)$.

Proof. We may assume $\dim V = n \geq 2$. Fix $R \in SL(V)$; then choose a hyperplane $W \subseteq V$ and $v \in V \backslash W$. If v and Rv are linearly independent then by

Proposition 3.2 there is a transvection T_1 with $T_1 Rv = v$. If v and Rv are linearly dependent first choose a transvection T_0 such that $T_0 Rv$ and v are linearly independent, then choose a transvection T'_1 with $T'_1 T_0 Rv = v$, and set $T_1 = T'_1 T_0$. In either case, then, $T_1 Rv = v$, with T_1 a product of transvections. Note that $v \notin T_1 RW$. If $T_1 RW = W$ set $T_2 = 1_V$. If $T_1 RW \neq W$ apply Proposition 3.3 to get a transvection T_2 with $T_2 T_1 RW = W$ and $T_2 v = v$. Set $R_1 = T_2 T_1 R$. Since $R_1 \in \mathrm{SL}(V)$ and $R_1 v = v$ it follows that $R_1 \mid W \in \mathrm{SL}(W)$. Now use induction on n. If $n = 2$, then $R_1 \mid W = 1_W$, so $R_1 = 1_V$ and $R = T_1^{-1} T_2^{-1}$. If $n > 2$, then by induction $R_1 \mid W$ is a product of transvections on W, each of which extends to a transvection fixing v on V by Exercise 3.3.2. Thus R_1 is the product of the extended transvections and so $R = T_1^{-1} T_2^{-1} R_1$, a product of transvections.

Theorem 3.5. If T_1 and T_2 are transvections on V, then T_1 and T_2 are conjugate in $\mathrm{GL}(V)$. If $n \geq 3$ they are conjugate in $\mathrm{SL}(V)$.

Proof. For $i = 1, 2$ let W_i be the fixed hyperplane of T_i, choose $x_i \in V \backslash W_i$, and say $w_i = T_i x_i - x_i \in W_i$. Choose bases $\{w_1, u_3, \ldots, u_n\}$ for W_1 and $\{w_2, v_3, \ldots, v_n\}$ for W_2. For each $a \neq 0$ in F define $S_a \in \mathrm{GL}(V)$ by means of $S_a x_1 = x_2$, $S_a w_1 = w_2$, $S_a u_i = v_i (3 \leq i \leq n-1)$, and $S_a u_n = a v_n$ (if $n \geq 3$). Then

$$\begin{aligned} S_a T_1 S_a^{-1} x_2 &= S_a T_1 x_1 = S_a(x_1 + w_1) = x_2 + w_2 = T_2 x_2, \\ S_a T_1 S_a^{-1} w_2 &= S_a T_1 w_1 = S_a w_1 = w_2 = T_2 w_2, \\ S_a T_1 S_a^{-1} v_i &= S_a T_1 u_i = S_a u_i = v_i = T_2 v_i, \qquad 3 \leq i \leq n-1, \end{aligned}$$

and

$$S_a T_1 S_a^{-1} v_n = a^{-1} S_a T_1 u_n = a^{-1} S_a u_n = v_n = T_2 v_n \qquad (\text{if } n \geq 3),$$

so $S_a T_1 S_a^{-1} = T_2$. If $n \geq 3$ we may set $b = (\det S_1)^{-1}$ and then $S_b \in \mathrm{SL}(V)$.

Theorem 3.6. Suppose $\dim V = 2$ and let $\{v_1, v_2\}$ be any basis for V. Each conjugacy class of transvections in $\mathrm{SL}(V)$ contains one whose matrix relative to $\{v_1, v_2\}$ is of the form $\left[\begin{smallmatrix} 1 & 0 \\ a & 1 \end{smallmatrix}\right]$, $a \in F^*$.

Proof. If T is a transvection with hyperplane W choose $v \in V \backslash W$ and set $w = Tv - v \in W$. Relative to the basis $\{v, w\}$ the matrix representing T is then $\left[\begin{smallmatrix} 1 & 0 \\ 1 & 1 \end{smallmatrix}\right]$. If M is the matrix representing T relative to $\{v_1, v_2\}$ there is a matrix $B \in \mathrm{GL}(2, F)$ with $BMB^{-1} = \left[\begin{smallmatrix} 1 & 0 \\ 1 & 1 \end{smallmatrix}\right]$. If $\det B = a^{-1}$ set $A = \left[\begin{smallmatrix} a & 0 \\ 0 & 1 \end{smallmatrix}\right]$. Then

$$BA \in \mathrm{SL}(2, F) \qquad \text{and} \qquad (BA)^{-1} M (BA) = A^{-1} \begin{bmatrix} 1 & 0 \\ 1 & 1 \end{bmatrix} A = \begin{bmatrix} 1 & 0 \\ a & 1 \end{bmatrix}.$$

Theorem 3.7. If $n \geq 3$ and $G = \mathrm{SL}(V)$, then $G' = G$.

Proof. By Theorems 3.4 and 3.5 it is sufficient to exhibit a transvection in G'. Choose a basis $\{v_1, v_2, \ldots, v_n\}$ for V and define T_1 and T_2 via

$$T_1: v_1 \mapsto v_1 - v_2,\ v_i \mapsto v_i \quad \text{if} \quad 2 \leq i \leq n,$$
$$T_2: v_1 \mapsto v_1,\ v_2 \mapsto v_2 - v_3,\ v_i \mapsto v_i, \quad 3 \leq i \leq n.$$

Clearly T_1 and T_2 are themselves transvections, and it is easily checked that

$$T_1 T_2 T_1^{-1} T_2^{-1}: v_1 \mapsto v_1 - v_3, v_i \to v_i, \qquad 2 \leq i \leq n,$$

so $T_1 T_2 T_1^{-1} T_2^{-1}$ is a transvection in G'.

Corollary. If $n \geq 3$, then $\mathrm{PSL}(V)$ is equal to its derived group.

Theorem 3.8. If $n = 2$, $|F| > 3$, and $G = \mathrm{SL}(V)$, then $G' = G$.

Proof. Choose a basis $\{v_1, v_2\}$ for V and choose $a \in F$, $a \notin \{0, \pm 1\}$. Define $S \in \mathrm{SL}(V)$ by means of S: $v_1 \mapsto a^{-1} v_1$, $v_2 \mapsto a v_2$, and for each $b \in F^*$ define $T_b \in \mathrm{SL}(V)$ by means of T_b: $v_1 \mapsto v_1 + b v_2$, $v_2 \mapsto v_2$. Then $S T_b S^{-1} T_b^{-1}$ is represented by the matrix

$$\begin{bmatrix} a^{-1} & 0 \\ 0 & a \end{bmatrix} \begin{bmatrix} 1 & 0 \\ b & 1 \end{bmatrix} \begin{bmatrix} a & 0 \\ 0 & a^{-1} \end{bmatrix} \begin{bmatrix} 1 & 0 \\ -b & 1 \end{bmatrix} = \begin{bmatrix} 1 & 0 \\ ba(a - a^{-1}) & 1 \end{bmatrix}.$$

The theorem follows from Theorems 3.4 and 3.6 since $b \in F^*$ is arbitrary.

Corollary. If $n = 2$ and $|F| > 3$, then $\mathrm{PSL}(V)$ is equal to its derived group.

Exercise 3.4. If $n \geq 3$ or $|F| > 3$ show that $\mathrm{GL}(n, F)$ has derived group $\mathrm{SL}(n, F)$. Determine the derived groups of $\mathrm{SL}(2, 2)$ and $\mathrm{SL}(2, 3)$.

The main result of this section is that, with two exceptions, $\mathrm{PSL}(V)$ is a simple group. We will need some auxiliary information about permutation groups.

Suppose G acts as a permutation group on a set S. A proper subset $B \subseteq S$, with $|B| > 1$, is called a *block* for G if either $xB = B$ or $xB \cap B = \varnothing$ for all $x \in G$. If G is transitive on S and has no blocks we say that G is *primitive* on S.

For example, the rotation group of a tetrahedron acts primitively on the vertices of the tetrahedron, whereas the dihedral group D_4 is transitive but not primitive on the vertices of a square.

Exercise 3.5. (1) If $|S| = p$, a prime, and G is transitive on S, show that G is primitive.

(2) If G is transitive on S show that G is primitive if and only if each stabilizer $\mathrm{Stab}_G(s)$, $s \in S$, is a maximal subgroup of G.

Proposition 3.9. If G is doubly transitive on S, then G is primitive.

Proof. Suppose $B \subseteq S$, $B \neq S$, and $|B| > 1$. Choose $a, b \in B$, $a \neq b$, and $c \in S \backslash B$, then choose $x \in G$ with $xa = a$ and $xb = c$. Then $a \in xB \cap B$ and $c \in xB \backslash B$ so B is not a block.

Proposition 3.10. Suppose G is faithful and primitive on S and $1 \neq N \lhd G$. Then N is transitive on S.

Proof. Since G is faithful and $N \neq 1$ there is an N-orbit $B \subseteq S$ with $|B| > 1$, say $B = \mathrm{Orb}_N(a) = Na$. If $x \in G$, then $xB = xNa = Nxa = \mathrm{Orb}_N(xa)$, so either $xB = B$ or $xB \cap B = \varnothing$. Thus $B = S$, since otherwise B would be a block for G, and so N is transitive.

Proposition 3.11. Suppose G acts on S, $H \leq G$ is transitive on S, and $s \in S$. Then $G = H \cdot \mathrm{Stab}_G(s)$.

Proof. If $x \in G$, then $xs = hs$ for some $h \in H$, so $h^{-1}x \in \mathrm{Stab}_G(s)$ and thus $x \in h \cdot \mathrm{Stab}_G(s) \subseteq H \cdot \mathrm{Stab}_G(s)$.

Theorem 3.12 (Iwasawa). Suppose G is faithful and primitive on S and $G = G'$. Fix $s \in S$ and set $H = \mathrm{Stab}_G(s)$. If there is a solvable subgroup $K \lhd H$ such that $G = \langle \bigcup \{K^x : x \in G\} \rangle$, then G is simple.

Proof. Suppose $1 \neq N \lhd G$. By Proposition 3.10 N is transitive on S and by Proposition 3.11 $G = NH$. Thus $NK \lhd NH = G$. But then $K^x \leq (NK)^x = NK$, all $x \in G$, so $\bigcup \{K^x : x \in G\} \subseteq NK$ and $G = NK$. Since K is solvable $K^{(m)} = 1$ for some integer m. It is easy to check inductively that $(NK)^{(k)} \leq N \cdot K^{(k)}$ for all k. Thus

$$G = G^{(m)} = (NK)^{(m)} \leq N \cdot K^{(m)} = N,$$

so $N = G$ and G is simple.

Proposition 3.13. If $n \geq 2$, then $\mathrm{PSL}(V)$ acts faithfully and doubly transitively on the projective space $P_{n-1}(V)$. In particular, it is primitive.

Proof. We observed earlier that $\mathrm{PSL}(V)$ is faithful on $P_{n-1}(V)$. It will suffice to show that $\mathrm{SL}(V)$ is doubly transitive. Take $[v_1] \neq [v_2]$ and $[w_1] \neq [w_2]$ in $P_{n-1}(V)$, so $\{v_1, v_2\}$ and $\{w_1, w_2\}$ are linearly independent sets in V. If $n = 2$ they are bases. If $n \geq 3$ set $V_1 = \langle v_1, v_2 \rangle$ and $V_2 = \langle w_1, w_2 \rangle$. Then either $V_1 = V_2 \neq V$ or else $V_1 \cup V_2$ is not a subspace, so $V_1 \cup V_2 \neq V$ in either case. If $v_3 \in V \backslash (V_1 \cup V_2)$, then both $\{v_1, v_2, v_3\}$ and $\{w_1, w_2, v_3\}$ are linearly independent. The argument can be repeated to obtain bases $\{v_1, v_2, v_3, \ldots, v_n\}$ and $\{w_1, w_2, v_3, \ldots, v_n\}$ for V. For any $b \in F^*$ define $T_b \in \mathrm{GL}(V)$ by means of $v_1 \mapsto bw_1$, $v_2 \mapsto w_2$, and $v_i \mapsto v_i$, $3 \leq i \leq n$. If $w_i = \sum_j a_{ji} v_j$, $i = 1, 2$, then det $T_b = b(a_{11}a_{22} - a_{21}a_{12})$. Choose b so that det $T_b = 1$. Then $T_b \in \mathrm{SL}(V)$ and T_b maps $[v_1]$ to $[w_1]$ and $[v_2]$ to $[w_2]$.

Proposition 3.14. If $0 \neq v \in V$ and $A = \mathrm{Stab}_{\mathrm{SL}(V)}([v])$, then A has a normal abelian subgroup B whose conjugates in $\mathrm{SL}(V)$ generate all of $\mathrm{SL}(V)$.

Proof. Choose a hyperplane W with $v \in V \backslash W$, so $V = W \oplus Fv$. If $S \in A$ and $w \in W$ write $Sw = S'w + a_w \cdot v$, with $S'w \in W$ and $a_w \in F$. It is easy to check that $S' \in \mathrm{GL}(W)$ and that $\phi: S \to S'$ is a homomorphism from A into $\mathrm{GL}(W)$. Thus $B = \ker \phi$ is a normal subgroup of A. Choose a basis $\{w_1, \ldots, w_{n-1}\}$ for W. For any $b \in F^*$ define $T_b \in A$ by means of T_b: $w_1 \mapsto w_1 + bv$, $w_i \mapsto w_i$ for $2 \leq i \leq n-1$, and $v \mapsto v$. Then T_b is a transvection (with hyperplane spanned by $\{w_2, \ldots, w_n, v\}$), and if $w \in W$, then $T_b w = w + bv$, so $T'_b = 1_W$ and $T_b \in B$. If $n = 2$ the matrix representing T_b relative to $\{w_1, v\}$ is $\left[\begin{smallmatrix} 1 & 0 \\ b & 1 \end{smallmatrix}\right]$, and b is arbitrary in F^*. It follows from Theorems 3.4, 3.5, and 3.6 that the conjugates of B generate $\mathrm{SL}(V)$.

If $S \in B$ then $S' = 1_w$, so $Sw_i = w_i + a_i v$, $a_i \in F$, all i. Thus if $S_1, S_2 \in B$, then their representing matrices relative to $\{w_1, \ldots, w_{n-1}, v\}$ have the partitioned form

$$\begin{bmatrix} I & 0 \\ u_1 & 1 \end{bmatrix} \quad \text{and} \quad \begin{bmatrix} I & 0 \\ u_2 & 1 \end{bmatrix}$$

Since

$$\begin{bmatrix} I & 0 \\ u_1 & 1 \end{bmatrix}\begin{bmatrix} I & 0 \\ u_2 & 1 \end{bmatrix} = \begin{bmatrix} I & 0 \\ u_1 + u_2 & 1 \end{bmatrix} = \begin{bmatrix} I & 0 \\ u_2 & 1 \end{bmatrix}\begin{bmatrix} I & 0 \\ u_1 & 1 \end{bmatrix}$$

it follows that B is abelian.

Theorem 3.15. Except for $\mathrm{PSL}(2,2)$ and $\mathrm{PSL}(2,3)$ every $\mathrm{PSL}(V)$ is a simple group.

Proof. Choose $v \neq 0$ in V and take $A = \mathrm{Stab}_{\mathrm{SL}}([v])$ and $B \lhd A$ as in Proposition 3.14. Write $Z = Z(\mathrm{SL}(V))$ and set $H = A/Z = \mathrm{Stab}_{\mathrm{PSL}}([v])$, $K = BZ/Z \lhd H$. Then K is solvable (in fact abelian) and the conjugates of K in $\mathrm{PSL}(V)$ generate $\mathrm{PSL}(V)$ by Proposition 3.14. The theorem follows from Proposition 3.13, the corollaries to Theorems 3.7 and 3.8, and Theorem 3.12.

Exercise 3.6. Use the action of PSL on projective space to show that $\mathrm{PSL}(2,2) \cong S_3$ and $\mathrm{PSL}(2,3) \cong A_4$, and conclude that the exceptions in Theorem 3.15 are not simple. Show also that $\mathrm{PSL}(2,4) \cong A_5$.

Note that $|\mathrm{PSL}(2, 5)| = 60$, so $\mathrm{PSL}(2, 5) \cong A_5$ by Proposition 4.5 of Chapter I, even though $|P_1(5)| = 6$.

Observe that $|\mathrm{PSL}(3,4)| = 20160 = |A_8|$, so $\mathrm{PSL}(3,4)$ and A_8 are simple groups having the same order. If $Z = Z[\mathrm{SL}(3,4)]$, then $|Z| = |F_4^*| = 3$. For any $S \in \mathrm{SL}(3,4)$ write $\bar{S}$ for the coset SZ that is its image in PSL.

Now let $\bar{S} \in \mathrm{PSL}(3,4)$ be any element of order 2. Then $S^2 \in Z$, so $S^6 = 1$. Replacing S by $T = S^3$ we have $\bar{S} = \bar{T}$ in PSL and $|T| = 2$ in SL. Let $W \subseteq V$

be the kernel of the linear transformation $1 + T$. Thus $\dim V = 3 = \dim W + \dim(1 + T)V$. Note that $v \in W$ if and only if $v + Tv = 0$, or $Tv = v$ (char $F = 2$), i.e., v is fixed by T. Thus $W \neq V$, since $T \neq 1$. Note also that $T(1 + T)v = (T + T^2)v = (T + 1)v$, so $(1 + T)v$ is fixed by T and hence $(1 + T)V \subseteq W$. Thus $\dim(1 + T)V \leq \dim W$ and it follows that $\dim(1 + T)V = 1$, $\dim W = 2$. Consequently W is a hyperplane and $Tv - v \in W$ for all $v \in V$, i.e., T is a transvection.

It follows from Theorem 3.5 that all elements of order 2 are conjugate in PSL(3, 4). The permutations (12)(34) and (12)(34)(56)(78) are not conjugate in S_8, hence certainly not in A_8, so A_8 has at least two conjugacy classes of elements of order 2. Thus PSL(3, 4) and A_8 are nonisomorphic simple groups having the same order.

Exercise 3.7. Show that PSL(4, 2) also has the same order as PSL(3, 4) and A_8. Show that PSL(4, 2) and PSL(3, 4) are not isomorphic.

(*Hint*: Consider

$$\begin{bmatrix} 1 & 0 & 0 & 0 \\ 0 & 1 & 0 & 0 \\ 0 & 0 & 1 & 0 \\ 0 & 0 & 1 & 1 \end{bmatrix} \quad \text{and} \quad \begin{bmatrix} 1 & 0 & 0 & 0 \\ 1 & 1 & 0 & 0 \\ 0 & 0 & 1 & 0 \\ 0 & 0 & 1 & 1 \end{bmatrix}.)$$

4. INTEGRAL DEPENDENCE AND DEDEKIND DOMAINS

All rings considered in this section will be commutative rings with 1 (proper ideals are obvious exceptions), and all modules will be assumed unitary. If R is a subring of S it will be assumed without further mention that $1_R = 1_S$.

If R is a subring of S, then $a \in S$ is *integral over* R if a is a root of some monic $f(x) \in R[x]$; the equation $f(a) = 0$ is called a relation of *integral dependence.* If every $a \in S$ is integral over R, then S is *integral over* R, or an *integral extension* of R. The concept generalizes the notion of algebraic field extensions to commutative rings.

Proposition 4.1. Suppose S is an integral domain that is integral over R. Then S is a field if and only if R is a field.

Proof. $\Rightarrow$: If $0 \neq a \in R$, then $a^{-1} \in S$ is integral over R, so there are $b_0, \ldots, b_{k-1} \in R$ with

$$b_0 + b_1 a^{-1} + \cdots + (a^{-1})^k = 0.$$

Multiply by a^{k-1} and observe that

$$a^{-1} = -(b_0 a^{k-1} + \cdots + b_{k-1}) \in R.$$

$\Leftarrow$: If $0 \neq a \in S$ choose $f(x)$ monic of minimal degree in $R[x]$ with $f(a) = 0$, say

$$f(x) = a_0 + a_1 x + \cdots + x^n.$$

Then $a_0 \neq 0$ (by minimality of degree) and

$$a_0 = -a(a_1 + a_2 a + \cdots + a^{n-1}) \in R \cap Sa.$$

Thus $a_0 = ba$ for some $b \in S$, and $1 = a_0^{-1} a_0 = (a_0^{-1} b)a$, so a has an inverse in S and S is a field.

In order to obtain a useful characterization of integrality we require the following technical proposition.

Proposition 4.2. Suppose that $A = (s_{ij})$ is an $n \times n$ matrix with entries from a ring S, that $u_1, u_2, \ldots, u_n \in S$, and that

$$\sum_{j=1}^{n} s_{ij} u_j = 0, \qquad 1 \leq i \leq n.$$

Then $(\det A)u_j = 0$, $\quad 1 \leq j \leq n$.

Proof. Fix j and denote by A_j the matrix obtained from A by multiplying column j by u_j. Thus $\det A_j = (\det A)u_j$. But $\det A_j$ is unchanged if we multiply column k of A_j by u_k and add to column j for each $k \neq j$. The resulting matrix has ij-entry $\sum_{k=1}^{n} s_{ik} u_k = 0$ for every i, so its determinant is 0, and hence $(\det A)u_j = 0$.

Proposition 4.3. Suppose R is a subring of S and $a \in S$. The following are equivalent:

(a) a is integral over R,
(b) $R[a]$ is a finitely generated R-submodule of S, and
(c) there is a finitely generated R-submodule M of S with $1 \in M$ and $aM \subseteq M$.

Proof. (a) $\Rightarrow$ (b): If $f(x) \in R[x]$ is monic of degree n with $f(a) = 0$, then a^n is an R-linear combination of $\{1, a, \ldots, a^{n-1}\}$, as are all higher powers of a. Thus $R[a] = R\langle 1, a, \ldots, a^{n-1}\rangle$.

(b) $\Rightarrow$ (c): Take $M = R[a]$.

(c) $\Rightarrow$ (a): Let $\{u_1, \ldots, u_n\} \subseteq S$ be a set of generators for M as an R-module. Multiply each u_i by a and we may write

$$\begin{aligned} au_1 &= s_{11}u_1 + \cdots + s_{1n}u_n, \\ &\vdots \\ au_n &= s_{n1}u_1 + \cdots + s_{nn}u_n, \end{aligned}$$

for suitable $s_{ij} \in R$. Thus

$$\sum_{j=1}^{n} (s_{ij} - a\delta_{ij})u_j = 0, \qquad 1 \leq i \leq n,$$

where δ_{ij} is the usual Kronecker delta. Denote by A the $n \times n$ matrix $(s_{ij} - a\delta_{ij})$; by Proposition 4.2 we have $(\det A)u_j = 0$ for all j. Write $1 = \sum_j b_j u_j$, $b_j \in R$; then

$$\det A = (\det A) \cdot 1 = \sum_j b_j(\det A)u_j = 0.$$

But then

$$f(x) = (-1)^n \det(s_{ij} - x\delta_{ij})$$

is a monic polynomial in $R[x]$ with $f(a) = 0$, so a is integral over R.

Corollary. If $a_1, a_2, \ldots, a_k \in S$ are integral over R, then $R[a_1, a_2, \ldots, a_k]$ is a finitely generated R-module.

Proof. Induction on k. Note that a_i is integral over $R[a_1, \ldots, a_{i-1}]$ since it is integral over R.

Exercise 4.1. Suppose R, S, and T are rings, with S integral over R and T integral over S. Show that T is integral over R. (See Proposition III.1.5.)

If R is a subring of S the *integral closure* of R in S, denoted $\mathrm{Int}_S(R)$, is the set of all elements $a \in S$ that are integral over R. Clearly $R \subseteq \mathrm{Int}_S(R)$. If $R = \mathrm{Int}_S(R)$ we say that R is *integrally closed* in S.

Proposition 4.4. If R is a subring of S, then $\mathrm{Int}_S(R)$ is a subring of S and $\mathrm{Int}_S(R)$ is integrally closed in S.

Proof. If a, $b \in \mathrm{Int}_S(R)$, then $M = R[a, b]$ is a finitely generated R-submodule of S containing $a - b$ and ab, so $a - b$ and ab are in $\mathrm{Int}_S(R)$ by Proposition 4.3(c). It follows that $\mathrm{Int}_S(R)$ is a ring, and $\mathrm{Int}_S(R)$ is its own integral closure in S by Exercise 4.1.

An integral domain R is called *integrally closed* if it is integrally closed as a subring of its field F of fractions. For example, $R = \mathbb{Z}$ is integrally closed. This fact is proved and generalized in the next proposition.

Proposition 4.5. If R is a UFD, then R is integrally closed.

Proof. Let F be the field of fractions of R and suppose $u \in F$ is integral over R. Write $u = a/b$ with $a, b \in R$. We may assume that $(a, b) = 1$ in R. Since u is integral we have

$$u^n = a^n/b^n = \sum_{i=0}^{n-1} r_i a^i/b^i$$

for some $n > 0$ in $\mathbb{Z}$ and $r_i \in R$. Multiply by b^n to obtain

$$a^n = \sum_{i=0}^{n-1} r_i a^i b^{n-i}.$$

Thus $b \mid a^n$, so $b \in U(R)$ since $(a, b) = (a^n, b) = 1$. But then $u = ab^{-1} \in R$, as required.

Proposition 4.6. Suppose R is an integral domain with field F of fractions, K is a field extension of F, and $S = \mathrm{Int}_K(R)$. If $a \in S$, with minimal polynomial $m(x)$ over F, then $m(x) \in S[x]$, and if R is integrally closed, then $m(x) \in R[x]$.

Proof. Let $L \supseteq K$ be a splitting field for $m(x)$ over K and let $a_1 = a, a_2, \ldots, a_k$ be all the roots of $m(x)$ in L. Since $a \in S$ we have $f(a) = 0$ for some monic $f(x) \in R[x]$. Thus $m(x) \mid f(x)$ in $F[x]$ by Proposition III.1.2, so $f(a_i) = 0$, $1 \le i \le k$, and all a_i are integral over R. Thus the coefficients of $m(x) = \prod_i (x - a_i)$ are integral over R (by Proposition 4.4), and they are in $F \subseteq K$, so they are in S, i.e., $m(x) \in S[x]$. If R is integrally closed, then $S \cap F = R$, so $m(x) \in R[x]$ in that case.

Corollary. If R is integrally closed and $a \in S$, then $\mathrm{Tr}_{K/F}(a) \in R$.

Proof. Recall that $\mathrm{Tr}_{F(a)/F}(a)$ is the negative of a coefficient of $m(x)$, as was observed in the proof of Theorem III.7.3, so $\mathrm{Tr}_{F(a)/F}(a) \in R$. Thus $\mathrm{Tr}_{K/F}(a) \in R$ by Proposition III.7.2

Proposition 4.7. Suppose R is an integral domain with field F of fractions, K is an algebraic extension field of F, and $S = \mathrm{Int}_K(R)$. If $u \in K^*$, then $ru \in S^*$ for some $r \in R$. As a consequence K is the field of fractions of S.

Proof. If denominators are cleared in the minimal polynomial for u over F we obtain

$$f(x) = a_0 + a_1 x + \cdots + a_n x^n \in R[x],$$

with $a_n \neq 0$ and $f(u) = 0$. Set

$$g(x) = a_n^{n-1} a_0 + a_n^{n-2} a_1 x + \cdots + a_{n-1} x^{n-1} + x^n \in R[x],$$

and observe that $g(a_n u) = a_n^{n-1} f(u) = 0$, so $a_n u \in S$. Set $r = a_n$.

Corollary. If $[K:F] = m$, finite, then there is a basis $\{v_1, \ldots, v_m\} \subseteq S$ for K over F.

Proof. Let $\{u_1, \ldots, u_m\}$ be any basis for K over F, choose $r_i \in R$ with $r_i u_i \in S^*$, set $r = \prod_i r_i$, and set $v_i = r u_i$ for each i.

Proposition 4.8. Suppose K is a finite separable extension of a field F and $\{u_1, \ldots, u_n\}$ is a basis for K over F. Then there is another basis $\{v_1, \ldots, v_n\}$ such that $\mathrm{Tr}_{K/F}(u_i v_j) = \delta_{ij}$ for all i and j.

Proof. Define $f_i\colon K \to F$, $1 \le i \le n$, by setting $f_i(a) = \mathrm{Tr}_{K/F}(u_i a)$ for all $a \in K$. Then $f_i \in \mathrm{Hom}_F(K, F)$ (the *dual space* of K), and it follows from the corollary to Proposition III.7.4 that $f_i \neq 0$. Thus $\dim_F(\ker f_i) = n - 1$. For each j set $V_j = \bigcap\{\ker f_i\colon i \neq j\}$. It is easy to verify by a dimension argument that $V_j \neq 0$ (in fact, $\dim V_j = 1$). Choose $v_j \neq 0$ in V_j. Let us show that $v_j \notin \ker f_j$. If $v_j \in \ker f_j$, then $\mathrm{Tr}(u_i v_j) = 0$ for all i. Choose $y \in K$ with $\mathrm{Tr}(y) = 1$ (the corollary to Proposition III.7.4) and write $yv_j^{-1} = \sum_i a_i u_i$, $a_i \in F$, so $y = \sum_i a_i u_i v_j$. Then

$$1 = \mathrm{Tr}(y) = \sum_i a_i \mathrm{Tr}(u_i v_j) = 0,$$

a contradiction. We may assume, then, that $f_j(v_j) = 1$, and hence $\mathrm{Tr}(u_i v_j) = \delta_{ij}$ for all i and j. If $\sum_j b_j v_j = 0$, $b_j \in F$, then

$$0 = f_i\left(\sum_j b_j v_j\right) = \sum_j b_j \mathrm{Tr}(u_i v_j) = b_i \qquad \text{for all } i,$$

so $\{v_j\}$ is a basis.

We remark that $\{v_i\}$ is called the *dual basis* for $\{u_i\}$ relative to the bilinear form $(x, y) \mapsto \mathrm{Tr}_{K/F}(xy)$ on K.

Theorem 4.9. Suppose R is an integrally closed domain with field F of fractions, K is a finite separable extension of F, and $S = \mathrm{Int}_K(R)$. There is an F-basis $\{v_1, \ldots, v_n\}$ for K such that the free R-module $R\langle v_1, \ldots, v_n\rangle$ contains S.

Proof. By the corollary to Proposition 4.7 there is a basis $\{u_1, \ldots, u_n\} \subseteq S$ for K over F, and by Proposition 4.8 there is a dual basis $\{v_1, \ldots, v_n\}$ with $\mathrm{Tr}_{K/F}(u_i v_j) = \delta_{ij}$ for all i and j. If $a \in S$ write $a = \sum_i b_i v_i$, with all $b_i \in F$. Then $au_i \in S$, and hence $\mathrm{Tr}_{K/F}(au_i) \in R$, all i, by the corollary to Proposition 4.6. But

$$\mathrm{Tr}_{K/F}(au_i) = \sum_j b_j \mathrm{Tr}(u_i v_j) = b_i, \qquad \text{all} \quad i,$$

so

$$a \in \sum_j Rv_j = R\langle v_1, \ldots, v_n\rangle.$$

Corollary 1. If R is Noetherian, then S is Noetherian.

Proof. The R-module $R\langle v_1, \ldots, v_n\rangle$ is Noetherian by Proposition IV.2.5, so S is a Noetherian R-module by Proposition IV.1.8. Ideals of S are R-submodules, so S satisfies the ACC for ideals, i.e., S is a Noetherian ring.

Corollary 2. If R is a PID, then S is a free R-module of rank $n = [K:F]$.

Proof. By Theorem IV.2.9 S is a free R-module of rank n or less. It follows from the corollary to Proposition 4.7 that rank $(S) = n$.

In the setting of Corollary 2 there is a free R-basis $\{w_1, \ldots, w_n\}$ for S. Such a basis is commonly called an *integral basis.*

It will prove useful to generalize the notion of ideal for integral domains. If R is an integral domain with field F of fractions, then a *fractional ideal* of R is a nonzero R-submodule M of F such that $rM \subseteq R$ for some $r \in R^*$. Note that then $I = rM$ is an R-submodule of R, i.e., an ideal in R. Basically, $M \subseteq F$ is a fractional ideal if $M \neq 0$ and we may simultaneously clear all denominators in M to obtain an ordinary ideal in R. For example, $M = \{3a/2 : a \in \mathbb{Z}\} \subseteq \mathbb{Q}$ is a fractional ideal of $\mathbb{Z}$, with $2M = (3) \subseteq \mathbb{Z}$. Clearly each ideal in R is a fractional ideal; when it is important to distinguish them from other fractional ideals ordinary ideals will be called *integral ideals.*

If $a \in F^*$, then $M = Ra$ is a fractional ideal of R; it is called the *principal fractional ideal* generated by a.

Suppose M and N are fractional ideals of R. Define

$$M + N = \{a + b : a \in M, b \in N\},$$

and

$$MN = \left\{\sum_{i=1}^{k} a_i b_i : a_i \in M, b_i \in N, 0 < k \in \mathbb{Z}\right\},$$

just as for integral ideals. Also, define

$$(M:N) = \{a \in F : aN \subseteq M\},$$

often called the *quotient* of M by N. If $MN = R$, then M is called *invertible.* If $MI = N$ for some integral ideal I in R we say that M *divides* N, and write $M \mid N$.

Let us write $\mathscr{F} = \mathscr{F}_R$ for the set of all fractional ideals of R, $\mathscr{I} = \mathscr{I}_R$ for the subset of invertible fractional ideals, and $\mathscr{P} = \mathscr{P}_R$ for the subset of principal fractional ideals.

The following exercise gathers together a number of elementary facts about fractional ideals.

Exercise 4.2. (1) If $M, N \in \mathscr{F}$ show that $M \cap N$, MN, $M + N$, and $(M:N) \in \mathscr{F}$.

(2) Show that $\mathscr{F}$ is a commutative monoid relative to multiplication, with identity element R.

(3) Show that $\mathscr{I}$ is a group, with identity R. If $M \in \mathscr{I}$ show that $M^{-1} = (R:M)$.

(4) Show that $\mathscr{P}$ is a subgroup of $\mathscr{I}$, with $(Ra)^{-1} = Ra^{-1}$ for each $a \in F^*$.

(5) If $M, N \in \mathscr{F}$ and $M \mid N$ show that $M \supseteq N$. Conclude that if $M \mid N$ and $N \mid M$ then $M = N$.

Proposition 4.10. If R is an integral domain and $M \in \mathscr{I}_R$, then M is a finitely generated R-module.

Proof. Write $1 \in R = MM^{-1}$ as $1 = \sum_{i=1}^{n} a_i b_i$, with $a_i \in M$, $b_i \in M^{-1}$. If $c \in M$, then $cb_i \in MM^{-1} = R$ for all i, so

$$c = c \cdot 1 = \sum_i (cb_i)a_i \in \sum_i Ra_i.$$

Thus $M = R\langle a_1, \ldots, a_n \rangle$.

Exercise 4.3. If R is a Noetherian domain and $M \in \mathscr{F}_R$ show that M is a finitely generated R-module. (*Hint*: Write $I = rM$, an integral ideal, for some $r \in R^*$ and show that I and M are isomorphic as R-modules).

Proposition 4.11. If every nonzero integral ideal in R is invertible, then $\mathscr{F} = \mathscr{I}$, i.e., every fractional ideal is invertible, and $\mathscr{F}$ is a group.

Proof. If $M \in \mathscr{F}$, then $I = rM$ is an integral ideal for some $r \in R^*$, so clearly $M = r^{-1}I$. Thus

$$M(rI^{-1}) = (r^{-1}I)(rI^{-1}) = R,$$

so $M^{-1} = rI^{-1}$ and $M \in \mathscr{I}$.

Proposition 4.12. Suppose R is an integral domain, I_1 and I_2 are nonzero integral ideals in R, and $I_1 I_2 = I \in \mathscr{I}_R$. Then also $I_1, I_2 \in \mathscr{I}_R$.

Proof. Say $I^{-1} = J$, so $JI_1I_2 = R$. Thus $JI_2 = I_1^{-1}$ and $JI_1 = I_2^{-1}$.

For the remainder of this section we will be concerned with writing ideals as products of prime ideals. Such products have been defined only for finitely many prime ideals, and that finiteness will be understood even when not explicitly mentioned. It will also be convenient to agree that the product of the empty collection of (nonzero) prime ideals is R. Of course, the product of a set $\{P\}$ containing just one prime ideal P is just P itself.

Proposition 4.13. Suppose R is an integral domain and $P_1, \ldots, P_m$ are invertible prime ideals in R, and set

$$I = \prod\{P_i : 1 \le i \le m\} \subseteq R.$$

If $Q_1, \ldots, Q_n$ are also prime ideals and $I = \prod\{Q_i : 1 \le i \le n\}$, then $m = n$ and $\{P_1, \ldots, P_m\} = \{Q_1, \ldots, Q_m\}$.

Proof. Induction on m. We may assume that P_1 is minimal in the set $\{P_i\}$, in the sense that if $P_j \subseteq P_1$, then $P_j = P_1$. We have $P_1 \supseteq I = \prod_i Q_i$, so $P_1 \supseteq Q_i$ for some i since P_1 is prime; we may assume $P_1 \supseteq Q_1$. By the same argument $Q_1 \supseteq P_j$ for some j, so $P_1 \supseteq Q_1 \supseteq P_j$ and hence $P_1 = Q_1 = P_j$ by the minimality of P_1. Multiplying by P_1^{-1} we obtain

$$\prod\{P_i : 2 \le i \le m\} = \prod\{Q_i : 2 \le i \le n\},$$

and the proposition follows by convention if $m = 1$, by induction if $m > 1$.

The point of Proposition 4.13 is that when factorization of an ideal as a product of invertible prime ideals occurs, it is unique.

Proposition 4.14. Suppose that R is an integral domain and that every proper ideal in R is a product of prime ideals. Then every invertible prime ideal P in R is maximal.

Proof. Take $a \in R\backslash P$ and set $I = P + Ra$. We must show that $I = R$, so suppose $I \neq R$. Set $J = P + Ra^2$ and write $I = P_1 \cdots P_k$, $J = Q_1 \cdots Q_m$ for prime ideals P_i, Q_j. Thus $P_i, Q_j \supseteq P$ for all i and j. Write $\bar{R}$ for R/P, $\bar{a} = a + P \in \bar{R}$, etc. Then $\bar{I} = \bar{P}_1 \cdots \bar{P}_k$ and

$$\bar{J} = \bar{Q}_1 \cdots \bar{Q}_m = \bar{I}^2 = \bar{P}_1^2 \cdots \bar{P}_k^2.$$

Since $\bar{I}$ and $\bar{J}$ are principal ideals in $\bar{R}$ they are invertible (Exercise 4.2.4), so all $\bar{P}_i$ and $\bar{Q}_j$ are invertible by Proposition 4.12. By Proposition 4.13 we may conclude that $m = 2k$, and by relabeling that $\bar{Q}_1 = \bar{Q}_2 = \bar{P}_1, \ldots, \bar{Q}_{2k-1} = \bar{Q}_{2k} = \bar{P}_k$. Since all contain P it follows that $P_i = Q_{2i} = Q_{2i-1}$ for each i, and hence $J = I^2$. Thus $P \subseteq J = (P + Ra)^2 \subseteq P^2 + Ra$. If $x \in P$ write $x = y + za$, with $y \in P^2$ and $z \in R$. Then $za = x - y \in P$, and $a \notin P$, so $z \in P$ and $x \in P^2 + Pa$. Thus $P \subseteq P^2 + Pa \subseteq P$ and $P = P^2 + Pa$. Multiply by P^{-1} to see that $R = P + Ra = I$ and the proof is complete.

Proposition 4.15. If R is a Noetherian domain, then every nonzero integral ideal in R contains a product of nonzero prime ideals.

Proof. Let $\mathscr{S}$ denote the set of nonzero integral ideals in R for which the conclusion fails. We must show that $\mathscr{S} = \varnothing$. If not there is a maximal element I in $\mathscr{S}$ (Proposition II.7.1). Since I cannot be prime there are $a, b \in R\backslash I$ with $ab \in I$. By maximality the ideals $(a) + I$ and $(b) + I$ are not in $\mathscr{S}$ so each of them contains a product of prime ideals, and consequently their product does likewise. But then

$$[(a) + I][(b) + I] \subseteq (ab) + aI + bI + I \subseteq I,$$

so I contains the same product, a contradiction.

If $F \subseteq \mathbb{C}$ is a finite extension of the rational field $\mathbb{Q}$ then F is called an *algebraic number field*. The integral closure R of $\mathbb{Z}$ in F is called the ring of *algebraic integers* in F. Thus an algebraic integer in F is any $a \in F$ that is a root of some monic $f(x) \in \mathbb{Z}[x]$. The rings R_m in Chapter II provide a class of examples of rings of algebraic integers. Note that in general F is the field of fractions of R by Proposition 4.7.

If R is the ring of algebraic integers in an algebraic number field F, then R is Noetherian by Corollary 1 to Theorem 4.9 and R is integrally closed by Proposition 4.4. In the next theorem we establish a third important property of R, and then use those three properties as defining properties for Dedekind domains.

Theorem 4.16. Suppose F is an algebraic number field and R is its ring of algebraic integers. If P is a nonzero prime ideal in R, then P is maximal.

Proof. Take $a \neq 0$ in P and let $m(x) = m_{a,\mathbb{Q}}(x)$ be its minimal polynomial over $\mathbb{Q}$. Then $m(x) \in \mathbb{Z}[x]$ by Proposition 4.6. If

$$m(x) = a_0 + a_1 x + \cdots + x^k,$$

then $a_0 \neq 0$, and

$$a_0 = -(a_1 a + \cdots + a^k) \in P,$$

so $P \cap \mathbb{Z} \neq 0$. It is easy to verify that $P \cap \mathbb{Z}$ is a prime ideal in $\mathbb{Z}$, so $P \cap \mathbb{Z} = p\mathbb{Z}$ for some prime $p \in \mathbb{Z}$. Thus

$$\mathbb{Z}/p\mathbb{Z} = \mathbb{Z}/P \cap \mathbb{Z} \cong (\mathbb{Z} + P)/P \subseteq R/P.$$

Since R is integral over $\mathbb{Z}$ it is immediate that R/P is integral over $(\mathbb{Z} + P)/P$. But $(\mathbb{Z} + P)/P \cong \mathbb{Z}_p$ is a field, so R/P is a field by Proposition 4.1, and hence P is a maximal ideal in R.

Exercise 4.4. (1) Show that the prime $p \in P \cap \mathbb{Z}$ in the proof above is unique.

(2) Show that R/P is a finite field (see Corollary 2 to Theorem 4.9).

A *Dedekind domain* is a Noetherian integral domain that is integrally closed, in which every nonzero prime ideal is maximal. Thus, for example, the ring of algebraic integers in an algebraic number field is a Dedekind domain. We shall see shortly that several alternative definitions are possible for Dedekind domains, some of them perhaps aesthetically more appealing than the one above. The definition above, however, tends to be more easily verified in specific cases.

Proposition 4.17. If R is a PID, then R is a Dedekind domain.

Proof. Propositions II.5.9, II.5.10, II.5.17, and Proposition 4.5.

Exercise 4.5. Show that R_{-5} is a Dedekind domain that is not a PID (see Exercise II.5.1). Show that the polynomial ring $\mathbb{Z}[x]$ is a Noetherian domain that is not a Dedekind domain.

Exercise 4.6. Suppose R is a Dedekind domain with field F of fractions, K is a finite separable extension of F, and $S = \mathrm{Int}_K(R)$. Show that S is also a Dedekind domain (see Proposition 4.4, Corollary 1 to Proposition 4.9, and the proof of Proposition 4.16).

Proposition 4.18. If R is a Dedekind domain and I is a proper integral ideal in R, then $(R:I) \neq R$ [clearly $(R:I) \supseteq R$].

Proof. By Proposition II.1.8 there is a maximal (and therefore prime) ideal P with $I \subseteq P \subseteq R$. Choose $a \in I^*$ and apply Proposition 4.15 to obtain nonzero prime (and therefore maximal) ideals $P_1, \ldots, P_k$ in R with $P_1 \cdots P_k \subseteq Ra \subseteq P$. We may assume that k is minimal in that respect. If P were different from all P_i we could choose $a_i \in P_i \backslash P$ for all i (since all P_i are maximal). But then $a_1 a_2 \cdots a_k \in P$, a contradiction since P is prime. Thus we may assume by relabeling that $P = P_1$, and hence $PP_2 \cdots P_k \subseteq Ra \subseteq P$. Since k is minimal we may choose $b \in P_2 \cdots P_k \backslash Ra$. If $c = b/a$, then $c \notin R$ (or else $b \in Ra$), but

$$cI = (b/a)I \subseteq (b/a)P \subseteq (1/a)PP_2 \cdots P_k \subseteq R,$$

so $c \in (R:I) \backslash R$, as required.

Proposition 4.19. If R is a Dedekind domain and $P \subseteq R$ is a nonzero prime ideal, then P is invertible, with $P^{-1} = (R:P)$.

Proof. Note that $P \subseteq (R:P)P \subseteq R$ by the definition of $(R:P)$. Since P is maximal we have either $(R:P)P = P$ or $(R:P)P = R$. Suppose $(R:P)P = P$. If $a \in (R:P)$, then $aP \subseteq P$, so $a^2P \subseteq aP \subseteq P$, and by induction $a^nP \subseteq P$ if $0 < n \in \mathbb{Z}$. Thus $ba^n \in P \subseteq R$ for any nonzero $b \in P$ and every n, so $R[a] = R\langle 1, a, a^2, \ldots \rangle$ is a fractional ideal of R. By Proposition 4.10 it is finitely generated. Thus a is integral over R by Proposition 4.3, and so $a \in R$ since R is integrally closed. But that means $(R:P) \subseteq R$, which contradicts Proposition 4.18. Thus $(R:P)P = R$, so $P \in \mathscr{I}$ and $P^{-1} = (R:P)$.

Theorem 4.20. If R is an integral domain the following are equivalent.

(a) R is a Dedekind domain.

(b) Every (integral) ideal I in R is a product of invertible prime ideals.

(c) Every nonzero (integral) ideal I in R is uniquely a product of prime ideals.

(d) Every fractional ideal M of R is invertible, i.e., $\mathscr{F}_R$ is a group with respect to multiplication.

Proof. (a) $\Rightarrow$ (b): If $0 \neq I \neq R$ apply Proposition 4.15 to obtain nonzero prime ideals $P_1, \ldots, P_k$ with $\prod_i P_i \subseteq I$. We may assume that k is minimal and apply induction on k. If $k = 1$, then $I = P_1$ since P_1 is maximal. Suppose then that $k > 1$ and assume that (b) holds for ideals containing products of fewer than k nonzero prime ideals. Choose a maximal (and hence prime) ideal P with $I \subseteq P$, so $\prod_i P_i \subseteq P$. Then some $P_i \subseteq P$, so we may assume $P_k \subseteq P$, and hence $P_k = P$ since P_k is maximal. By Proposition 4.19 P is invertible. Multiply by P^{-1} to see that $P_1 \cdots P_{k-1} \subseteq IP^{-1}$. Note that $IP^{-1} \neq R$ for otherwise $I = P$, contradicting the minimality of $k > 1$. By the induction hypothesis there are prime ideals $Q_1, \ldots, Q_m$ with $IP^{-1} = Q_1 \cdots Q_m$, and hence $I = Q_1 \cdots Q_m P$.

(b) ⇒ (c): Apply Proposition 4.13.

(c) ⇒ (d): It will suffice, by Proposition 4.11 and (c), to show that every nonzero prime ideal P is invertible. Choose $a \in P$, $a \neq 0$, and write Ra as a product of prime ideals, say $P_1 \cdots P_k = Ra \subseteq P$. Then Ra is invertible by Exercise 4.2.4 and each P_i is invertible by Proposition 4.12. Since P is prime $P \supseteq P_i$, so $P = P_i$ since P_i is maximal (Proposition 4.14). Thus P is invertible.

(d) ⇒ (a): If I is a nonzero (integral) ideal in R, then I is finitely generated by Proposition 4.10, so R is Noetherian by Proposition II.7.2. We must show that R is integrally closed (in its field F of fractions). Take $a = b/c \in F$, with $b, c \in R$, integral over R; say a is a root of

$$f(x) = a_0 + a_1 x + \cdots + x^k \in R[x].$$

Let M be the fractional ideal of R generated by $\{1, a, \ldots, a^{k-1}\}$. Then

$$a^k = -(a_0 + a_1 a + \cdots + a_{k-1} a^{k-1}) \in M.$$

It follows easily that

$$M^2 = (R + Ra + \cdots + Ra^{k-1})(R + Ra + \cdots + Ra^{k-1}) = M,$$

so $M = MM^{-1} = R$, and hence $a \in R$, as required. Suppose finally that $P \neq 0$ is a prime ideal in R, and take $a \in R \backslash P$. Set $J = P + Ra$ and set $M = J^{-1}P$. Then $JM = JJ^{-1}P = P$, so $M \mid P$ and hence $M \supseteq P$. If $b \in M$, then $ab \in JM = P$, and $a \notin P$, so $b \in P$. Thus $M = P$ or $J^{-1}P = P$. Multiply by P^{-1} to see that $J^{-1} = R$ and hence $J = R$. Thus P is maximal and the proof is complete.

There are several other characterizations of Dedekind domains. Two that are well known involve the notions of localization (see Section 2 of Chapter II) and of projective modules (see Exercise IV.7.46).

Exercise 4.7. Suppose R is a Dedekind domain with group $\mathscr{F}$ of fractional ideals. Show that $\mathscr{F}$ is a free abelian group with basis the set of nonzero prime ideals of R.

Theorem 4.21. A Dedekind domain R is a UFD if and only if it is a PID.

Proof. Suppose R is a UFD.It will be sufficient to show that each nonzero prime ideal P in R is principal. Take $a \in P$, $a \neq 0$, and write $a = \prod\{p_i^{e_i} : 1 \leq e_i \in \mathbb{Z}\}$, where the p_i are distinct primes. Then $p_i \in P$ for some i, and hence $Rp_i \subseteq P$. Since p_i is prime Rp_i is a prime ideal, so it is maximal, and $p = Rp_i$ is principal.

Exercise 4.8. If R is a Dedekind domain and $M, N \in \mathscr{F}_R$ show that $M \mid N$ if and only if $M \supseteq N$.

If R is a Dedekind domain and $M_1, M_2 \in \mathscr{F}_R$, then a *greatest common divisor* (GCD) for M_1 and M_2 is any $M \in \mathscr{F}$ such that $M \mid M_1$, $M \mid M_2$, and if $N \in \mathscr{F}$ with $N \mid M_1$, $N \mid M_2$, then $N \mid M$. A *least common multiple* (LCM) for M_1 and M_2 is any $M \in \mathscr{F}$ such that $M_1 \mid M$, $M_2 \mid M$, and if $N \in \mathscr{F}$ with $M_1 \mid N$, $M_2 \mid N$, then $M \mid N$. There are obvious extensions to three or more fractional ideals. If $\mathrm{GCD}(M_1, M_2) = R$ we say that M_1 and M_2 are *relatively prime.*

Exercise 4.9. Suppose R is a Dedekind domain.

(1) If $M_1, M_2 \in \mathscr{F}$ show that $\mathrm{GCD}(M_1, M_2) = M_1 + M_2$ and $\mathrm{LCM}(M_1, M_2) = M_1 \cap M_2$. Conclude in particular that the GCD and LCM exist and are unique.

(2) If I and J are proper nonzero integral ideals write

$$I = \prod\{P_i^{a_i} : 0 \le a_i \in \mathbb{Z}, 1 \le i \le k\}$$

and

$$J = \prod\{P_i^{b_i} : 0 \le b_i \in \mathbb{Z}, 1 \le i \le k\},$$

where the P_i are distinct prime ideals and P_i^0 is understood to be R. Set $c_i = \min\{a_i, b_i\}$ and $d_i = \max\{a_i, b_i\}$, for each i. Show that

$$\mathrm{GCD}(I, J) = \prod\{P_i^{c_i} : 1 \le i \le k\}$$

and

$$\mathrm{LCM}(I, J) = \prod\{P_i^{d_i} : 1 \le i \le k\}.$$

Conclude that $IJ = (I + J)(I \cap J)$ and that $IJ = I \cap J$ if and only if I and J are relatively prime.

The quotient group $\mathscr{C} = \mathscr{C}_R = \mathscr{F}/\mathscr{P}$ is called the *ideal class group* of a Dedekind domain R (or of its field F of fractions). Note that $|\mathscr{C}| = 1$ if and only if R is a PID (and hence a UFD), so $|\mathscr{C}|$ can be viewed as a measure of deviation from unique factorization for elements of R. It can be shown that $|\mathscr{C}| = h$ is finite for algebraic number fields; the *class number* h is an important numerical invariant in algebraic number theory.

Historically the notion of ideal was introduced by Dedekind, following ideas of Kummer, to provide a replacement for unique factorization of elements when rings of algebraic integers are not UFD's.

The next proposition will show that, in a rather different sense, ideals in Dedekind domains are never very far from being principal (see the corollary).

Proposition 4.22. Suppose R is a Dedekind domain and I, J are proper nonzero integral ideals in R. Then there is an integral ideal L with $J + L = R$ (i.e., J and L are relatively prime) and $IL \in \mathscr{P}_R$.

Proof. Write

$$I = \prod\{P_i^{a_i}: 0 \leq a_i \in \mathbb{Z}, 1 \leq i \leq k\}$$

and

$$J = \prod\{P_i^{b_i}: 0 \leq b_i \in \mathbb{Z}, 1 \leq i \leq k\}$$

with distinct prime ideals P_i. For each i choose $a_i \in P_i^{a_i} \backslash P_i^{a_i+1}$ (this is possible because of unique factorization). The ideals $P_i^{a_i+1}$ are pairwise relatively prime so we may apply the Chinese Remainder Theorem (II.6.2) to obtain $a \in R$ with $a \equiv a_i (\text{mod } P_i^{a_i+1})$ for all i. Thus $a \in P_i^{a_i} \backslash P_i^{a_i+1}$, all i, so $a \in \bigcap_i P_i^{a_i} = \prod_i P_i^{a_i} = I$. It follows that $P_i^{a_i} \mid Ra$ for all i. Also $P_i^{a_i} \mid IJ$ for all i. Since $P_i^{a_i+1} \nmid Ra$ it follows from Exercise 4.9 that

$$\text{GCD}(Ra, IJ) = Ra + IJ = \prod_i P_i^{a_i} = I.$$

If we write $Ra = \prod_i P_i^{e_i}$, then $e_i \geq a_i$ for all i since $I \mid Ra$. Set $L = \prod_i P_i^{e_i - a_i}$, so that L is an integral ideal with $IL = Ra \in \mathscr{P}$. Also $I = Ra + IJ = IL + IJ$. Multiply by I^{-1} to see that $\text{GCD}(L, J) = L + J = R$, so L and J are relatively prime.

Corollary. Each integral ideal I in R requires at most two generators.

Proof. We may assume $0 \neq I \neq R$. Choose $a \in I$, $a \neq 0$, and apply the proposition to I and $J = Ra$ to obtain an integral ideal L with $IL = Rb$ for some $b \in R^*$ and $Ra + L = R$. Then $Ra + Rb = Ra + IL \subseteq I$. Write $1 = ra + c$, with $r \in R$ and $c \in L$. If $d \in I$, then

$$d = dra + dc \in Ra + IL = Ra + Rb.$$

Thus $I = (a, b)$.

We end this section with a detailed look at an easy numerical example. Take $R = R_{-5}$ and $F = \mathbb{Q}(\sqrt{-5})$ as in Section 5 of Chapter II. Then R is not a UFD, and $6 = 2 \cdot 3 = (1 + \sqrt{-5})(1 - \sqrt{-5})$ is an example of different factorizations into nonassociated primes. Set $I_1 = (2)$, $I_2 = (3)$, $J_1 = (1 + \sqrt{-5})$, and $J_2 = (1 - \sqrt{-5})$, so $(6) = I_1 I_2 = J_1 J_2$. Set

$$P_1 = \text{GCD}(I_1, J_1) = I_1 + J_1 = (2, 1 + \sqrt{-5}).$$

Let us see that P_1 is a prime ideal. (Why is $P_1 \neq R$?) Suppose $(x + y\sqrt{-5}) \times (u + v\sqrt{-5}) \in P_1$, i.e.,

$$(xu - 5yv) + (xv + yu)\sqrt{-5} = 2a + b + b\sqrt{-5} \qquad \text{for some} \quad a, b \in \mathbb{Z}.$$

Thus $xv + yu = b$ and

$$xu - 5yv = 2a + b = 2a + xv + yu,$$

so

$$u(x-y)-v(x+5y)=2a\equiv 0 \pmod 2,$$

and therefore

$$u(x-y)-v(x-y)=(u-v)(x-y)\equiv 0 \pmod 2,$$

so we may assume $u \equiv v (\mathrm{mod}\ 2)$. Writing $u = v + 2c$ we have $u + v\sqrt{-5} = 2c + v(1+\sqrt{-5}) \in P_1$, and P_1 is prime.

Exercise 4.10. Set $P_2 = I_1 + J_2$, $P_3 = I_2 + J_1$, $P_4 = I_2 + J_2$, and show that each is a proper prime ideal.

Continuing with the example it is clear that $I_1 \subseteq P_1 \cap P_2 = P_1P_2$. If $\alpha \in P_1P_2$ there are $a_i, b_i, c_i, d_i \in \mathbb{Z}$ with

$$\begin{aligned}\alpha &= \sum_i [2a_i + b_i(1+\sqrt{-5})][2c_i + d_i(1-\sqrt{-5})]\\ &\in (2) + \sum_i b_i d_i (1+\sqrt{-5})(1-\sqrt{-5})\\ &\subseteq (2) + \sum_i b_i d_i (6) \subseteq (2) + (6) \subseteq (2) = I_1,\end{aligned}$$

so $I_1 = P_1P_2$. Similarly, $I_2 = P_3P_4$, $J_1 = P_1P_3$, and $J_2 = P_2P_4$. Thus

$$(6) = I_1I_2 = P_1P_2P_3P_4 = J_1J_2 = P_1P_3P_2P_4,$$

and uniqueness of factorization is restored at the level of ideals.

5. TRANSCENDENTAL FIELD EXTENSIONS

Throughout this section F will denote a field and $K \supseteq F$ an extension field. A set $S \subseteq K$ is *algebraically dependent* over F if there are distinct elements $a_1, a_2, \ldots, a_k \in S$ and a nonzero polynomial

$$f(x_1, x_2, \ldots, x_k) \in F[x_1, x_2, \ldots, x_k]$$

with $f(a_1, a_2, \ldots, a_k) = 0$. Otherwise S is *algebraically independent* over F. Note that a set $S = \{a\} \subseteq K$ with just one element is algebraically dependent if a is algebraic over F; it is algebraically independent if a is transcendental over F. In general an algebraically independent set $S \subseteq K$ is called a *transcendence set* over F.

Exercise 5.1. Show that $S \subseteq K$ is algebraically dependent over F if and only if there is some $a \in S$ that is algebraic over $F(S \setminus \{a\})$.

If K is not algebraic over F, then K is called a *transcendental* extension. If $K = F(S)$ for some transcendence set S over F, then K is called a *purely transcendental* extension of F. The simplest example of a purely transcendental extension is a field $K = F(x)$ of rational functions in an indeterminate x over F, or more generally $K = F(x_1, x_2, \ldots, x_n)$ with distinct indeterminates $x_1, \ldots, x_n$ over F.

Exercise 5.2. Show that the real field $\mathbb{R}$ is a transcendental extension of the rational field $\mathbb{Q}$, but that it is not purely transcendental.

A transcendence set $B \subseteq K$ over F is called a *transcendence basis* if it is maximal with respect to set inclusion. By an easy Zorn's Lemma argument any transcendence set $S \subseteq K$ is contained in a transcendence basis B. In particular K *has* a transcendence basis over F. Note that K is algebraic over F if and only if the empty set $\varnothing$ is the only transcendence basis for K over F.

Proposition 5.1. Suppose $S \subseteq K$ is a transcendence set and $a \in K \setminus S$. Then $S \cup \{a\}$ is algebraically dependent over F if and only if a is algebraic over $F(S)$.

Proof. $\Rightarrow$: There is a finite (possibly empty) set $\{b_1, \ldots, b_k\} \subseteq S$ and a nonzero polynomial

$$f(x_0, x_1, \ldots, x_k) \in F[x_0, x_1, \ldots, x_k]$$

with $f(a, b_1, \ldots, b_k) = 0$. Note that x_0 actually occurs in $f(X)$, since otherwise there would be a nontrivial algebraic dependence relation on S. Define

$$g(x_0) = f(x_0, b_1, \ldots, b_k) \in F(S)[x_0]$$

and observe that $g(x_0) \neq 0$ but $g(a) = 0$, so a is algebraic over $F(S)$.

$\Leftarrow$: If $g(a) = 0$ for a nonzero polynomial $g(x)$ over $F(S)$, then we may clear denominators in the coefficients of $g(x)$ and assume that $g(x) \in F[b_1, \ldots, b_k][x]$ for some $\{b_1, \ldots, b_k\} \subseteq S$. But then the equation $g(a) = 0$ is in fact a nontrivial algebraic dependence relation over F for $\{a, b_1, \ldots, b_k\}$, and hence for $S \cup \{a\}$.

Corollary. If S is a transcendence set for K over F, then S is a transcendence basis if and only if K is algebraic over $F(S)$.

Exercise 5.3. If $S \subseteq K$ and K is algebraic over $F(S)$ show that there is a transcendence basis B for K over F with $B \subseteq S$.

If $S \subseteq K$ write

$$\mathfrak{a}(S) = \mathfrak{a}_{K,F}(S) = \{a \in K : a \text{ is algebraic over } F(S)\},$$

commonly called the *algebraic closure* of $F(S)$ in K. The following observations are easy consequences of the definition of $\mathfrak{a}(S)$:

(i) $S \subseteq \mathfrak{a}(S)$;
(ii) if $S \subseteq T \subseteq K$, then $\mathfrak{a}(S) \subseteq \mathfrak{a}(T)$;
(iii) if $a \in \mathfrak{a}(S)$, then $a \in \mathfrak{a}(S')$ for some finite set $S' \subseteq S$; and
(iv) $\mathfrak{a}(\mathfrak{a}(S)) = \mathfrak{a}(S)$

[(iv) follows from Proposition III.1.5]. These four facts, together with the result of the next proposition, show that $\mathfrak{a}$ determines a "dependence relation" as defined abstractly in Section 1of Chapter I of Zariski and Samuel [41].

Proposition 5.2. Suppose $S \subseteq K$, $a, b \in K$, and $b \in \mathfrak{a}(S \cup \{a\})\backslash\mathfrak{a}(S)$. Then $a \in \mathfrak{a}(S \cup \{b\})$.

Proof. Set $L = F(S)$, so b is transcendental over L but algebraic over $L(a)$. Then $\{a, b\}$ is algebraically dependent over L by Proposition 5.1. Choose $f(x_1, x_2) \neq 0$ in $L[x_1, x_2]$ with $f(a, b) = 0$, and note that x_1 must occur in $f(x_1, x_2)$, for otherwise b would be algebraic over L. Thus

$$0 \neq g(x_1) = f(x_1, b) \in L(b)[x_1] = F(S \cup \{b\})[x_1],$$

with $g(a) = 0$, so $a \in \mathfrak{a}(S \cup \{b\})$.

Theorem 5.3. If A and B are transcendence bases for K over F, then $|A| = |B|$.

Proof. We may assume that $0 < |A| \leq |B|$. Suppose first that $A = \{a_1, a_2, \ldots, a_n\}$ is finite. Note that $B \nsubseteq \mathfrak{a}(A\backslash\{a_1\})$, e.g., since $a_1 \in \mathfrak{a}(B)$ but $a_1 \notin \mathfrak{a}(A\backslash\{a_1\})$. Choose $b_1 \in B\backslash\mathfrak{a}(A\backslash\{a_1\})$ and observe that $A_1 = \{b_1, a_2, \ldots, a_n\}$ is a transcendence set by Proposition 5.1. Also $a_1 \in \mathfrak{a}(A_1)$ by Proposition 5.2, and it follows that A_1 is a transcendence basis. Replacing A by A_1 begins an induction. Suppose now that $1 < k \leq n$ and that $b_1, b_2, \ldots, b_{k-1} \in B$ have been chosen so that

$$A_{k-1} = \{b_1, \ldots, b_{k-1}, a_k, \ldots, a_n\}$$

is a transcendence basis. Then $B \nsubseteq \mathfrak{a}(A_{k-1}\backslash\{a_k\})$ and we may choose $b_k \in B\backslash\mathfrak{a}(A_{k-1}\backslash\{a_k\})$. It follows as above that

$$A_k = \{b_1, \ldots, b_k, a_{k+1}, \ldots, a_n\}$$

is a transcendence basis. Take $k = n$, then $A_n \subseteq B$ is a transcendence basis, so $A_n = B$ and $|B| = |A_n| = |A|$.

Suppose then that A is infinite. For each $a \in A$ there is a finite set $B_a \subseteq B$ with $a \in \mathfrak{a}(B_a)$. Clearly $B = \bigcup\{B_a : a \in A\}$, for if $\bigcup\{B_a : a \in A\} = C \neq B$, then $B \subseteq K = \mathfrak{a}(A) \subseteq \mathfrak{a}(C)$, contradicting the algebraic independence of B. Thus

$$|B| = |\bigcup\{B_a : a \in A\}| \leq \aleph_0|A| = |A|,$$

and hence $|B| = |A|$.

The cardinality of any transcendence basis for K over F is called the *transcendence degree* of K over F; it will be denoted by $TD(K:F)$.

Exercise 5.4. Show that $TD(\mathbb{R}:\mathbb{Q}) = TD(\mathbb{C}:\mathbb{Q}) = |\mathbb{R}|$ (see Exercise III.1.4).

Proposition 5.4. If $F \subseteq L \subseteq K$, then $TD(K:F) = TD(K:L) + TD(L:F)$.

Proof. Let A be a transcendence basis for L over F and B a transcendence basis for K over L; clearly $A \cap B = \varnothing$. Then L is algebraic over $F(A)$, so $L(B)$ is algebraic over $F(A)(B) = F(A \cup B)$; hence K, being algebraic over $L(B)$, is algebraic over $F(A \cup B)$ by Proposition III.1.5. It will suffice, then, to show that $A \cup B$ is algebraically independent over F. Suppose $0 \neq f(X, Y) \in F[X, Y]$ gives a dependence relation $f(a_1, \ldots, a_m, b_1, \ldots, b_n) = 0$, with $a_i \in A$, $b_j \in B$. Then $f(a_1, \ldots, a_m, b_1, \ldots, b_n)$ can be viewed as a polynomial evaluated at $b_1, \ldots, b_n$ and having coefficients of the form

$$g(a_1, \ldots, a_m) \in F[a_1, \ldots, a_m] \subseteq F(A).$$

All the coefficients must be 0 since B is a transcendence set over $L \supseteq F(A)$. But then each such $g(X) = 0$ in $F[X]$ since A is a transcendence set over F, and hence $f(X, Y) = 0$ in $F[X, Y]$. Thus $A \cup B$ is a transcendence basis for K over F, and

$$TD(K:F) = |A \cup B| = |A| + |B| = TD(L:F) + TD(K:L).$$

Exercise 5.5 (D. Gay).

(1) Suppose $S, T \subseteq K$ are transcendence sets over F and $\sigma: S \to T$ is a 1–1 correspondence. Show that there is a (unique) F-isomorphism $\tau: F(S) \to F(T)$ with $\tau | S = \sigma$.

(2) Let B be a transcendence basis for $K = \mathbb{C}$ over $F = \mathbb{Q}$ and choose $A \subseteq B$ with $|A| = |B \backslash A|$. For each $S \subseteq A$ apply (1) to obtain an isomorphism $\tau_S: \mathbb{Q}(B) \to \mathbb{Q}(B \backslash S)$. Extend τ_S to an isomorphism between $\mathfrak{a}(B) = \mathbb{C}$ and $\mathfrak{a}(B \backslash S)$ (see Theorem III.1.11).

(3) Conclude that $\mathbb{C}$ has $2^{|\mathbb{C}|}$ proper subfields isomorphic with itself, and furthermore $2^{|\mathbb{C}|}$ subfields $L \cong \mathbb{R}$ with $[\mathbb{C}:L]$ infinite and $L \nsubseteq \mathbb{R}$ (see Exercise III.8.13 and note that there is no monomorphism $\theta: \mathbb{R} \to \mathbb{R}$ except $\theta = 1_{\mathbb{R}}$).

If K is a purely transcendental extension of F with $TD(K:F) = 1$ we may as well take $K = F(x)$, the field of rational functions in an indeterminate x over F. The transcendence basis most ready at hand is $B = \{x\}$. If $\alpha \in K^*$ write $\alpha = f(x)/g(x)$, with $f(x)$ and $g(x)$ relatively prime in $F[x]$, and define the *degree* of α to be $\deg \alpha = \max\{\deg f(x), \deg g(x)\}$.

Proposition 5.5. If $K = F(x)$ and $\alpha \in K \backslash F$, then α is transcendental over F, and $[K:F(\alpha)] = \deg \alpha$.

Proof. Say $\deg \alpha = n > 0$, and write $\alpha = f(x)/g(x)$, with

$$f(x) = a_0 + a_1 x + \cdots + a_n x^n,$$
$$g(x) = b_0 + b_1 x + \cdots + b_n x^n$$

in $F[x]$, and at least one of a_n, b_n nonzero. Let y be an indeterminate over K and set

$$h(y) = h_\alpha(y) = \alpha g(y) - f(y) \in F[\alpha][y] \subseteq K[y].$$

Then $h(y)$ has leading coefficient $\alpha b_n - a_n$, so $\deg h(y) = n$, and $h(x) = 0$, so x is algebraic of degree n or less over $F(\alpha)$. Thus α must be transcendental over F, for otherwise x would be algebraic over F. If $h_\alpha(y)$ were reducible in $F[\alpha][y]$ it would also be reducible in $F[\alpha][y] = F[\alpha, y]$, by Proposition II.5.15. Since $h_\alpha(y)$ is of degree 1 in α any factorization would take the form $h_\alpha(y) = u(y)v(\alpha, y)$, where $u(y)$ is of degree 0 in α, i.e., $u(y) \in F[y]$, and $v(\alpha, y) \in F[\alpha, y]$ is of degree 1 in α. If we apply the homomorphism $\phi: F[\alpha, y] \to F[y]$ determined by $\phi(\alpha) = 0$, $\phi(y) = y$, to the equation $\alpha g(y) - f(y) = u(y)v(\alpha, y)$ we obtain $f(y) = -u(y)v(0, y)$, so $u(y) \mid f(y)$ in $F[y]$. But then also

$$u(y) \mid \alpha g(y) = h_\alpha(y) + f(y),$$

and consequently $u(y) \mid g(y)$. Since $(f(y), g(y)) = 1$ in $F[y]$ we conclude that $\deg u(y) = 0$ and $h_\alpha(y)$ is irreducible over $F(\alpha)$. Since $K = F(x) = F(\alpha)(x)$ we conclude further that $[K:F(\alpha)] = n = \deg \alpha$.

Corollary 1. The minimal polynomial $m(y)$ for x over $F(\alpha)$ is an $F(\alpha)$-multiple of $\alpha g(y) - f(y)$.

Corollary 2. If $K = F(x)$ and $\alpha \in K \setminus F$, then $K = F(\alpha)$ if and only if $\deg \alpha = 1$, i.e., $\alpha = (ax + b)/(cx + d)$, with $a, b, c, d \in F$ and $ad \neq bc$.

Proof. Since $[K:F(\alpha)] = \deg \alpha$ we have $K = F(\alpha)$ if and only if $\deg \alpha = 1$, hence $\alpha = (ax + b)/(cx + d)$, as indicated. If $ad = bc$, then either $\alpha = a/c$ or $\alpha = b/d$, both in F.

The next theorem requires a bit of information from the beginning of Section 3, above.

Theorem 5.6. If $K = F(x)$, with x transcendental over F, then the Galois group $G = G(K:F)$ is isomorphic with the projective general linear group $\mathrm{PGL}(2, F)$.

Proof. Any $\phi \in G$ must carry the primitive element x to another primitive element α, hence $\alpha(x) = (ax + b)/(cx + d)$ for some $a, b, c, d \in F$ with $ad \neq bc$, by Corollary 2 above. Conversely, defining $\phi(x) = (ax + b)/(cx + d)$ completely determines $\phi \in G$ since $K = F(x)$. Define a map $f: \mathrm{GL}(2, F) \to G$ via

$f: \left[\begin{smallmatrix} a & b \\ c & d \end{smallmatrix}\right] \mapsto \phi$, with $\phi(x) = (ax + b)/(cx + d)$. It is a routine matter to check that f is an epimorphism. If $A = \left[\begin{smallmatrix} a & b \\ c & d \end{smallmatrix}\right] \in \mathrm{GL}(2, F)$ and $f(A) = 1 \in G$, then $(ax + b)/(cx + d) = x$, or $ax + b = cx^2 + dx$, so $b = c = 0$ and $a = d$. Thus

$$\ker f = \{a1 : a \in F^*\} = Z[\mathrm{GL}(2, F)]$$

and

$$G \cong \mathrm{GL}(2, F)/Z[\mathrm{GL}(2, F)] = \mathrm{PGL}(2, F).$$

It will be convenient for the proof of the next theorem to introduce some temporary notation. If $f(x, y) \in F[x, y]$ is viewed as a polynomial in x with coefficients in $F[y]$ we will write $f(x, y)$ as $f_y(x)$, and similarly $f_x(y)$ will denote $f(x, y)$ viewed as polynomial in y with coefficients in $F[x]$. Thus, for example, if

$$f(x, y) = xy - xy^3 + x^3y + x^4y^2,$$

then

$$f_y(x) = (y - y^3)x + yx^3 + y^2x^4 \quad \text{and} \quad \deg f_y(x) = 4;$$
$$f_x(y) = (x + x^3)y + x^4y^2 - xy^3 \quad \text{and} \quad \deg f_x(y) = 3.$$

Similar remarks apply to rational functions $g(x, y) \in F(x, y)$.

Theorem 5.7 (Luroth's Theorem). Suppose $K = F(x)$, with x transcendental over F, and $L \neq F$ is an intermediate field, $F \subseteq L \subseteq K$. Then $L = F(\gamma)$ for some $\gamma \in K$ that is transcendental over F.

Proof. If $\beta \in L \backslash F$, then x is algebraic over $F(\beta) \subseteq L$ by Proposition 5.5, so x is algebraic over L. Let

$$m_x(y) = a_0 + a_1 y + \cdots + y^n$$

be the minimal polynomial of x over L, so $[K:L] = n$ [since $K = L(x)$]. At least one of the coefficients $a_i = a_i(x)$ is not in F since x is transcendental over F, say $a_i = \gamma \in L \backslash F$. Write $\gamma = f(x)/g(x)$ with $f(x)$ and $g(x)$ relatively prime in $F[x]$, and say $\deg \gamma = k$. Then $[K:F(\gamma)] = k$ by Proposition 5.5, and $k \geq n$ since $F(\gamma) \subseteq L \subseteq K$. The theorem will be proved if we show that $k = n$. We may assume that all coefficients $a_j = a_j(x)$ are reduced to lowest terms in $F(x)$; then multiply through $m_x(y)$ by the least common multiple $b_n = b_n(x) \in F[x]$ of their denominators and cancel the denominators. Thus we replace $m_x(y)$ by

$$u_x(y) = b_0 + b_1 x + \cdots + b_n x^n,$$

with each $b_j = b_j(x) \in F[x]$ and hence $u(x, y) \in F[x, y]$. It is easily verified that $u_x(y)$ is primitive over $F[x]$. Since $\gamma = a_i = b_i/b_n = f(x)/g(x)$ it is clear that $\deg u_y(x) \geq k$.

If we set $h_x(y) = \gamma g(y) - f(y) \in L[y]$, then $h_x(x) = 0$, so $m_x(y) \mid h_x(y)$ in $L[y]$, say

$$m_x(y)p_x(y) = \gamma g(y) - f(y) = [f(x)/g(x)]g(y) - f(y)$$

with $p_x(y) \in L[y]$. Set

$$r(x, y) = f(x)g(y) - f(y)g(x) \in F[x, y]$$

and note that $\deg r_x(y) = deg\, r_y(x) = k$. Also

$$m_x(y)p_x(y)g(x) = f(x)g(y) - f(y)g(x) = r_x(y), \qquad (*)$$

so if the left-hand side is viewed as a polynomial in y over $F(x)$, then all denominators of coefficients must cancel and, since $u_x(y)$ is primitive, we may rewrite (*) as $u(x, y)q(x, y) = r(x, y)$ for some $q(x, y) \in F[x, y]$. Observe now that

$$k = \deg r_y(x) = \deg u_y(x) + \deg q_y(x) \geq k + \deg q_y(x),$$

so $\deg q_y(x) = 0$ and $q(x, y) = q(y) \in F[y]$ (also $\deg u_y(x) = k$). Since $q(y)$ is primitive (its nonzero coefficients are units in $F[x]$) we have $u_x(y)q(y)$ primitive by Gauss's Lemma (Theorem II.5.13), and hence $r_x(y) = u_x(y)q(y)$ is primitive over $F[x]$. But $r(x, y) = -r(y, x)$ so it follows that $r_y(x) = u_y(x)q(y)$ is primitive over $F[y]$ and consequently $q(y)$ must be a constant, $q(y) = q \in F^*$. Finally, then, $n = \deg u_x(y) = \deg r_x(y) = k$.

Luroth's Theorem is important in the study of algebraic curves. Loosely speaking its significance there is as follows: If an irreducible curve can be parametrized (finitely many points excepted) by rational functions then it can be so parametrized in a 1–1 fashion (see Walker [38, Chapter V, Section 7]).

There is an analog of Luroth's Theorem for purely transcendental extensions $K = F(x, y)$ of transcendence degree 2, due to Castelnuovo and Zariski. It requires, however, that F be algebraically closed and applies only to intermediate fields L such that K is separable over L. Beyond a few counterexamples very little is known for transcendence degree 3 or higher.

Suppose $F = F_q$, a finite field, and set $K = F(x)$, x an indeterminate. Then the Galois group $G = G(K:F)$ is the group of linear fractional transformations $(ax + b)/(cx + d)$ as in the proof of Theorem 5.6; $G \cong \mathrm{PGL}(2, F)$ and $|G| = (q - 1)q(q + 1) = q^3 - q$ by Proposition 3.1. Set $L = \mathscr{F}G$, the fixed field of G in K. Then K is Galois over L by definition, and $[K:L] = |G| = q^3 - q$ by the Fundamental Theorem of Galois Theory (III.2.10).

By Luroth's Theorem L is of the form $F(\gamma)$ for some $\gamma \in K$ that is transcendental over F. In order to determine a choice for γ it will be convenient to describe first a set of generators for G. For each $a \in F^*$ define $\rho_a \in G$ by setting $\rho_a(x) = ax$, and for each $b \in F$ define $\tau_b \in G$ by setting $\tau_b(x) = x + a$. Define $\sigma \in G$ by means of $\sigma(x) = 1/x$.

Exercise 5.6. Suppose $\phi \in G$.

(1) If $\phi(x) = ax + b$ show that $\phi = \tau_b \rho_a$.

(2) If $\phi(x) = (ax + b)/(cx + d)$, with $c \neq 0$, show that $\phi = \tau_r \rho_s \sigma \tau_d \rho_c$, where $r = a/c$ and $s = (bc - ad)/c$.

(3) Conclude that $G = \langle \rho_a, \tau_b, \sigma : a \in F^*, b \in F \rangle$.

Now set

$$\gamma = (x^q - x)^{q^2+1}/(x^{q^2} - x)^{q+1} \in F(x) = K.$$

Recall that

$$x^q - x = \prod\{x - a : a \in F\},$$

and if F_1 is a degree 2 extension of F, then $|F_1| = q^2$ and $x^{q^2} - x = \prod\{x - a : a \in F_1\}$ (see Theorem III.3.11). Thus if we divide $x^{q^2} - x$ by $x^q - x$, then the quotient, which is easily checked to be $\sum_{i=0}^{q} x^{i(q-1)}$, has as roots the elements of $F_1 \backslash F$, so it is relatively prime to $x^q - x$. Thus we may write

$$\gamma = (x^q - x)^{q^2-q} \bigg/ \left(\sum_{i=0}^{q} x^{i(q-1)}\right)^{q+1}$$

in lowest terms, and $\deg \gamma = q^3 - q = |G|$. Thus $[K:L] = [K:F(\gamma)]$ by Proposition 5.5, so it will suffice to show that $\gamma \in L$ in order to conclude that $L = F(\gamma)$.

If $b \in F$, then

$$\tau_b(x^q - x) = (x + b)^q - (x + b) = x^q - x + b^q - b = x^q - x,$$

and likewise $\tau_b(x^{q^2} - x) = x^{q^2} - x$, so $\tau_b(\gamma) = \gamma$. If $a \in F^*$ then

$$\rho_a(x^q - x) = a^q x^q - ax = ax^q - ax = a(x^q - x),$$

and

$$\rho_a(x^{q^2} - x) = a(x^{q^2} - x),$$

so $\rho_a(\gamma) = \gamma$. Finally,

$$\sigma(x^q - x) = 1/x^q - 1/x = (x - x^q)/x^{q+1}$$

and

$$\sigma(x^{q^2} - x) = (x - x^{q^2})/x^{q^2+1},$$

so $\sigma(\gamma) = \gamma$. Thus $\phi(\gamma) = \gamma$ for all $\phi \in G$ by Exercise 5.6(3), so $\gamma \in L = \mathscr{F}G$ and $L = F(\gamma)$.

Exercise 5.7. Take $F = \mathbb{Z}_2$, $K = F(x)$, $G = G(K:F)$, and $L = \mathscr{F}G$, so $L = F(\gamma)$, with

$$\gamma = (x^2 - x)^5/(x^4 - x)^3 = (x^2 - x)^2/(1 + x + x^2)^3.$$

Exhibit primitive elements (over F) for each of the six intermediate fields M, $L \subseteq M \subseteq K$.

6. VALUATIONS AND p-ADIC NUMBERS

A *Diophantine equation* is not actually a specific type of equation, it is rather an equation for which only certain types of solutions are of interest. For example, if

$$f(x_1, x_2, \ldots, x_k) \in \mathbb{Z}[x_1, x_2, \ldots, x_k],$$

then the equation $f(x_1, \ldots, x_k) = 0$ is called a Diophantine equation if the only solutions of interest are solutions $x_1 = a_1, \ldots, x_k = a_k$, with all $a_i \in \mathbb{Z}$.

For a concrete example consider $x^2 - 7y^2 - 11 = 0$. Viewed as an equation involving real variables x, y its graph is a hyperbola and the equation has infinitely many solutions. The corresponding Diophantine problem asks whether there are any points on that hyperbola having integers for both coordinates. That might appear to be a more difficult question. If there were an integral solution, however, it would also be a solution to the congruence $x^2 - 7y^2 \equiv 11 \pmod 4$, or equivalently $x^2 + y^2 \equiv 3 \pmod 4$. Since the square of any integer is $\equiv 0$ or $1 \pmod 4$ the congruence has no solution in $\mathbb{Z}$ and, a priori, the Diophantine equation has no solutions.

The example suggests some necessary conditions for the existence of solutions to certain Diophantine equations: if $f(x_1, \ldots, x_k) = 0$ is to have integral solutions it is necessary that the congruences $f(x_1, \ldots, x_k) \equiv 0 \pmod n$ have solutions for all moduli n.

Another necessary condition is even more obvious: if the equation has integral solutions it has real solutions. For example, $x^2 + y^2 + 1 = 0$ has no solutions in integers.

Surprisingly, perhaps, there are some equations for which the necessary conditions above are, collectively, also sufficient. An important example is afforded by pure quadratic equations

$$f(x_1, \ldots, x_k) = \sum_{i,j} a_{ij} x_i x_j = 0,$$

with all $a_{ij} \in \mathbb{Z}$. If that equation has a nontrivial real solution and if there is a nontrivial solution to $f(x_1, \ldots, x_k) \equiv 0 \pmod{p^m}$ for all prime powers p^m $(0 < m \in \mathbb{Z})$, then there is a nontrivial solution in integers. When the necessary conditions are also sufficient it is commonly said that the "Hasse Principle" applies.

Around 1900 K. Hensel constructed a field $\hat{\mathbb{Q}}_p \supseteq \mathbb{Q}$ for each prime $p \in \mathbb{Z}$, the field of *p-adic numbers*. Each $\hat{\mathbb{Q}}_p$ has a subring $\hat{R}_p \supseteq \mathbb{Z}$ of *p-adic integers*, with the following property. If $f(X) \in \mathbb{Z}[x_1, \ldots, x_k]$, then $f(X) = 0$ has a

p-adic integer solution if and only if all the congruences $f(X) \equiv 0 \pmod{p^m}$, $0 < m \in \mathbb{Z}$, have solutions. Furthermore, each $\hat{\mathbb{Q}}_p$ is a complete metric space, as is $\mathbb{R} \supseteq \mathbb{Q}$, and methods of analysis are available.

It is convenient to define the p-adic fields, and many other important fields, in terms of completions of fields with valuations.

A function ϕ from a field F to the real field $\mathbb{R}$ is called a *valuation* of F if

(i) $\phi(ab) = \phi(a)\phi(b)$ for all $a, b \in F$,
(ii) $\phi(0) = 0$, $\phi(a) > 0$ for all $a \in F^*$, and
(iii) there exists $c \geq 1$ in $\mathbb{R}$ such that if $\phi(a) \leq 1$, $a \in F$, then $\phi(a + 1) \leq c$.

Note that ϕ: $F^* \to \mathbb{R}^*$ is a homomorphism. If $\phi(a) = 1$ for all $a \in F^*$, then ϕ is called the *trivial valuation* on F.

Terminology pertaining to valuations is not universally agreed upon. A valuation as defined above is often call a *real*, or *rank* 1, valuation, and sometimes is called an *absolute value*. There is a more general notion of valuation, in which the positive real numbers are replaced by any ordered group.

Exercise 6.1. If ϕ is a valuation on F and $a^k = 1$ in F, $0 < k \in \mathbb{Z}$, show that $\phi(a) = 1$. Conclude that if F is a finite field, then F can have only the trivial valuation.

Probably the most familiar example of a nontrivial valuation is the ordinary absolute value function ϕ_∞ on $F = \mathbb{Q}$ (or on any $F \subseteq \mathbb{C}$), i.e., $\phi_\infty(a) = |a|$, all $a \in F$.

For another class of examples take $F = \mathbb{Q}$ and let $p \in \mathbb{Z}$ be a fixed (positive) prime. If $r \in \mathbb{Q}^*$ write $r = p^{v_p(r)}a/b$, with $v_p(r) \in \mathbb{Z}$, $a, b \in \mathbb{Z}$, and $(a, p) = (b, p) = 1$ (note that

$$r = \pm\prod\{p^{v_p(r)}\text{: all primes } p\}).$$

Define $\phi_p(r) = p^{-v_p(r)}$, and, of course, $\phi_p(0) = 0$. Then ϕ_p is a valuation, called the *p-adic valuation* of $\mathbb{Q}$.

Exercise 6.2. Show that each p-adic valuation ϕ_p on $\mathbb{Q}$ is in fact a valuation, and that the constant c in (iii) of the definition can be taken equal to 1.

For another example suppose K is a field and let $F = K(x)$, the field of rational functions over K. Suppose $\alpha = f(x)/g(x) \in F$, with $f(x), g(x) \in K[x]$. If $\deg f(x) = m$ and $\deg g(x) = n$ define $\phi(\alpha) = 2^{m-n}$. Then ϕ is a valuation on F. Further examples can be obtained by fixing an irreducible (i.e., prime) $p(x) \in K[x]$ and proceeding by analogy with the definition of the p-adic valuation on $\mathbb{Q}$.

If ϕ_1, ϕ_2 are nontrivial valuations on F we say that ϕ_1 is *equivalent* with ϕ_2, and write $\phi_1 \sim \phi_2$, if $\phi_2(a) < 1$ for all $a \in F$ with $\phi_1(a) < 1$. Note that if ϕ_1 is a nontrivial valuation and $0 < r \in \mathbb{R}$, and if we define $\phi_2: F \to \mathbb{R}$ by means of $\phi_2(a) = \phi_1(a)^r$, i.e., $\phi_2 = \phi_1^r$, then ϕ_2 is also a valuation and $\phi_1 \sim \phi_2$. We agree that the trivial valuation is equivalent only to itself.

Proposition 6.1. Suppose ϕ_1, ϕ_2 are valuations on F, $\phi_1 \sim \phi_2$, and $a \in F$. Then

$$\phi_1(a) < 1 \quad \text{if and only if} \quad \phi_2(a) < 1,$$
$$\phi_1(a) = 1 \quad \text{if and only if} \quad \phi_2(a) = 1,$$

and

$$\phi_1(a) > 1 \quad \text{if and only if} \quad \phi_2(a) > 1.$$

Proof. We may assume that ϕ_1 and ϕ_2 are nontrivial. Note that $\phi_i(a) < 1$ ($a \in F^*$) if and only if $\phi_i(a^{-1}) > 1$. A moment's thought shows that it will suffice to show that if $\phi_1(a) = 1$, then $\phi_2(a) = 1$. Choose $b \in F^*$ with $\phi_1(b) < 1$. If $0 < n \in \mathbb{Z}$, then $\phi_1(a^n b) = \phi_1(b) < 1$, so

$$\phi_2(a^n b) = \phi_2(a)^n \phi_2(b) < 1,$$

or $\phi_2(a) < \phi_2(b)^{-1/n}$. Thus

$$\phi_2(a) \leq \lim_{n \to \infty} \phi_2(b)^{-1/n} = 1.$$

By the same argument $\phi_2(a^{-1}) \leq 1$, so $\phi_2(a) \geq 1$ and hence $\phi_2(a) = 1$.

Corollary. Equivalence of valuations is an equivalence relation, and the valuations on F are partitioned into equivalence classes.

Proposition 6.2. If ϕ_1, ϕ_2 are valuations on F and $\phi_1 \sim \phi_2$, then $\phi_2 = \phi_1^r$ for some positive $r \in \mathbb{R}$.

Proof. If both are trivial take $r = 1$. Otherwise take $a \in F^*$ with $\phi_1(a) < 1$, so also $\phi_2(a) < 1$. Set $r = \log \phi_2(a)/\log \phi_1(a) \in \mathbb{R}$ so $\phi_2(a) = \phi_1(a)^r$. Take any $b \in F^*$ with $\phi_1(b) < 1$, and choose m, n positive in $\mathbb{Z}$ with $m/n > \log \phi_1(b)/\log \phi_1(a)$. Then $\log \phi_1(a^m) < \log \phi_1(b^n)$, so $\phi_1(a^m) < \phi_1(b^n)$. It follows that $\phi_2(a^m) < \phi_2(b^n)$ and, reversing the argument, that $m/n > \log \phi_2(b)/\log \phi_2(a)$. Since m and n are arbitrary it follows that

$$\log \phi_1(b)/\log \phi_1(a) \geq \log \phi_2(b)/\log \phi_2(a).$$

The reverse inequality follows symmetrically, so they are equal and hence

$$\log \phi_2(b)/\log \phi_1(b) = \log \phi_2(a)/\log \phi_1(a) = r.$$

Thus $\phi_2 = \phi_1^r$.

A valuation ϕ on F satisfies the *triangle inequality* if $\phi(a + b) \leq \phi(a) + \phi(b)$ for all $a, b \in F$.

Proposition 6.3. Suppose ϕ is a valuation on F whose constant c [(iii) in the definition] can be taken to be 2, i.e., if $\phi(a) \leq 1$, then $\phi(a + 1) \leq 2$. Then ϕ satisfies the triangle inequality.

Proof. Take $a, b \in F$, say with $\phi(a) \leq \phi(b) \neq 0$. Then $\phi(a/b) \leq 1$ so

$$\phi(a/b + 1) = \phi[(a + b)/b] \leq 2$$

or

$$\phi(a + b) \leq 2\phi(b) = 2\max\{\phi(a), \phi(b)\}.$$

An easy induction gives

$$\phi\left(\sum_{i=1}^{2^k} a_i\right) \leq 2^k \max\{\phi(a_i): 1 \leq i \leq 2^k\}$$

for all $a_i \in F$ if $1 \leq k \in \mathbb{Z}$. If $1 \leq n \in \mathbb{Z}$ choose $k \in \mathbb{Z}$ with $n \leq 2^k < 2n$. Then for any $a_1, \ldots, a_n \in F$ we may add enough 0's to $\sum_{i=1}^n a_i$ to have 2^k summands and conclude that

$$\phi\left(\sum_{i=1}^{n} a_i\right) \leq 2^k \max\{\phi(a_i): 1 \leq i \leq n\} \leq 2n \max\{\phi(a_i): 1 \leq i \leq n\}.$$

As a special case we see that $\phi(n \cdot 1) \leq 2n$ for all positive $n \in \mathbb{Z}$. Furthermore,

$$\phi\left(\sum_{i=1}^{n} a_i\right) \leq 2n \sum_{i=1}^{n} \phi(a_i), \qquad \text{all} \quad a_i \in F.$$

Apply that inequality and the Binomial Theorem to obtain

$$\begin{aligned}\phi((a + b)^n) = \phi\left(\sum_{i=0}^{n} \binom{n}{i} a^{n-i} b^i\right) &\leq 2(n + 1) \sum_{i=0}^{n} \phi\binom{n}{i} \phi(a)^{n-i} \phi(b)^i \\ &\leq 4(n + 1) \sum_{i=0}^{n} \binom{n}{i} \phi(a)^{n-i} \phi(b)^i = 4(n + 1)[\phi(a) + \phi(b)]^n.\end{aligned}$$

Thus

$$\phi(a + b) \leq (4n + 4)^{1/n}[\phi(a) + \phi(b)].$$

Since $\lim_{n \to \infty} (4n + 4)^{1/n} = 1$ we conclude that $\phi(a + b) \leq \phi(a) + \phi(b)$.

Corollary. Every valuation ϕ_1 is equivalent with a valuation ϕ_2 that satisfies the triangle inequality.

Proof. Suppose ϕ_1 has constant $c > 2$, set $r = \log 2/\log c$, and set $\phi_2 = \phi_1^r \sim \phi_1$. Then if $\phi_2(a) \leq 1$ we have $\phi_2(a + 1) = \phi_1(a + 1)^r \leq c^r = 2$, and ϕ_2 satisfies the triangle inequality.

If the constant c in the definition of a valuation can be taken to be 1, then the triangle inequality can be strengthened considerably. If $\phi(a+1) \leq 1$ whenever $\phi(a) \leq 1$, then ϕ is called a *nonarchimedean* valuation. Otherwise ϕ is *archimedean*. Note that the ordinary absolute value ϕ_∞ on $\mathbb{Q}$ is archimedean.

Exercise 6.3. Show that the p-adic valuations ϕ_p on $\mathbb{Q}$ are all nonarchimedean.

Proposition 6.4. Suppose ϕ is a valuation on F. Then the following are equivalent:

(a) ϕ is nonarchimedean;
(b) $\{\phi(n \cdot 1): n \in \mathbb{Z}\}$ is bounded in $\mathbb{R}$;
(c) $\phi(a+b) \leq \max\{\phi(a), \phi(b)\}$, all $a, b \in F$.

Proof. (a)$\Rightarrow$(b): Since $\phi(1) = 1$ we have $\phi(2 \cdot 1) = \phi(1+1) \leq 1$, and inductively $\phi(n \cdot 1) \leq 1$ if $1 \leq n \in \mathbb{Z}$. Since $\phi(-n \cdot 1) = \phi(n \cdot 1)$ we conclude $\phi(n \cdot 1) \leq 1$ for all $n \in \mathbb{Z}$.

(b)$\Rightarrow$(c): By the corollary to Proposition 6.3 we may assume that ϕ satisfies the triangle inequality. Say $\phi(n \cdot 1) \leq M$ for all $n \in \mathbb{Z}$ and take $a, b \in F$. If $0 < n \in \mathbb{Z}$, then

$$\phi(a+b)^n = \phi\left(\sum_{k=0}^{n} \binom{n}{k} a^{n-k} b^k\right) \leq M \sum_{k=0}^{n} \phi(a)^{n-k}\phi(b)^k$$
$$\leq M(n+1)\max\{\phi(a)^n, \phi(b)^n\},$$

and so

$$\phi(a+b) \leq [M(n+1)]^{1/n} \max\{\phi(a), \phi(b)\}.$$

Since

$$\lim_{n\to\infty} [M(n+1)]^{1/n} = 1$$

we conclude that $\phi(a+b) \leq \max\{\phi(a), \phi(b)\}$.

(c)$\Rightarrow$(a): If $a \in F$ and $\phi(a) \leq 1$, then

$$\phi(a+1) \leq \max\{\phi(a), \phi(1)\} = 1$$

so ϕ is nonarchimedean.

Corollary. If $\operatorname{char} F = p$ and ϕ is a valuation on F, then ϕ is nonarchimedean.

Remark: The proof of Proposition 6.4 shows that if ϕ is nonarchimedean, then $\phi(n \cdot 1) \leq 1$ for all $n \in \mathbb{Z}$.

Exercise 6.4. If ϕ is a nonarchimedean valuation on $F, a, b \in F$, and $\phi(a) < \phi(b)$ show that $\phi(a+b) = \phi(b)$. Conclude that if $\phi(a_i) \leq \phi(a_1)$

for $2 \leq i \leq k$ but $\phi(a_1 + a_2 + \cdots + a_k) < \phi(a_1)$, then $\phi(a_i) = \phi(a_1)$ for some $i \geq 2$.

The next two propositions, due to A. Ostrowski, characterize (up to equivalence) all valuations on $\mathbb{Q}$.

Proposition 6.5. If ϕ is an archimedean valuation on $\mathbb{Q}$, then $\phi \sim \phi_\infty$, the ordinary absolute value.

Proof. (Artin [4]). We may assume that ϕ satisfies the triangle inequality. Take $m, n > 1$ in $\mathbb{Z}$. It is an easy consequence of the division algorithm that we may write m as a polynomial in n, $m = a_0 + a_1 n + \cdots + a_k n^k$, with $a_i \in \mathbb{Z}$, $0 \leq a_i < n$, and $a_k \neq 0$. Thus

$$\phi(m) \leq \sum_{i=0}^{k} a_i \phi(n^i) < n \sum_{i=0}^{k} \phi(n^i) \leq n(k+1)\max\{1, \phi(n)^k\}$$

(consider the two possibilities $\phi(n) \leq 1$ and $\phi(n) > 1$). Note that $k \leq \log m/\log n$ since $m \geq n^k$. Thus

$$\phi(m) < n(\log m/\log n + 1) \cdot \max\{1, \phi(n)^{\log m/\log n}\}.$$

For any positive integer j we may replace m, above, by m^j and conclude that

$$\phi(m)^j < n(j\log m/\log n + 1)\max\{1, \phi(n)^{j\log m/\log n}\},$$

and hence

$$\phi(m) < [n(j\log m/\log n + 1)]^{1/j}\max\{1, \phi(n)^{\log m/\log n}\}.$$

Since $\lim_{j\to\infty}(j\alpha + \beta)^{1/j} = 1$ for any $\alpha, \beta \in \mathbb{R}$ with $\alpha > 0$, we conclude that

$$\phi(m) \leq \max\{1, \phi(n)^{\log m/\log n}\}.$$

Since ϕ is archimedean we may assume that $\phi(m) > 1$, so $1 < \phi(m) \leq \phi(n)^{\log m/\log n}$. It follows that $\phi(n) > 1$ for all $n > 1$ in $\mathbb{Z}$. We may thus reverse the roles of m and n above and conclude that

$$\phi(n) \leq \phi(m)^{\log n/\log m} \leq \phi(n),$$

and consequently

$$\phi(n)^{1/\log n} = \phi(m)^{1/\log m} = e^r$$

for some positive $r \in \mathbb{R}$, independent of the choice of $m, n > 1$ in $\mathbb{Z}$. Taking logarithms we see that $\log \phi(n) = r\log n = \log n^r$, and hence $\phi(n) = n^r = |n|^r$. Since $\phi(-a) = \phi(a)$ and $\phi(a/b) = \phi(a)/\phi(b)$ it follows that $\phi = \phi_\infty^r \sim \phi_\infty$.

Proposition 6.6. If ϕ is a nontrivial nonarchimedean valuation on $\mathbb{Q}$, then $\phi \sim \phi_p$ for some prime p.

Proof. Set $P = \{a \in \mathbb{Z}: \phi(a) < 1\}$, $P \neq \{0\}$ since ϕ is nontrivial. If $a, b \in P$ and $n \in \mathbb{Z}$, then

$$\phi(a - b) \leq \max\{\phi(a), \phi(b)\} < 1$$

and

$$\phi(na) = \phi(n)\phi(a) \leq \phi(a) < 1,$$

so P is an ideal in $\mathbb{Z}$. If $m, n \in \mathbb{Z}\backslash P$, then $\phi(mn) = \phi(m)\phi(n) \geq 1$, so $mn \notin P$ and P is a prime ideal. Thus $P = (p)$ for some prime $p \in \mathbb{Z}$. In particular $\phi(p) < 1$, so $\phi(p) = p^{-r}$ for some $r > 0$ in $\mathbb{R}$. If $n \in \mathbb{Z}^*$ write $n = p^k m$, $p \nmid m$. Then $m \notin P$, so $\phi(m) = 1$ and $\phi(n) = p^{-kr} = \phi_p(n)^r$. It follows that $\phi = \phi_p^r \sim \phi_p$.

If ϕ is a valuation on a field F, then ϕ determines a topology on F if we take as a neighborhood basis at each $a \in F$ the sets

$$B_\varepsilon(a) = \{b \in F: \phi(b - a) < \varepsilon\}, \qquad 0 < \varepsilon \in \mathbb{R}.$$

It is clear from Proposition 6.2 that equivalent valuations determine the same topology. Since we may assume, by Proposition 6.3, corollary, that ϕ satisfies the triangle inequality, it follows easily that F is in fact a metric space, with metric $\delta = \delta_\phi$ defined by $\delta(a, b) = \phi(a - b)$.

As usual a sequence $a_1, a_2, a_3, \ldots$ in F *converges* to $a \in F$ if $\delta(a_n, a) = \phi(a_n - a) \to 0$ as $n \to \infty$. The sequence $\{a_n\}$ is a *Cauchy sequence* if for any $\varepsilon > 0$ in $\mathbb{R}$ there is some $N \in \mathbb{Z}$ such that $\phi(a_m - a_n) < \varepsilon$ if $m, n \geq N$. Clearly a convergent sequence is a Cauchy sequence. If every Cauchy sequence converges in F, then F is called *complete* (relative to ϕ). Note for example that F is complete if ϕ is trivial.

Exercise 6.5. Show that a Cauchy sequence $\{a_n\}$ is bounded, i.e., $\phi(a_n) \leq M$ for some $M \in \mathbb{R}$, all n.

If ϕ is a valuation on F, then a *completion* of F is a field extension $K \supseteq F$ with a valuation θ on K satisfying (i) $\theta | F = \phi$, (ii) K is complete, and (iii) F is dense in K, i.e., if $a \in K$ there is a sequence $\{a_n\}$ in F with $\theta(a_n - a) \to 0$. It is a well-known fact from topology that every metric space has a completion; the construction is in fact a bit easier in the present setting.

If ϕ is a valuation on F denote by $\mathscr{C}$ the set of all Cauchy sequences $\{a_n\} \subseteq F$. If $\{a_n\}$, $\{b_n\} \in \mathscr{C}$ define $\{a_n\} + \{b_n\} = \{a_n + b_n\}$ and $\{a_n\}\{b_n\} = \{a_n b_n\}$.

Exercise 6.6. Show that $\mathscr{C}$ is a commutative ring with 1 relative to the operations above (for multiplication see Exercise 6.5).

Let $\mathscr{N} = \{\{a_n\} \in \mathscr{C}: a_n \to 0\}$. It is easily checked that $\mathscr{N}$ is an ideal since Cauchy sequences are bounded. Suppose I is an ideal of $\mathscr{C}$ with $\mathscr{N} \subseteq I$, $I \neq \mathscr{N}$, and choose $\{a_n\} \in I \backslash \mathscr{N}$. Then $\{a_n\}$ is eventually bounded away from

0, say $\phi(a_n) \geq \varepsilon > 0$ for all $n \geq N$. Define $b_i = 1$ for $i < N$ and $b_i = -a_i$ for $i \geq N$, then set $c_i = a_i + b_i$. Clearly $\{c_n\} \in \mathscr{N}$; let us see that $\{b_n\}$ is a unit in $\mathscr{C}$. If $m, n \geq N$, then

$$\phi(1/b_n - 1/b_m) = [1/\phi(a_m a_n)]\phi(a_m - a_n) \leq (1/\varepsilon^2)\phi(a_m - a_n),$$

which tends to 0 as $m, n \to \infty$. Thus $\{1/b_n\} \in \mathscr{C}$ and $\{b_n\} \in U(\mathscr{C})$. Since $\{c_n\} \in \mathscr{N}$ and $\{a_n\} \in I$ we have $\{b_n\} \in I$, so $I = \mathscr{C}$ and $\mathscr{N}$ is a maximal ideal.

As a result $K = \mathscr{C}/\mathscr{N}$ is a field. If $a \in F$ the constant sequence $\{a\}$ is is $\mathscr{C}$, and $a \mapsto \{a\} + \mathscr{N}$ is a monomorphism from F into K, so we may identify F with its image and agree that $F \subseteq K$. If $\{a_n\} \in \mathscr{C}$, then $\{\phi(a_n)\}$ is a Cauchy sequence in $\mathbb{R}$, which is complete, so suppose $\phi(a_n) \to r \in \mathbb{R}$. If $\{a_n\} + \mathscr{N} = \{b_n\} + \mathscr{N}$ in K, then $\{a_n - b_n\} \in \mathscr{N}$ and it follows that $\phi(b_n) \to r$ as well. We may thus define $\theta(\{a_n\} + \mathscr{N}) = r$ and thereby obtain a well-defined function θ from K to $\mathbb{R}$.

Exercise 6.7. Verify that $\theta: K \to \mathbb{R}$, as defined above, is a valuation on K, and that $\theta | F = \phi$.

Theorem 6.7. The field K, above, is a completion of F relative to ϕ.

Proof. It remains to be shown that K is complete and that F is dense in K. We may assume that θ satisfies the triangle inequality. It will be convenient to establish the denseness first. If $\alpha = \{a_n\} + \mathscr{N} \in K$, then for each m we have $\theta(a_m - \alpha) = \lim_{k \to \infty} \phi(a_m - a_k)$, and so

$$\lim_{m \to \infty} \theta(a_m - \alpha) = \lim_{m \to \infty} \lim_{k \to \infty} \phi(a_m - a_k) = 0,$$

since $\{a_n\} \in \mathscr{C}$, i.e., $\{a_n\}$ is a sequence in F that converges to $\alpha \in K$. Next let $\{\alpha_n\}$ be a θ-Cauchy sequence in K, with $\alpha_n = \{a_k^{(n)} + \mathscr{N}\}$, $a_k^{(n)} \in F$, $k = 1, 2, \ldots$. For each n we may choose $b_n \in F$ with $\theta(b_n - \alpha_n) < 2^{-n}$, by the denseness of F in K. Then

$$\phi(b_n - b_m) \leq \phi(b_n - \alpha_n) + \phi(\alpha_n - \alpha_m) + \phi(\alpha_m - b_m),$$

which converges to 0 as $m, n \to \infty$, so $\{b_n\} \in \mathscr{C}$. Set $\beta = \{b_n\} + \mathscr{N} \in K$, and observe that

$$\theta(\alpha_n - \beta) = \theta(\{a_k^{(n)} - b_k\} + \mathscr{N}) = \lim_k \phi(a_k^{(n)} - b_k).$$

Thus

$$\lim_n \theta(\alpha_n - \beta) = \lim_n \lim_k \phi(a_k^{(n)} - b_k) \leq \lim_n \lim_k [\phi(a_k^{(n)} - b_n) + \phi(b_n - b_k)]$$

$$= \lim_n \theta(\alpha_n - b_n) + \lim_n \theta(b_n - \beta).$$

Since $\theta(\alpha_n - b_n) < 2^{-n}$ and $b_n \to \beta$ by the proof above of denseness, we conclude that $\alpha_n \to \beta$ and K is complete.

Exercise 6.8. Show that the completion K is unique in the following sense. If L, with valuation ψ, is another completion of F relative to ϕ, then there is an F-isomorphism $f: K \to L$ with $\theta = \psi f$. (*Hint*: For any Cauchy sequence $\{a_n\} \subseteq F$ define $f(\lim_n a_n) = \lim_n a_n \in L$.)

If K is the completion of F relative to ϕ, then the extended valuation (θ above) is usually also denoted by ϕ.

We remark that the completion of $\mathbb{Q}$ relative to ϕ_∞ is (isomorphic with) $\mathbb{R}$; in fact, the completion process above is in essence one of the standard methods for constructing $\mathbb{R}$ from $\mathbb{Q}$.

Suppose now that ϕ is a nonarchimedean valuation on a field F. Define

$$R = R_\phi = \{a \in F : \phi(a) \leq 1\},$$
$$P = P_\phi = \{a \in F : \phi(a) < 1\},$$

and

$$U = U_\phi = \{a \in F : \phi(a) = 1\}$$

and note that any valuation equivalent with ϕ determines the same sets.

Since ϕ is nonarchimedean it is immediate that R is a subring of F; it is called the *ring of integers* at ϕ, or the *valuation ring* at ϕ. If we write $R^{-1} = \{r^{-1} : r \in R^*\}$, then $F = R \cup R^{-1}$, and in particular F is the field of fractions of R.

Proposition 6.8. If ϕ is nonarchimedean on F with valuation ring R, then $U = U_\phi$ is the group of units of R and $P = P_\phi$ is the unique maximal ideal in R.

Proof. If $u \in U$, then $\phi(u) = 1$, so $\phi(u^{-1}) = 1$ and $u^{-1} \in R$, i.e., $u \in U(R)$, and conversely. Since ϕ is nonarchimedean it is clear that P is an ideal in R (see the proof of Proposition 6.6) and P is maximal and unique since R is the disjoint union of P and U.

Since P is maximal in R the quotient R/P is a field $\bar{F} = \bar{F}_\phi$, called the *residue class field* of F at ϕ.

Exercise 6.9. If ϕ_p is the p-adic valuation on $\mathbb{Q}$ describe R, P, and U explicitly and show that $\bar{F} \cong \mathbb{Z}_p$.

Proposition 6.9. Suppose ϕ is a nonarchimedean valuation on F and K is a completion of F, with ϕ extended to K. Then F and K have isomorphic residue class fields.

Proof. Write R_F, R_K for the valuation rings in F and in K, etc. Then $R_F = R_K \cap F$ and $P_F = P_K \cap F$. If $a \in R_K$, then by denseness $\phi(a - b) < 1$

for some $b \in F$, so $a - b \in P_K$ and hence $b \in R_K \cap F = R_F$. Thus $a \in b + P_K \subseteq R_F + R_K$, so $R_K \subseteq R_F + P_K$ and hence $R_K = R_F + P_K$. As a result

$$\bar{K} = R_K/P_K = (R_F + P_K)/P_K \cong R_F/R_F \cap P_K = R_F/P_F = \bar{F}.$$

For any valuation ϕ of a field F the image $\phi(F^*)$ is a multiplicative subgroup of $\{r \in \mathbb{R}: r > 0\}$, called the *value group* of ϕ. If the value group is cyclic, then ϕ is called a *discrete* valuation.

If ϕ is discrete, nonarchimedean, and nontrivial, then one of the two generators of the value group is its largest element less than 1, call it r_0. Take $\pi \in F$ with $\phi(\pi) = r_0$, so $\pi \in P$. If $a \in P$, then $\phi(a) \leq \phi(\pi)$, so $\phi(a\pi^{-1}) \leq 1$ and $a\pi^{-1} \in R$, or $a \in R\pi$. It follows that $P = R\pi$ is a principal ideal.

Exercise 6.10. Show that each p-adic valuation ϕ_p on $\mathbb{Q}$ is discrete, and that π can be taken to be p. Show that ϕ_∞ is not discrete.

Proposition 6.10. Suppose ϕ is nonarchimedean on F and K is the ϕ-completion of F. Then the value group of K is equal to that of F. In particular ϕ remains discrete on K if it is discrete on F.

Proof. Take $a \in K^*$. Since F is dense $\phi(a - b) < \phi(a)$ for some $b \in F$. Thus

$$\phi(b) = \phi(a + (b - a)) = \max\{\phi(a), \phi(b - a)\} = \phi(a)$$

by Exercise 6.4, so $\phi(a) \in \phi(F^*)$.

Suppose now that F is a field that is complete relative to a nontrivial nonarchimedean valuation ϕ. As usual an infinite series $\sum_{k=1}^{\infty} a_k$, with each $a_k \in F$, is defined to be the sequence $\{b_n\}$ of partial sums, $b_n = \sum_{k=1}^{n} a_k$. If $b_n \to a \in F$, then a is called the *sum* of the series, and we write $a = \sum_{k=1}^{\infty} a_k$. Since F is complete $\sum_{k=1}^{\infty} a_k$ converges if and only if $\{b_n\}$ is a Cauchy sequence, i.e., $\phi(a_m + \cdots + a_n) \to 0$ as $m, n \to \infty$. Since ϕ is nonarchimedean

$$\phi(a_m + \cdots + a_n) \leq \max\{\phi(a_k): m \leq k \leq n\},$$

so clearly $\sum_{k=1}^{\infty} a_k$ converges if and only if $\phi(a_k) \to 0$ as $k \to \infty$.

Suppose further that ϕ is discrete, with π a generator for the maximal ideal P in the valuation ring R. By Proposition 6.2 we may assume that $\phi(\pi) = p^{-1}$ for some prime $p \in \mathbb{Z}$. Choose a representative element a of R from each element (coset) in the residue class field $\bar{F} = R/P$, taking $a = 0$ in the coset P itself. Denote by Δ the resulting set of coset representatives so each $r \in R$ can be written uniquely as $r = a + b$, with $a \in \Delta$, $b \in P$. Note that $\phi(a) = 1$ for every $a \neq 0$ in Δ, since $a \in R \setminus P = U$.

Theorem 6.11. In the setting above every element $a \in F^*$ has a unique representation as a convergent "Laurent" series $a = \sum_{k=m}^{\infty} a_k \pi^k$, where $m \in \mathbb{Z}$

(possibly $m < 0$), each $a_k \in \Delta$, $a_m \neq 0$, and $\phi(a) = p^{-m}$. Conversely every series of that form converges to an element $a \in F^*$ with $\phi(a) = p^{-m}$. We have

$$R = \left\{ \sum_{k=0}^{\infty} a_k \pi^k : a_k \in \Delta \right\},$$

$$P = \left\{ \sum_{k=1}^{\infty} a_k \pi^k : a_k \in \Delta \right\},$$

and

$$U = \left\{ \sum_{k=0}^{\infty} a_k \pi^k : a_k \in \Delta \text{ and } a_0 \neq 0 \right\}.$$

Proof. We prove the converse first. Any series $\sum_{k=m}^{\infty} a_k \pi^k$, $a_k \in \Delta$, converges since $\phi(a_k \pi^k) = \phi(a_k) p^{-k} \to 0$, so write $a = \sum_{k=m}^{\infty} a_k \pi^k$. Then

$$\phi(a) = \lim_{n \to \infty} \phi\left(\sum_{k=m}^{n} a_k \pi^k \right) \leq \lim_{n \to \infty} \max\{\phi(a_k \pi^k) : m \leq k \leq n\}$$

$$= \lim_{n \to \infty} \max\{p^{-k} : m \leq n \leq k\} = p^{-m}.$$

Since $a_m \neq 0$ we have $a = a_m \pi^m + b$, where $b = \sum_{k=m+1}^{\infty} a_k \pi^k$, so

$$\phi(b) \leq p^{-(m+1)} < \phi(a_m \pi^m) = p^{-m},$$

hence $\phi(a) = p^{-m}$ by Exercise 6.4. Suppose $\sum_{k=m}^{\infty} c_k \pi^k$ is another such series, that $c_k = a_k$ for $m \leq k < n$, but $c_n \neq a_n$. Then $c - a = \sum_{k=n}^{\infty} (c_k - a_k) \pi^k$. Each coefficient $c_k - a_k$ is either 0 (if $c_k = a_k$) or else $c_k - a_k \in R \backslash P = U$, so $\phi(c_k - a_k) = 1$. It follows as above that $\phi(c - a) = p^{-n} \neq 0$, so $c \neq a$ and series representations are unique.

Now let us see inductively that every $a \in F^*$ has such a series representation. Say $\phi(a) = p^{-m}$, $m \in \mathbb{Z}$. Then $\phi(a) = \phi(\pi^m)$, so $\phi(a\pi^{-m}) = 1$ and $a\pi^{-m} \in U$. Choose $a_m \in \Delta$ with $a\pi^{-m} - a_m \in P$, so that

$$\phi(a\pi^{-m} - a_m) = \phi(a - a_m \pi^m) \phi(\pi^{-m}) < 1$$

or

$$\phi(a - a_m \pi^m) < \phi(\pi^m) = p^{-m}.$$

By induction we may write

$$a - a_m \pi^m = \sum_{k=m+1}^{\infty} a_k \pi^k$$

(possibly $a_{m+1} = 0$), and hence $a = \sum_{k=m}^{\infty} a_k \pi^k$. The descriptions of R, P, and U follow immediately.

If $p > 0$ is a prime in $\mathbb{Z}$, then the completion $\hat{\mathbb{Q}}_p$ of $\mathbb{Q}$ relative to the p-adic valuation ϕ_p is called the field of *p-adic numbers*. We will write $\hat{R}_p$ for its valuation ring $R_{\hat{\mathbb{Q}}_p}$ of p-adic integers and write $\hat{U}_p$ for $U(\hat{R}_p)$.

By Exercise 6.9 and Proposition 6.9 the residue class field of $\hat{\mathbb{Q}}_p$ is $\mathbb{Z}_p$, so it is natural to take $\Delta = \Delta_p = \{0, 1, \ldots, p-1\}$. Thus each $a \in \hat{\mathbb{Q}}_p^*$ has a unique expansion as a ϕ_p-convergent series $a = \sum_{k=m}^{\infty} a_k p^k$, with $m \in \mathbb{Z}$, $\phi(a) = p^{-m}$, and $a_k \in \Delta$. That series is called the *canonical representation* of $a \in \hat{\mathbb{Q}}_p^*$. If $m \neq 0$ it is often convenient to "factor out" p^m, relabel the coefficients $(a_k \mapsto a_{k-m})$, and write $a = p^m \sum_{k=0}^{\infty} a_k p^k$, with $u = \sum_{k=0}^{\infty} a_k p^k \in \hat{U}_p$.

Sums and products of p-adic numbers in canonical form can be carried out as usual for convergent series, except that coefficients should be reduced (by division) when necessary to integers between 0 and $p-1$, the quotients being carried as contributions to coefficients of higher powers of p.

In many respects the canonical representation of p-adic numbers is analogous to the decimal representation of real numbers. We may observe, for example, that a p-adic number a is rational if and only if its canonical expansion is ultimately periodic. In order to sketch a proof we may clearly assume that $a \in \hat{U}_p$, since multiplication by p^m will not alter periodicity.

Suppose first that the sequence of coefficients of $a \in \hat{U}_p$ has the form $b_0, \ldots, b_{j-1}, \overline{c_0, \ldots, c_{k-1}}$, the bar denoting periodic repetition. Set $b = \sum_{i=0}^{j-1} b_i p^i \in \mathbb{Q}$ and $c = \sum_{i=0}^{k-1} c_i p^i \in \mathbb{Q}$, so $a = b + p^j c(1 + p^k + p^{2k} + \cdots)$. The geometric series converges since $\phi_p(p^{nk}) = p^{-nk} \underset{n}{\to} 0$, so $a = b + p^j c/(1 - p^k) \in \mathbb{Q}$.

The proof of the converse amounts basically to representing $a \in \mathbb{Q}^*$ in the form just obtained above. Again take a to be a p-adic unit, so $a = d/e$, with d, e prime to p in $\mathbb{Z}$ and $e > 0$. Take k to be the order of p when viewed as an element of $U(\mathbb{Z}_e)$, so $p^k \equiv 1 (\operatorname{mod} e)$, and write $1 - p^k = ef$, $f \in \mathbb{Z}$. Thus $a = g/(1 - p^k)$, where $g = df$ is prime to p. Choose $j \in \mathbb{Z}$, minimal, for which $|g| < p^j$ (one exception: if $g = -1$ take $j = 0$). Let $c \in \mathbb{Z}$ satisfy $cp^j \equiv g(\operatorname{mod} p^k - 1)$ and set $b = (g - cp^j)/(1 - p^k) \in \mathbb{Z}$. If $a > 0$, hence $g < 0$, choose c with $0 \leq c \leq p^k - 2$; if $a < 0$ and $g > 0$ choose c with $1 \leq c \leq p^k - 1$. In either case it follows easily that $0 \leq b < p^j$. Note that $g = (1 - p^k)b + p^j c$, so $a = b + p^j c/(1 - p^k)$. If we write $b = \sum_{i=0}^{j-1} b_i p^i$ and $c = \sum_{j=0}^{k-1} c_i p^i$, canonically, and $1/(1 - p^k) = \sum_{i=0}^{\infty} p^{ki}$ we obtain the canonical form of a and see that its sequence of coefficients is $b_0, \ldots, b_{j-1}, \overline{c_0, \ldots, c_{k-1}}$.

An example may be illuminating. Let us express $a = \frac{2}{15}$ canonically in $\hat{\mathbb{Q}}_7$. The order of $7 \operatorname{mod} 15$ is $k = 4$, and $1 - 7^4 = -2400 = 15(-160)$, so $a = -320/(1 - 7^4)$, $g = -320$. Thus $7^2 < |g| < 7^3 = 343$ and $j = 3$. We may solve $343x \equiv 1(\operatorname{mod} 2400)$ (e.g., by using the Euclidean algorithm) and find that $x_0 = 7$ is a solution. Thus $c = 160 \equiv -320 \cdot 7 (\operatorname{mod} 2400)$ is a solution to $cp^j \equiv g(\operatorname{mod} p^k - 1)$, with $0 \leq c \leq p^k - 2$. We set

$$b = (-320 - 160 \cdot 343)/(-2400) = 23$$

and obtain

$$a = \tfrac{2}{15} = 23 + 7^3 \cdot 160/(1 - 7^4).$$

Write $23 = 2 + 3 \cdot 7$ and $160 = 6 + 1 \cdot 7 + 3 \cdot 7^2$, so

$$\begin{aligned}\tfrac{2}{15} &= (2 + 3 \cdot 7) + 7^3(6 + 1 \cdot 7 + 3 \cdot 7^2)(1 + 7^4 + 7^8 + \cdots)\\ &= 2 + 3 \cdot 7 + 0 \cdot 7^2 + 6 \cdot 7^3 + 1 \cdot 7^4 + 3 \cdot 7^5 + 0 \cdot 7^6 + 6 \cdot 7^7 + \cdots\end{aligned}$$

and the sequence of coefficients is $2, 3, 0, \overline{6, 1, 3, 0}$.

Exercise 6.11. (1) Find the canonical expansion of $a = \frac{2}{15}$ in $\hat{\mathbb{Q}}_{11}$, $\hat{\mathbb{Q}}_{13}$, and $\hat{\mathbb{Q}}_5$.

(2) Suppose $a \in \hat{U}_p$ has sequence $\overline{5, 1}$ of coefficients. Determine a if $p = 7$ and if $p = 11$.

(3) Suppose $0 > a \in \mathbb{Z} \cap \hat{U}_p$. Choose j minimal in $\mathbb{Z}$ with $a + p^j > 0$, and write

$$a + p^j = a_0 + a_1 p + \cdots + a_{j-1} p^{j-1}, \qquad a_i \in \Delta_p.$$

Show that the canonical representation of a in $\hat{\mathbb{Q}}_p$ has coefficient sequence $a_0, a_1, \ldots, a_{j-1}, \overline{p - 1}$.

Proposition 6.12. The ring $\hat{R}_p$ of p-adic integers is compact (with the topology it inherits as a subspace of the metric space $\hat{\mathbb{Q}}_p$).

Proof. For $0 \le k \in \mathbb{Z}$ set $Y_k = \Delta_p = \{0, 1, \ldots, p - 1\}$ with the discrete topology, and set $X = \prod\{Y_k : 0 \le k \in \mathbb{Z}\}$, the Cartesian product with the product topology. Then X is compact by Tychonov's Theorem (e.g., see Kelley [21]). Define f: $X \to \hat{R}_p$ by setting $f(a_0, a_1, a_2, \ldots) = \sum_{k=0}^{\infty} a_k p^k$. Then f is a 1–1 map from X onto $\hat{R}_p$ by Theorem 6.11, and it will suffice to show that f is continuous. Take $\alpha = (a_0, a_1, \ldots) \in X$, so $f(\alpha) = a \in \hat{R}_p$. As a basic neighborhood of a we may take $N = \{b \in \hat{R}_p : \phi_p(b - a) < p^{-k}\}$, with $k \ge 0$ in $\mathbb{Z}$. If we set

$$V = \{a_0\} \times \{a_1\} \times \cdots \times \{a_k\} \times \prod\{Y_j : j \ge k + 1\},$$

then V is open in X, $\alpha \in V$, and $f(V) \subseteq N$, so f is continuous and $\hat{R}_p$ is compact.

We remark that the map f in the proof above is also open, so X and $\hat{R}_p$ are in fact homeomorphic. It follows in particular that $\hat{R}_p$ is totally disconnected.

Exercise 6.12. Show that $\hat{\mathbb{Q}}_p$ is locally compact. (*Hint*: $\hat{\mathbb{Q}}_p = \bigcup\{p^k \hat{R}_p : k \in \mathbb{Z}\}$.)

Theorem 6.13. If $f(x) \in \mathbb{Z}[x]$, then the congruences $f(x) \equiv 0 (\operatorname{mod} p^m)$ have solutions for all $m \ge 1$ if and only if the equation $f(x) = 0$ has a solution in $\hat{R}_p$.

Proof. $\Leftarrow$: Suppose $a = \sum_{k=0}^{\infty} a_k p^k \in \hat{R}_p$ is a solution to $f(x) = 0$. For each n set $b_n = \sum_{k=0}^{n} a_k p^k \in \mathbb{Z}$, so $\lim_n b_n = a$ in $\hat{R}_p$. Thus $f(b_n) \in \mathbb{Z}$ and $f(b_n) \to f(a) = 0$ (since polynomials are continuous), i.e. $\phi_p(f(b_n)) \to 0$, which means that high powers of p divide $f(b_n)$ for large n. Now, given $m \geq 1$ choose n large enough so that $\phi_p(f(b_n)) < p^{-m}$, i.e., $p^m \mid f(b_n)$, or $f(b_n) \equiv 0(\text{mod } p^m)$.

$\Rightarrow$: Choose $c_m \in \mathbb{Z}$, $1 \leq m \in \mathbb{Z}$, with $f(c_m) \equiv 0(\text{mod } p^m)$. Since all $c_m \in \hat{R}_p$, which is compact (Proposition 6.12), there is a subsequence $\{c_{m_i}\}$ that converges, say $c_{m_i} \to a \in \hat{R}_p$ as $i \to \infty$. By continuity $f(c_{m_i}) \to f(a)$. But $p^{m_i} \mid f(c_{m_i})$; hence $\phi_p(f(c_{m_i})) \leq p^{-m_i}$, and $m_i \to \infty$, so $f(a) = \lim_i f(c_{m_i}) = 0$.

With not much more effort the polynomial $f(x)$ in Theorem 6.13 can be replaced by $f(x_1, \ldots, x_r) \in \mathbb{Z}[x_1, \ldots, x_r]$, with the same conclusion (see Borevich and Shafarevich [6, p. 41]).

Exercise 6.13. Show that if $f(x) = x^2 - a \in \mathbb{Z}[x]$, then $f(x)$ has a root in $\mathbb{Z}$ if and only if it has a root in $\mathbb{R}$ and in $\hat{R}_p$ for every prime p.

We conclude this section with a brief discussion of an approximation theorem that has come to play a central role in algebraic number theory.

Proposition 6.14. Suppose $\phi_1, \phi_2, \ldots, \phi_k$ are inequivalent nontrivial valuations on a field F. Then there is an element $a \in F$ with $\phi_1(a) > 1$, $\phi_i(a) < 1$ for $2 \leq i \leq k$.

Proof. Induction on k. The case $k = 1$ is trivial but the induction step will also require the result for $k = 2$. Since ϕ_1 and ϕ_2 are inequivalent we may choose $b, c \in F$ with $\phi_1(b) > 1$, $\phi_2(b) \leq 1$, $\phi_1(c) \geq 1$, and $\phi_2(c) < 1$. Set $a = bc$; then $\phi_1(a) > 1$ and $\phi_2(a) < 1$.

Suppose then that $k \geq 3$ and assume the result for $k - 1$ or fewer valuations. Choose b, $c \in F$ with $\phi_1(b) > 1$, $\phi_i(b) < 1$ for $2 \leq i \leq k - 1$ and $\phi_1(c) > 1$, $\phi_k(c) < 1$. If $\phi_k(b) \leq 1$ set $a_n = b^n c$, $1 \leq n \in \mathbb{Z}$. Then $\phi_1(a_n) > 1$ and $\phi_k(a_n) < 1$ for all n, and if $2 \leq i \leq k - 1$, then $\phi_i(a_n) = \phi_i(b)^n \phi_i(c) < 1$ for large enough n. Thus we may take $a = a_n$ for some n. If $\phi_k(b) > 1$, then instead set $a_n = b^n c/(1 + b^n)$, $1 \leq n \in \mathbb{Z}$. Then $\phi_1(a_n) \underset{n}{\to} \phi_1(c)$ and $\phi_k(a_n) \underset{n}{\to} \phi_k(c)$, so $\phi_1(a_n) > 1$ and $\phi_k(a_n) < 1$ for large enough n. If $2 \leq i \leq k - 1$, then $\phi_i(a_n) \underset{n}{\to} 0$, and again we may take $a = a_n$ for some n.

Corollary. Suppose $\phi_1, \ldots, \phi_k$ are inequivalent nontrivial valuations on F and $0 < \varepsilon \in \mathbb{R}$. Then for some $a \in F$ we have $\phi_1(a - 1) < \varepsilon$, $\phi_i(a) < \varepsilon$ for $2 \leq i \leq k$.

Proof. Apply the proposition to get $b \in F$ with $\phi_1(b) > 1$ and $\phi_i(b) < 1$ for $2 \leq i \leq k$. Set $a_n = b^n/(1 + b^n)$, $1 \leq n \in \mathbb{Z}$, and take $a = a_n$ for sufficiently large n.

Theorem 6.15 (The Artin–Whaples Approximation Theorem). Suppose that $\phi_1, \ldots, \phi_k$ are inequivalent nontrivial valuations on F, that $0 < \varepsilon \in \mathbb{R}$, and that $a_1, \ldots, a_k \in F$. Then there is an element $a \in F$ for which $\phi_i(a - a_i) < \varepsilon$, $1 \le i \le k$.

Proof. We may assume that each ϕ_i satisfies the triangle inequality. Set

$$M = \max\{\phi_i(a_j): 1 \le i, j \le k\}.$$

Apply the corollary, above, to each ϕ_i in turn and obtain $b_1, \ldots, b_k \in F$ with

$$\phi_i(b_i - 1) < \varepsilon/kM, \qquad \phi_j(b_i) < \varepsilon/kM \qquad \text{if} \quad j \ne i.$$

Set $a = \sum_{j=1}^k a_j b_j$. Then

$$\begin{aligned} \phi_i(a - a_i) &= \phi_i\left(\sum_j a_j b_j - a_i\right) \\ &\le \sum\{\phi_i(a_j)\phi_i(b_j): j \ne i\} + \phi_i(a_i)\phi_i(b_i - 1) \\ &< \frac{(k-1)\varepsilon M}{kM} + \frac{\varepsilon M}{kM} = \varepsilon. \end{aligned}$$

Exercise 6.14. If $F = \mathbb{Q}, a_1, \ldots, a_k \in \mathbb{Z}$, and $\phi_{p_1}, \ldots, \phi_{p_k}$ are distinct p_i-adic valuations show that the conclusion of the Artin–Whaples Theorem follows from the Chinese Remainder Theorem.

Exercise 6.15. Let ϕ_2, ϕ_3, ϕ_5 be the p-adic valuations for $p = 2, 3, 5$ on $\mathbb{Q}$ and set $a_2 = a_3 = 4$, $a_5 = 6$. Find $a \in \mathbb{Q}$ with $\phi_p(a - a_p) < 0.01$, $p = 2, 3, 5$.

Exercise 6.16. If $\phi_1, \ldots, \phi_k$ are inequivalent nontrivial valuations on F show that $\prod\{\phi_i(a): 1 \le i \le k\} = 1$ cannot be true for all $a \in F^*$. On the other hand let ϕ_p be the p-adic valuation on $\mathbb{Q}$ for each prime $p > 0$ and show that $\phi_\infty(a)\prod\{\phi_p(a): \text{all } p\} = 1$ for all $a \in \mathbb{Q}^*$.

7. REAL FIELDS AND STURM'S THEOREM

The essence of the usual order relationship on $R = \mathbb{Z}$, $\mathbb{Q}$, or $\mathbb{R}$ can be captured as follows. There is a subset P of R (the set of positive elements) such that

(i) R is the disjoint union of P, $\{0\}$, and $-P = \{-x: x \in P\}$ and
(ii) if $x, y \in P$, then $x + y \in P$ and $xy \in P$.

Then $x > y$ if and only if $x - y \in P$.

Suppose now that R is any commutative ring. If there is a subset P of R satisfying (i) and (ii) above, then R is said to be *ordered* (relative to P), the elements of P are called *positive* and the elements of $-P$ are called *negative.* We write $x > y$ (or $y < x$) if $x - y \in P$. Thus $P = \{x \in R : x > 0\}$ and $-P = \{x \in R : x < 0\}$. We write $x \geq y$ (or $y \leq x$) to mean that $x > y$ or $x = y$, as usual.

Note that if $x > 0$ and $y < 0$ in an ordered ring R, then $x, -y \in P$, so $x(-y) = -xy \in P$ and hence $xy < 0$. Similarly if $x < 0$ and $y < 0$, then $xy > 0$. In particular R can have no zero divisors, so every ordered ring is an integral domain.

For an example, take $R = \mathbb{Q}[x]$ and agree that $f(x) \in P$ if and only if

$$f(x) = a_0 + a_1 x + \cdots + a_n x^n$$

with $a_n > 0$ in $\mathbb{Q}$. With this order we have, for example, $2 + x > 1 - x - 3x^3$.

Exercise 7.1. Verify the following facts in an ordered ring R.

(1) If $x > y$ and $y > z$, then $x > z$.
(2) If $x > y$ and $u \geq v$, then $x + u > y + v$.
(3) If $x > y$ and $z > 0$, then $xz > yz$.
(4) If $0 < x \in U(R)$, then $0 < x^{-1}$.
(5) If $x > y > 0$ in $U(R)$, then $0 < x^{-1} < y^{-1}$.
(6) If $r_1, r_2, \ldots, r_k \in R$, then $\sum\{r_i^2 : 1 \leq i \leq k\} \geq 0$.

Clearly any subring S of an ordered ring R inherits an order from R if the set of positive elements in S is taken to be $P \cap S$. We say that the resulting order on S is *induced* from the order on R.

In the opposite direction suppose an ordered ring R is a subring of a ring K. The order on R is said to be *extended* to K if K has an order that induces the original order on R. Thus, for example, the usual order on $\mathbb{Q}$ induces the usual order on $\mathbb{Z}$, and extends to the usual order on $\mathbb{R}$.

Proposition 7.1. Suppose R is an ordered ring and $F \supseteq R$ is its field of fractions. Then $\operatorname{char}(F) = 0$ and there is a unique extension of the order on R to an order on F.

Proof. Let us first extend the order. Set

$$P_F = \left\{x \in F : x = \frac{a}{b} \text{ with } a > 0 \text{ and } b > 0 \text{ in } R\right\},$$

i.e., $x > 0$ in F if and only if x can be expressed as a quotient of positive elements from R. It is easy to check then that F is ordered, and if $a > 0$ in R, then $a = a^2/a > 0$ in F, so the order on F extends that on R. Uniqueness is clear from the definition of order and Exercise 7.1.4. Finally note that $1 = 1^2 > 0$, so $n \cdot 1 = 1^2 + 1^2 + \cdots + 1^2 > 0$ if n is any positive integer, and hence $\operatorname{char}(F) = 0$.

Corollary. The only possible order on $\mathbb{Q}$ is the usual order.

Proof. The order on $\mathbb{Z}$ is unique since its positive elements are just 1, $2 = 1 + 1$, $3 = 1 + 1 + 1$, etc.

The order on $R = \mathbb{Q}[x]$ given as an example above extends uniquely to an order on the field $\mathbb{Q}(x)$ of rational functions, whereby $f(x)/g(x) > 0$ in $\mathbb{Q}(x)$ if and only if $f(a)/g(a) > 0$ in $\mathbb{Q}$ for all sufficiently large positive $a \in \mathbb{Q}$.

Exercise 7.2. Not all ordered rings have unique orders. For example, $R = \mathbb{Q}(\sqrt{2})$ inherits the usual order from $\mathbb{R}$, so that P consists of all $a + b\sqrt{2}$, a, $b \in \mathbb{Q}$, that are positive real numbers. Show that $P_1 = \{a + b\sqrt{2} : a - b\sqrt{2} \in P\}$ determines a different order on $\mathbb{Q}(\sqrt{2})$.

A field F is called *formally real* if -1 is not a sum of squares of elements of F, or equivalently if

$$\sum\{a_i^2 : 1 \leq i \leq k\} = 0, \qquad a_i \in F,$$

is possible only if every $a_i = 0$. Note that in any field of characteristic $p > 0$ we have $\sum\{1^2 : 1 \leq i \leq p\} = 0$, so a formally real field must have characteristic 0.

Proposition 7.2. Suppose F is a field and $a, b \in F$ are sums of squares in F, say $a = \sum_{i=1}^k a_i^2$ and $b = \sum_{j=1}^m b_j^2$. Then ab is a sum of squares in F, as is b^{-1} if $b \neq 0$.

Proof. Clearly $ab = \sum_{i,j}(a_i b_j)^2$ and

$$b^{-1} = b(b^{-1})^2 = \sum_{j=1}^{m} (b_j b^{-1})^2.$$

A formally real field F is called *real closed* if the only formally real algebraic extension of F is F itself. Note for example that $F = \mathbb{R}$ is real closed since its only nontrivial algebraic extension is (isomorphic with) $\mathbb{C}$, which is not formally real.

Proposition 7.3. If F is a formally real field, then F has an algebraic extension K that is real closed.

Proof. Zorn's Lemma.

Proposition 7.4. Suppose F is a real closed field and $a = \sum_{i=1}^k a_i^2$, $a_i \in F$. Then $a = b^2$ for some $b \in F$.

Proof. Suppose not. Then $x^2 - a$ is irreducible in $F[x]$. Let c be a root in a splitting field $K = F(c)$. Since F is real closed K is not formally real, and we may write $-1 = \sum_i (b_i + cc_i)^2$, with $b_i, c_i \in F$. Thus

$$-1 = \sum_i b_i^2 + a\sum_i c_i^2 + 2c\sum_i b_i c_i,$$

and we see that $\sum_i b_i c_i = 0$ since $c \notin F$. Thus

$$-1 = \sum_i b_i^2 + \left(\sum_j a_j^2\right)\left(\sum_i c_i^2\right),$$

a sum of squares by Proposition 7.2, contradicting the formal reality of F.

Proposition 7.5. If F is a real closed field and $0 \neq a \in F$, then either a or $-a$ is a square in F (but not both).

Proof. If a is not a square, then just as in the proof of Proposition 7.4 we may write $-1 = \sum_i b_i^2 + a\sum_i c_i^2$, with $b_i, c_i \in F$ and $\sum_i c_i^2 \neq 0$. Solving for $-a$ we find

$$-a = \left(1^2 + \sum_i b_i^2\right) \Big/ \sum_i c_i^2,$$

so $-a$ is a sum of squares by Proposition 7.2. Thus $-a$ is a square by Proposition 7.4. If both a and $-a$ were squares, then $-1 = -a/a$ would be a square.

Theorem 7.6. A real closed field F is ordered, with the nonzero squares as positive elements, and the order is unique.

Proof. If P denotes the set of nonzero squares in F, then P determines an order by Propositions 7.4 and 7.5. The order is unique since squares must be positive for any order.

Corollary. Every formally real field can be ordered.

Proof. Proposition 7.3.

The next theorem generalizes to all real closed fields F an important property of the real field $\mathbb{R}$. In the case of $F = \mathbb{R}$ it is a consequence of the Intermediate Value Theorem of calculus.

Theorem 7.7. Suppose F is a real closed field and $f(x) \in F[x]$ has odd degree. Then $f(x)$ has a root in F.

Proof. Suppose the theorem to be false and choose a monic polynomial $f(x) \in F[x]$ of minimal odd degree n not having a root in F. Clearly $n \geq 3$. Note that $f(x)$ must be irreducible by the minimality of n. Let a be a root of $f(x)$ in an extension field. Then $K = F(a)$ is not formally real and we may write $-1 = \sum_i f_i(a)^2$, where each $f_i(x) \in F[x]$, $\deg f_i(x) < n = [K:F]$, and at least one $f_i(x)$ is nonconstant. But then $g(x) = 1 + \sum_i f_i(x)^2$ is a multiple, in $F[x]$, of the minimal polynomial $f(x)$ of a. Write $g(x) = f(x)h(x)$, $h(x) \in F[x]$. Note that $\deg g(x)$ is even and less than $2n - 1$, and

$$\deg g(x) = \deg f(x) + \deg h(x) = n + \deg h(x),$$

so deg $h(x)$ is odd and less than $n - 1$. Consequently $h(x)$ has a root $b \in F$, and b is a root of $g(x)$. But then $-1 = \sum_i f_i(b)^2$ in F, a contradiction.

Theorem 7.8. If F is a real closed field and i is a root for $x^2 + 1 \in F[x]$, then $K = F(i)$ is algebraically closed.

Proof. Note that if $a + bi \in K$, with $a, b \in F$, then $a^2 + b^2 \geq 0$; hence $a^2 + b^2$ is a square in F. Say $a^2 + b^2 = c^2$, with $c \geq 0$. Then $c^2 \geq a^2$, and so $c \geq \pm a$. Thus $(a + c)/2$ and $(-a + c)/2$ are squares in F, say of u and v. If $b \geq 0$ take $u \geq 0$ and $v \geq 0$; if $b < 0$ take $u > 0$ and $v < 0$. Then it is easy to check that $(u + vi)^2 = a + bi$, so every element of K is a square. It follows from the classical quadratic formula then that there are no irreducible polynomials of degree 2 in $K[x]$, so K has no extension fields of degree 2. That fact, together with Theorem 7.7, allows the proof of the Fundamental Theorem of Algebra (Theorem III.3.9), with $\mathbb{R}$ and $\mathbb{C}$ replaced by F and K, to be repeated verbatim in the present context.

Theorem 7.9 (The Intermediate Value Theorem). Suppose F is a real closed field and $f(x) \in F[x]$. If $a, b \in F$ with $f(a) < 0 < f(b)$, then $f(c) = 0$ for some $c \in F$ between a and b.

Proof. By Theorem 7.8 each irreducible factor of $f(x)$ in $F[x]$ has degree 1 or 2. Any irreducible quadratic factor $x^2 + rx + s$ can be written as $(x + r/2)^2 + (s - r^2/4)$, with $s - r^2/4$ positive (i.e., a square). Thus $x^2 + rx + s > 0$ for all values assigned to x (from F). If no linear factor of $f(x)$ changed signs from a to b, then $f(x)$ could not change signs, so there is a factor $ux + v$ of $f(x)$ with $ua + v < 0$ and $ub + v > 0$. If $u > 0$, then $a < -v/u < b$, and if $u < 0$, then $b < -v/u < a$. In either case $c = -v/u$ is a root of $f(x)$ between a and b.

Suppose F is any field and $f(x) \in F[x]$ has positive degree. Apply the Euclidean algorithm (Exercise II.8.26) to $f_0(x) = f(x)$ and $f_1(x) = f'(x)$, with one small variation: at each stage *subtract* the negative of the usual remainder. This process determines the *standard sequence* $f_0(x), f_1(x), \ldots, f_k(x)$ for $f(x)$. It satisfies the following relations:

$$
\begin{aligned}
f_0(x) &= f(x), \qquad f_1(x) = f'(x); \\
f_0(x) &= f_1(x)q_1(x) - f_2(x), \qquad \deg f_2(x) < \deg f_1(x); \\
f_1(x) &= f_2(x)q_2(x) - f_3(x), \qquad \deg f_3(x) < \deg f_2(x); \\
&\vdots \\
f_{k-2}(x) &= f_{k-1}(x)q_{k-1}(x) - f_k(x), \qquad \deg f_k(x) < \deg f_{k-1}(x); \\
f_{k-1}(x) &= f_k(x)q_k(x).
\end{aligned}
$$

Thus

$$f_k(x) = \mathrm{GCD}(f(x), f'(x)) \qquad \text{and} \qquad f_k(x) \mid f_i(x), \qquad 0 \leq i \leq k.$$

Next set $g_i(x) = f_i(x)/f_k(x)$, $0 \leq i \leq k$, and call the sequence $g_0(x)$, $g_1(x), \ldots, g_k(x) \in F[x]$ the *Sturm sequence* for $f(x)$.

Observe that $g_k(x) = 1$, that $\mathrm{GCD}(g_0(x), g_1(x)) = 1$, and that

$$g_{i-1}(x) = g_i(x)q_i(x) - g_{i+1}(x) \qquad \text{for} \qquad 1 \leq i \leq k-1.$$

Observe also that $f_0(x)$ and $g_0(x)$ have exactly the same set of roots in F, but that $g_0(x)$ has no repeated roots (see Proposition III.3.2). It is not generally true that $g_1(x) = g_0'(x)$.

Proposition 7.10. Suppose F is a real closed field, $f(x) \in F[x]$ has positive degree, and $g_0(x), \ldots, g_k(x) = 1$ is the Sturm sequence for $f(x)$. Choose $a, b \in F$ with $a < b$ such that $f(a) \neq 0$ and $f(b) \neq 0$. Then

(1) if $a \leq c \leq b$ in F and $g_i(c) = 0$ for some i, $0 < i < k$, then $g_{i-1}(c)$ and $g_{i+1}(c)$ differ in sign, i.e., $g_{i-1}(c)g_{i+1}(c) < 0$;

(2) if $a < c < b$ and $f(c) = 0$, then there are $c_1, c_2 \in F$, with $c_1 < c < c_2$, such that $g_0(u)g_1(u) < 0$ if $c_1 < u < c$ and $g_0(u)g_1(u) > 0$ if $c < u < c_2$.

Proof. (1) Since $g_{i-1}(x) = g_i(x)q_i(x) - g_{i+1}(x)$ and $g_i(c) = 0$ we have $g_{i-1}(c) = -g_{i+1}(c)$, so either they differ in sign or both are 0. But $g_{i+1}(c) = 0$ entails that $g_j(c) = 0$ for all $j \geq i$, contradicting the fact that $g_k(c) = 1$.

(2) Say that c is a root of $f(x)$ with multiplicity m, so $f(x) = (x-c)^m q(x)$, $q(c) \neq 0$. Thus

$$f_1(x) = f'(x) = (x-c)^m q'(x) + m(x-c)^{m-1}q(x)$$

and

$$f_k(x) = (f(x), f'(x)) = (x-c)^{m-1}d(x), \qquad \text{with} \quad d(c) \neq 0.$$

We may write $q(x) = d(x)G(x)$ and $q'(x) = d(x)H(x)$, with $H(c) \neq 0$. Consequently $g_0(x) = (x-c)G(x)$ and $g_1(x) = (x-c)H(x) + mG(x)$, and in particular $g_1(c) \neq 0$. Choose $c_1 < c$ and $c_2 > c$ such that if $c_1 < u < c_2$, then $g_1(u) \neq 0$ and $G(u) \neq 0$. (Why is this possible?) Note that then, by the Intermediate Value Theorem (7.9), $g_1(u)G(u) > 0$ if $c_1 < u < c_2$, since $g_1(c) = mG(c)$ and hence $g_1(c)G(c) > 0$. But then

$$g_0(u)g_1(u) = (u-c)g_1(u)G(u)$$

is < 0 if $c_1 < u < c$ and is > 0 if $c < u < c_2$.

Suppose F is an ordered field and let $f_0(x), f_1(x), \ldots, f_k(x)$ be any sequence in $F[x]$. For any $a \in F$ define the *variation* $V_a(f)$ of the sequence at a to be the number of changes from positive to negative (ignoring 0's) in the sequence $f_0(a), f_1(a), \ldots, f_k(a)$. Thus $V_a(f)$ is an integer between 0 and k.

For example, if $F = \mathbb{Q}$ and $k = 4$ and the sequence $\{f_i(a)\}$ is $1, -1, 0, 1, -2$, then $V_a(f) = 3$; if $\{f_i(b)\}$ is $1, 0, 1, 2, -3$, then $V_b(f) = 1$.

For the proof of the following theorem it will be convenient to use the usual notation from analysis for intervals in an ordered field F. Thus

$$[a, b] = \{c \in F : a \leq c \leq b\},$$
$$(a, b] = \{c \in F : a < c \leq b\},$$

etc.

Theorem 7.11 (Sturm). Suppose F is a real closed field and $f(x) \in F[x]$ has positive degree. Suppose $a, b \in F$ with $a < b$ and $f(a) \neq 0$, $f(b) \neq 0$. If $f_0(x), f_1(x), \ldots, f_k(x)$ is the standard sequence for $f(x)$, then $V_a(f) - V_b(f)$ is the number of distinct roots c of $f(x)$ in the interval (a, b).

Proof. Note that $f_k(a) \neq 0$ and $f_k(b) \neq 0$ since $f_k(x) \mid f(x)$, and recall that $g_i(x) = f_i(x)/f_k(x)$ for $g_i(x)$ in the Sturm sequence for $f(x)$. It follows that $V_a(f) = V_a(g)$ and $V_b(f) = V_b(g)$. Thus it will suffice to show that $V_a(g) - V_b(g)$ is the number of roots (automatically distinct) of $g_0(x)$ between a and b in F.

Choose elements a_i, $0 \leq i \leq m$, in $[a, b]$, with

$$a = a_0 < a_1 < a_2 < \cdots < a_m = b,$$

so that all roots of all $g_j(x)$ in the Sturm sequence that lie in $[a, b]$ are included among the a_i. Consequently no $g_j(x)$ has a root in any of the open intervals (a_i, a_{i+1}), $0 \leq i \leq m - 1$. Take $c_1 \in (a_0, a_1)$. Each $g_j(x)$ remains either always positive or always negative throughout (a_0, a_1) by the Intermediate Value Theorem, so if no $g_j(a_0)$ is 0 we have $V_{a_0}(g) = V_{c_1}(g)$. On the other hand, if $g_j(a_0) = 0$ for some j, $i < j < k$, then $g_{j-1}(a_0)$ and $g_{j+1}(a_0)$ differ in sign by (1) of Proposition 7.10; hence $g_{j-1}(c_1)$ and $g_{j+1}(c_1)$ must also differ in sign by the Intermediate Value Theorem. Thus each of $g_{j-1}(a_0)$, 0, $g_{j+1}(a_0)$ and $g_{j-1}(c_1)$, $g_j(c)$, $g_{j+1}(c_1)$ contribute one sign change in the determination of $V_{a_0}(g)$ and $V_{c_1}(g)$, and again $V_{a_0}(g) = V_{c_1}(g)$. Similarly, if $c_m \in (a_{m-1}, a_m)$, then $V_{c_m}(g) = V_{a_m}(g)$.

Next choose $c_i \in (a_{i-1}, a_i)$ and $c_{i+1} \in (a_i, a_{i+1})$, $1 < i < m - 1$. Suppose first that $f(a_i) \neq 0$, and so $g_0(a_i) \neq 0$. Then the arguments in the preceding paragraph apply to (a_i, a_{i+1}) in place of (a_0, a_1), and to (a_{i-1}, a_i) in place of (a_{m-1}, a_m), and we conclude that $V_{c_{i+1}}(g) = V_{a_i}(g) = V_{c_i}(g)$. Suppose then that $f(a_i) = g_0(a_i) = 0$. By (2) of Proposition 7.10 (and the Intermediate Value Theorem again) we see that $g_0(c_i)$ and $g_1(c_i)$ differ in sign, but $g_0(c_{i+1})$ and $g_1(c_{i+1})$ have the same sign. Other than that, the arguments used above show the same number of sign changes in $g_{j-1}(c_i)$, $g_j(c_i)$, $g_{j+1}(c_i)$ as in $g_{j-1}(c_{i+1})$, $g_j(c_{i+1})$, $g_{j+1}(c_{i+1})$ for $j > 1$. Thus $V_{c_i}(g) - V_{c_{i+1}}(g) = 1$ at each i for which $f(a_i) = g_0(a_i) = 0$.

Now write $V_a(g) - V_b(g)$ as a telescoping sum,

$$V_a(g) - V_b(g) = V_a(g) - V_{c_1}(g) + \sum_{i=1}^{m-1} (V_{c_i}(g) - V_{c_{i+1}}(g)) + V_{c_m}(g) - V_b(g)$$

$$= \sum_{i=1}^{m-1} (V_{c_i}(g) - V_{c_{i+1}}(g)).$$

Each $V_{c_i}(g) - V_{c_{i+1}}(g)$ contributes 1 if $g_0(a_i) = 0$ and contributes 0 otherwise. Thus $V_a(g) - V_b(g)$ is the number of roots of $g_0(x)$ in $[a, b]$.

It is worth noticing that sign changes in the sequence $f_0(c), f_1(c), \ldots, f_k(c)$ are unaffected when some or all of the $f_i(x)$ are replaced by various positive multiples of themselves. Thus at each stage of the Euclidean algorithm we are free to replace $f_i(x)$ by a positive multiple and proceed to the next stage. This simple observation can simplify the arithmetic considerably. By a slight abuse of language we will continue to call any resulting sequence the standard sequence for $f(x)$.

For an example take $F = \mathbb{R}$ and $f(x) = f_0(x) = x^4 - x - 1$. Then $f_1(x) = 4x^3 - 1$, $f_2(x) = 3x + 4$, and $f_3(x) = 1$. We have

a	$f_0(a)$	$f_1(a)$	$f_2(a)$	$f_3(a)$	$V_a(f)$
$-\infty$	+	−	−	+	2
−1	1	−5	1	1	2
0	−1	−1	4	1	1
1	−1	3	7	1	1
2	13	31	10	1	0
$+\infty$	+	+	+	+	0

It follows that $f(x)$ has just two real roots [neither is repeated since $(f(x), f'(x)) = 1$], one in $(-1, 0)$ and one in $(1, 2)$.

For a second example take $f(x) = 36x^2 - 36x + 5$. Then $f_1(x) = 2x - 1$ and $f_2(x) = 1$. Thus $V_0(f) = 2$ and $V_1(f) = 0$, and $f(x)$ has two real roots between 0 and 1. This example shows that Sturm's Theorem is an improvement over the Intermediate Value Theorem, which would not detect any roots between 0 and 1.

Exercises

1. Show that $f(x) = x^3 - 3x - 1$ has three real roots and isolate them within intervals of length 1.
2. Show that $f(x) = x^4 + 4x^3 - 12x + 9$ has no real roots.
3. Show that $f(x) = x^4 - 6x^2 - 4x + 2$ has four real roots and isolate the roots within four pairwise disjoint intervals.

4. If $f(x) = x^3 + px + q \in \mathbb{R}[x]$, with $p \neq 0$, set $D = -4p^3 - 27q^2$. Show that if $D < 0$, then $f(x)$ has one real and two nonreal roots; if $D > 0$, then $f(x)$ has three real roots; and if $D = 0$, then $x = -3q/2p$ is a root with multiplicity 2.

5. If

$$f(x) = a_0 + a_1 x + \cdots + a_{n-1} x^{n-1} + x^n \in \mathbb{R}[x]$$

show that all real roots of $f(x)$ lie in the interval $[-b, b]$, where

$$b = \max\{1, |a_0| + |a_1| + \cdots + |a_{n-1}|\}.$$

(*Hint*: If $|u| > b$ show that $|f(u)| > 0$.) Generalize from $\mathbb{R}$ to any ordered field F.

8. REPRESENTATIONS AND CHARACTERS OF FINITE GROUPS

We assume throughout this section that G is a finite group and V is a finite-dimensional vector space over the complex field $\mathbb{C}$.

A *representation* of G is a homomorphism T from G to the group $\mathrm{GL}(V)$ of all invertible linear transformations of V. A basis $\{v_1, \ldots, v_n\}$ for V provides an isomorphism between $\mathrm{GL}(V)$ and the group $\mathrm{GL}(n, \mathbb{C})$ of all invertible $n \times n$ complex matrices, and a corresponding homomorphism $\hat{T}$ from G to $\mathrm{GL}(n, \mathbb{C})$. The homomorphism $\hat{T}$ is called a *matrix representation* of G. Thus each $T(x)$, $x \in G$, is a linear transformation of V, and $\hat{T}(x)$ is a matrix that represents $T(x)$ relative to the chosen basis $\{v_1, \ldots, v_n\}$.

If $\dim V = n$ we say that the representation T (or $\hat{T}$) has degree n, and write $\deg T = \deg \hat{T} = n$. A representation is called *faithful* if it is 1–1.

Examples

1. If G is any finite group and V any complex vector space, then the *trivial representation* is defined by $T(x) = 1_V$, all $x \in G$. It is faithful only when $G = 1$.

2. Suppose $G = \langle x \mid x^m = 1 \rangle$, $\zeta \in \mathbb{C}$ is a primitive mth root of unity, $V = \mathbb{C}$, and $\hat{T}(x^i) = \zeta^i$, $0 \leq i \leq m - 1$. Then T is a faithful representation.

3. Suppose G is a subgroup of the symmetric group S_m and set $H = G \cap A_m$. Set $V = \mathbb{C}$ and define

$$\hat{T}(x) = \begin{cases} 1 & \text{if } x \in H, \\ -1 & \text{if } x \in G \backslash H. \end{cases}$$

Then T is a representation of G.

4. Suppose G acts as a permutation group on a set $S = \{s_1, \ldots, s_n\}$. Let V be a vector space with basis $\{v_1, \ldots, v_n\}$ and define $T\colon G \to \mathrm{GL}(V)$ by means of

$T(x)\colon v_i \mapsto v_j$ if and only if $xs_i = s_j$. Then T is called a *permutation representation* of G [note that each $\hat{T}(x)$ is a permutation matrix, i.e., each row and column of $\hat{T}(x)$ has exactly one 1, the other entries are 0's].

5. In the example above suppose $S = G$ and the action is left multiplication. The resulting permutation representation is called the *left regular permutation representation* of G. It will be denoted in general by $R = R_G$.

Exercise 8.1. (1) Write out the permutation matrix representation of the dihedral group D_4 corresponding to its action on the vectices of a square.

(2) Write out the left regular matrix representation if $G = \mathbb{Z}_4$, $\mathbb{Z}_2 \times \mathbb{Z}_2$, S_3, or D_4.

Suppose T and S are representations of G on vector spaces V and W, respectively. Then T and S are called *equivalent* if there is an isomorphism $\theta\colon V \to W$ such that the diagram

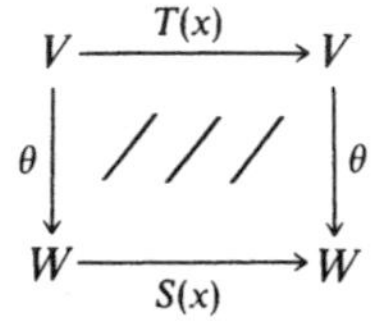

is commutative for all $x \in G$, i.e., $T(x) = \theta^{-1}S(x)\theta$ for all $x \in G$. For corresponding matrix representations $\hat{T}$ and $\hat{S}$ this means there is an invertible matrix M for which $\hat{T}(x) = M^{-1}\hat{S}(x)M$, all $x \in G$, a simultaneous similarity transformation.

We write $S \sim T$ (and $\hat{S} \sim \hat{T}$) if S and T are equivalent.

If T is a representation of G on V and W is a subspace of V we say that W is *T-invariant* if $T(x)W \subseteq W$ for all $x \in G$. If the only T-invariant subspaces of V are 0 and V, then T is called *irreducible;* otherwise T is *reducible.*

Suppose T is reducible, with W a T-invariant subspace, $0 \neq W \neq V$. If a basis for W is enlarged to a basis for V, then each $\hat{T}(x)$ has the partitioned form

$$\hat{T}(x) = \left[\begin{array}{c|c} A(x) & C(x) \\ \hline 0 & B(x) \end{array}\right].$$

It is easily checked that A and B are each matrix representations of G, with respective degrees $\dim W$ and $\dim(V/W)$.

If T and S are representations of G on vector spaces V and W define their *direct sum* $T \oplus S$ on $V \oplus W$ by means of

$$(T \oplus S)(x)\colon (v, w) \mapsto (T(x)v, S(x)w).$$

Clearly $T \oplus S$ is also a representation of G. If $\{v_i\}$ is a basis for V and $\{w_j\}$ is a basis for W, then $\{(v_i, 0)\} \cup \{(0, w_j)\}$ is a basis for $V \oplus W$ and the matrix representation $(T \oplus S)\hat{}$ takes the partitioned form

$$(T \oplus S)\hat{}(x) = \left[\begin{array}{c|c} \hat{T}(x) & 0 \\ \hline 0 & \hat{S}(x) \end{array}\right],$$

all $x \in G$. The definition extends in an obvious fashion to the direct sum of any finite number of representations.

Theorem 8.1 (Maschke). Every representation T of a finite group G is equivalent with a direct sum of irreducible representations.

Proof. Induction on $n = \deg T$. If $n = 1$, then T is irreducible and there is nothing to prove. Assume that $n \geq 2$ and that the theorem holds for representations of degree less than n. For a suitable basis

$$\hat{T}(xy) = \begin{bmatrix} A(xy) & C(xy) \\ 0 & B(xy) \end{bmatrix} = \hat{T}(x)\hat{T}(y) = \begin{bmatrix} A(x) & C(x) \\ 0 & B(x) \end{bmatrix}\begin{bmatrix} A(y) & C(y) \\ 0 & B(y) \end{bmatrix},$$

and so

$$C(xy) = A(x)C(y) + C(x)B(y), \qquad \text{all} \quad x, y \in G.$$

Thus

$$C(xy)B((xy)^{-1})B(x) = C(xy)B(y^{-1}) = A(x)C(y)B(y^{-1}) + C(x),$$

and so

$$\sum\{C(xy)B((xy)^{-1}) : y \in G\}B(x) = A(x)\sum\{C(y)B(y^{-1}) : y \in G\} + |G|C(x)$$

for all $x \in G$.

Set

$$D = |G|^{-1}\sum\{C(y)B(y^{-1}) : y \in G\}.$$

Then the equation above becomes

$$|G|DB(x) = |G|A(x)D + |G|C(x),$$

or

$$DB(x) = A(x)D + C(x), \qquad \text{all} \quad x \in G.$$

Observe now that

$$\begin{bmatrix} I & D \\ 0 & I \end{bmatrix}\begin{bmatrix} A(x) & 0 \\ 0 & B(x) \end{bmatrix} = \begin{bmatrix} A(x) & DB(x) \\ 0 & B(x) \end{bmatrix}$$

$$= \begin{bmatrix} A(x) & A(x)D + C(x) \\ 0 & B(x) \end{bmatrix} = \begin{bmatrix} A(x) & C(x) \\ 0 & B(x) \end{bmatrix}\begin{bmatrix} I & D \\ 0 & I \end{bmatrix}.$$

It follows that $\hat{T} \sim A \oplus B$, and the induction hypothesis can be applied to A and B.

Proposition 8.2 (Schur's Lemma). Suppose T and S are irreducible representations of G on V and W, respectively. Suppose $\theta\colon W \to V$ is a linear transformation with $\theta S(x) = T(x)\theta$ for all $x \in G$. Then either $\theta = 0$ or else θ is an isomorphism and consequently $T \sim S$.

Proof. Suppose $\theta \neq 0$. Set $W_1 = \ker\theta$ and $V_1 = \operatorname{Im}\theta$. If $w \in W_1$ and $x \in G$, then $0 = T(x)\theta w = \theta S(x)w$, so $S(x)w \in \ker\theta = W_1$, i.e., W_1 is S-invariant. Thus $W_1 = 0$ ($W_1 \neq W$ since $\theta \neq 0$), and so θ is 1–1. If $v \in V_1$ write $v = \theta u, u \in W$. If $x \in G$, then

$$T(x)v = T(x)\theta u = \theta S(x)u \in \operatorname{Im}\theta = V_1,$$

so V_1 is T-invariant. Thus $V_1 = V$ and θ is onto.

If T is a representation of G on V define the *centralizer* of T to be the algebra $C(T)$ of all linear transformations $A\colon V \to V$ such that $T(x)A = AT(x)$ for all $x \in G$.

Proposition 8.3. Suppose T is a representation of G. (i) If $C(T)$ is a division algebra, then T is irreducible. (ii) If T is irreducible, then $C(T) = \{a1 : a \in \mathbb{C}\} \cong \mathbb{C}$.

Proof. (i) If T is reducible we may assume, by Maschke's Theorem (8.1), that $T = T_1 \oplus T_2$ acting on $V = V_1 \oplus V_2$, with both $V_i \neq 0$. Let P be the corresponding projection onto V_1. Then P is not invertible, since $\ker P = V_2 \neq 0$. If $v \in V$ write $v = v_1 + v_2, v_i \in V_i$. Then

$$PT(x)v = P\big(T_1(x)v_1 + T_2(x)v_2\big) = T_1(x)v_1 = T(x)Pv,$$

so $P \in C(T)$ and $C(T)$ is not a division algebra.

(ii) If $A \in C(T)$ let $a \in \mathbb{C}$ be an eigenvalue for A. Then $(A - a1)T(x) = T(x)(A - a1)$, all $x \in G$, and $A - a1$ is singular so it cannot be an isomorphism. Thus $A - a1 = 0$ by Schur's Lemma (8.2), and $A = a1$.

Corollary. If G is abelian and T is an irreducible representation of G on V, then $\deg T = 1$.

Proof. Since G is abelian each $T(x), x \in G$, is in $C(G)$, and so $T(x) = a_x \cdot 1$, $a_x \in \mathbb{C}$. But then every subspace of V is invariant, so T can be irreducible only if $\dim V = 1$.

Proposition 8.4 (Schur). Suppose T and S are inequivalent irreducible representations of G, with $\deg T = n$, and that bases are chosen so that $\hat{T}(x) = (t_{ij}(x))$, $\hat{S}(x) = (s_{ij}(x))$, all $x \in G$. Then

(i) $\sum\{s_{ij}(x)t_{km}(x^{-1}) : x \in G\} = 0$ and

(ii) $\sum\{t_{ij}(x)t_{km}(x^{-1}) : x \in G\} = \delta_{im}\delta_{jk}|G|/n$,

all i, j, k, and m (as usual $\delta_{jk} = 1$ if $j = k$, $= 0$ if $j \neq k$).

Proof. (i) For any $\mathbb{C}$-matrix M of appropriate size set

$$L = \sum\{\hat{S}(x)M\hat{T}(x^{-1}):x \in G\},$$

and observe that

$$\hat{S}(y)L = \sum_x \hat{S}(y)\hat{S}(x)M\hat{T}(x^{-1})\hat{T}(y^{-1})\hat{T}(y)$$

$$= \sum_x \hat{S}(yx)M\hat{T}((yx)^{-1})\hat{T}(y) = L\hat{T}(y),$$

all $y \in G$. By Schur's Lemma (8.2) $L = 0$. Now choose M to be E_{jk} (see p. 181). Then the *im*-entry of L is

$$\sum_x s_{ij}(x)t_{km}(x^{-1}) = 0.$$

(ii) This time set

$$L = \sum\{\hat{T}(x)M\hat{T}(x^{-1}):x \in G\},$$

and see as above that $L \in C(\hat{T})$. Thus if $M = E_{jk}$ we have by Proposition 8.3 that $L = a_{jk}I$ for some $a_{jk} \in \mathbb{C}$. The *im*-entry of L is then

$$a_{jk}\delta_{im} = \sum_x t_{ij}(x)t_{km}(x^{-1}) = \sum_x t_{ij}(x^{-1})t_{km}(x)$$

$$= \sum_x t_{km}(x)t_{ij}(x^{-1}) = a_{mi}\delta_{kj}.$$

Choose $i = m$ and $j \neq k$ to see that $a_{jk} = 0$ if $j \neq k$. Next choose $i = m$ and $j = k$ to see that $a_{ii} = a_{jj} = a$ (say) for all i and j. Thus

$$a = \sum_x t_{ij}(x)t_{ji}(x^{-1}) \qquad \text{for all} \quad i, j.$$

Sum over j to obtain

$$na = \sum_x \sum_j t_{ij}(x)t_{ji}(x^{-1}) = \sum_x 1 = |G|,$$

since $\sum_j t_{ij}(x)t_{ji}(x^{-1})$ is the *ii*-entry of $\hat{T}(x)\hat{T}(x^{-1}) = I$, and so $a = |G|/n$. Finally, we see that

$$\sum_x t_{ij}(x)t_{km}(x^{-1}) = a_{jk}\delta_{im} = a\delta_{jk}\delta_{im} = \delta_{im}\delta_{jk}|G|/n.$$

Exercise 8.2. Suppose $T_1, T_2, \ldots, T_k$ are mutually inequivalent irreducible representations of G, with $\hat{T}_m(x) = (t_{ij}^{(m)}(x))$, $1 \leq m \leq k$. Use Proposition 8.4 to show that the functions $t_{ij}^{(m)}$, all m, i, and j, are $\mathbb{C}$-linearly independent. If $n_m = \deg T_m$, $1 \leq m \leq k$, conclude that $\sum_{m=1}^k n_m^2 \leq |G|$, and hence conclude that there can be no more than $|G|$ mutually inequivalent representations of G. (The key observation is that $|G|$ is the dimension of the space of all $\mathbb{C}$-valued functions on G.)

Recall that if $A = (a_{ij})$ is an $n \times n$ matrix, then the *trace of* A is $\operatorname{tr} A = \sum_{i=1}^{n} a_{ii}$. It is easy to check that if B is also $n \times n$, then $\operatorname{tr}(AB) = \operatorname{tr}(BA)$, and consequently if C is $n \times n$ and invertible then $\operatorname{tr}(C^{-1}AC) = \operatorname{tr}(A)$. As a result, if $T: V \to V$ is a linear transformation represented by a matrix A relative to some basis of V, we may define the trace of T to be $\operatorname{tr} T = \operatorname{tr} A$.

If T is a representation of G on V define its *character* $\chi = \chi_T$, a function from G to $\mathbb{C}$, by means of $\chi(x) = \operatorname{tr} T(x)$, all $x \in G$. Observe that $\chi(1) = \operatorname{tr} 1_V = \dim V = \deg T$; $\chi(1)$ is called the *degree* of χ. We say that a character $\chi = \chi_T$ is *reducible* or *irreducible* according as T is reducible or irreducible. If T is faithful, then χ is called faithful.

The character χ of a representation T of degree 1 coincides with the (1×1) matrix representation $\hat{T}$. Thus characters of degree 1, which are called *linear* characters, are just homomorphisms from G to the multiplicative group $\mathbb{C}^*$. The character of the trivial degree one representation is called the *principal* character and is often denoted 1_G. Thus $1_G(x) = 1$, all $x \in G$.

Exercise 8.3. Suppose G is a permutation group acting on a set S, let T be the resulting permutation representation of G, and let $\chi = \chi_T$. If $x \in G$ show that $\chi(x)$ is the number of fixed points of x in S, i.e., of $s \in S$ for which $xs = s$.

Proposition 8.5. If T and S are equivalent representations of G, then $\chi_T = \chi_S$.

Proof. For appropriate choices of bases we have $\hat{T}(x) = \hat{S}(x)$ for all $x \in G$.

Proposition 8.6. Characters are class functions on G, i.e., they are constant on conjugacy classes.

Proof. Say $\chi = \chi_T$ and take $x, y \in G$. Then

$$\chi(y^{-1}xy) = \operatorname{tr}\big(T(y^{-1}xy)\big) = \operatorname{tr}\big(T(y)^{-1}T(x)T(y)\big) = \operatorname{tr} T(x) = \chi(x).$$

Proposition 8.7. If T and S are representations of G, then $\chi_{T \oplus S} = \chi_T + \chi_S$.

Proof. Obvious.

Corollary. If T is a representation of G with character χ, then there are irreducible characters $\chi_1, \chi_2, \ldots, \chi_r$ (not necessarily distinct) such that $\chi = \chi_1 + \chi_2 + \cdots + \chi_r$.

Proof. By Maschke's Theorem (8.1) there are irreducible representations $T_1, \ldots, T_r$ with $T \sim T_1 \oplus T_2 \oplus \cdots \oplus T_r$. Take traces.

Proposition 8.8. If T is a representation of G with character χ and $x \in G$, then $\chi(x^{-1}) = \overline{\chi(x)}$, the complex conjugate of $\chi(x)$.

Proof. Set $H = \langle x \rangle$. Then $T|H$ is a representation of H. By Maschke's Theorem and the corollary to Proposition 8.3 (applied to $T|H$) there is a basis relative to which

$$\hat{T}(x) = \begin{bmatrix} a_1 & & 0 \\ & \ddots & \\ 0 & & a_n \end{bmatrix},$$

a diagonal matrix. If $|x| = m$, then

$$I = \hat{T}(x^m) = \hat{T}(x)^m = \begin{bmatrix} a_1^m & & 0 \\ & \ddots & \\ 0 & & a_n^m \end{bmatrix},$$

so each a_i is an mth root of unity, and hence $a_i^{-1} = \bar{a}_i$. But then

$$\hat{T}(x^{-1}) = \begin{bmatrix} a_1^{-1} & & 0 \\ & \ddots & \\ 0 & & a_n^{-1} \end{bmatrix} = \begin{bmatrix} \bar{a}_1 & & 0 \\ & \ddots & \\ 0 & & \bar{a}_0 \end{bmatrix},$$

and consequently $\chi(x^{-1}) = \operatorname{tr} \hat{T}(x^{-1}) = \sum_1^n \bar{a}_i = \overline{\chi(x)}$.

Exercise 8.4. Suppose $\hat{T}$ is a matrix representation of G. Define its *contragredient* matrix representation $\hat{T}^*$ by setting $\hat{T}^*(x) = \hat{T}(x^{-1})^t$, all $x \in G$. Show that $\hat{T}^*$ is a representation. If χ is the character of $\hat{T}$ show that $\hat{T}^*$ has character $\bar{\chi}$. Show that $\hat{T}^*$ is irreducible if and only if $\hat{T}$ is irreducible.

If ϕ, θ are functions from G to $\mathbb{C}$ define their *inner product* to be

$$(\phi, \theta) = |G|^{-1} \sum \{\phi(x)\theta(x^{-1}) : x \in G\}.$$

Then the inner product is easily seen to be a symmetric bilinear form on the vector space of all functions from G to $\mathbb{C}$. Note that if ϕ and θ are characters of G, then (ϕ, θ) is real (use Proposition 8.8), and for any character χ we have

$$(\chi, \chi) = |G|^{-1} \sum \{|\chi(x)|^2 : x \in G\} > 0.$$

Theorem 8.9 (The First Orthogonality Relation). Suppose $\chi_1, \chi_2, \ldots, \chi_k$ are all the irreducible characters of G. Then $(\chi_i, \chi_j) = \delta_{ij}$.

Proof. Take $i \neq j$ and let T and S be representation whose characters are χ_i and χ_j, respectively. Choose bases and say $\hat{T}(x) = (t_{ij}(x))$, $\hat{S}(x) = (s_{ij}(x))$, all $x \in G$. Then

$$(\chi_i, \chi_j) = |G|^{-1} \sum_x \chi_i(x)\chi_j(x^{-1}) = |G|^{-1} \sum_x \sum_{i,j} t_{ii}(x) s_{jj}(x^{-1}) = 0$$

by Proposition 8.4(i), since S and T are inequivalent by Proposition 8.5. On the other hand,

$$\begin{aligned}(\chi_i, \chi_i) &= |G|^{-1} \sum_x \sum_{i,j} t_{ii}(x) t_{jj}(x^{-1}) \\ &= |G|^{-1} \sum_{i,j} \delta_{ij} |G|/\chi_i(1) \\ &= |G|^{-1} \chi_i(1) |G|/\chi_i(1) = 1\end{aligned}$$

by Proposition 8.4(ii).

Corollary 1. The characters $\chi_1, \chi_2, \ldots, \chi_k$ are linearly independent.

Proof. Suppose $\sum_{i=1}^k a_i \chi_i = 0$, $a_i \in \mathbb{C}$. Then

$$0 = (0, \chi_j) = \left(\sum_i a_i \chi_i, \chi_j\right) = \sum_i a_i (\chi_i, \chi_j) = a_j \qquad \text{for all } j.$$

We observed in the corollary to Proposition 8.7 that a character χ of G can be written as a sum of irreducible characters. Thus if $\chi_1, \chi_2, \ldots, \chi_k$ are all the irreducible characters of G we may write

$$\chi = n_1 \chi_1 + n_2 \chi_2 + \cdots + n_k \chi_k,$$

where each n_i is a nonnegative integer, the *multiplicity* of χ_i as a *constituent* of χ. By Corollary 1 above the coefficients n_i are uniquely determined by χ.

We draw two more corollaries from Theorem 8.9.

Corollary 2. If χ is a character of G, then $\chi = \sum_{i=1}^k (\chi, \chi_i)\chi_i$, the coefficient (χ, χ_i) being the multiplicity n_i of χ_i as a constituent of χ.

Proof. We may write $\chi = \sum_j n_j \chi_j$, and so $(\chi, \chi_i) = \sum_j n_j (\chi_j, \chi_i) = n_i$.

Corollary 3. If $\chi = \sum_i n_i \chi_i$ and $\psi = \sum_j m_j \chi_j$ are two characters of G, then $(\chi, \psi) = \sum_{i=1}^k n_i m_i$. In particular, $(\chi, \chi) = \sum_i n_i^2$.

The next result is really just a further corollary of Theorem 8.9, but it is of considerable importance as a test for irreducibility so we list it as a separate theorem.

Theorem 8.10. A character χ of G is irreducible if and only if $(\chi, \chi) = 1$.

Proof. If χ is irreducible, then $(\chi, \chi) = 1$ by Theorem 8.9. If χ is reducible write $\chi = \sum_i n_i \chi_i$, with either some $n_i > 1$ or else at least two nonzero multiplicities. Thus $(\chi, \chi) = \sum_i n_i^2 > 1$ by Corollary 3 of Theorem 8.9.

Exercise 8.5. If T and S are two representations of G having the same character χ show that $T \sim S$.

Proposition 8.11. Denote by ρ the character of the left regular permutation representation R of G. Then $\rho(1) = |G|$ and $\rho(x) = 0$ if $1 \neq x \in G$. If $\chi_1, \ldots, \chi_k$ are all the irreducible characters of G, then $\rho = \sum_{i=1}^k \chi_i(1)\chi_i$.

Proof. Since R has degree $|G|$ and $\hat{R}(1) = I$ it is clear that $\rho(1) = |G|$. If $1 \neq x \in G$, then $xy \neq y$ for each $y \in G$, and consequently every diagonal entry of $\hat{R}(x)$ is 0. Thus $\rho(x) = 0$. By Corollary 2 of Theorem 8.9 the multiplicity of χ_i in ρ is

$$(\rho, \chi_i) = |G|^{-1} \sum_x \rho(x)\chi_i(x^{-1}) = \chi_i(1).$$

Corollary. $\sum_{i=1}^k \chi_i(1)^2 = |G|$.

Proof. $\rho(1) = \sum_{i=1}^k \chi_i(1)\chi_i(1) = |G|$.

Proposition 8.12 (Burnside's Orbit Formula). Suppose G acts on a finite set S and let θ denote the character of the permutation representation. Then the number of G-orbits in S is $(\theta, 1_G)$.

Proof. See Theorem 2.2 above in this chapter. Also see Exercise 8.3.

Let $\chi_1 = 1_G, \chi_2, \ldots, \chi_k$ be the distinct irreducible characters of G, and let $K_1 = \{1\}, K_2, \ldots, K_m$ be the conjugacy classes of G. For $1 \leq i \leq k$ and $1 \leq j \leq m$ define $\omega_{ij} = |K_j|\chi_i(K_j)/\chi_i(1)$.

Proposition 8.13. Suppose T_i is an irreducible representation with character χ_i and K_j is a conjugacy class of G. Set $M_{ij} = \sum\{T_i(x): x \in K_j\}$. Then $M_{ij} = \omega_{ij}1 \in C(T_i)$.

Proof. If $y \in G$, then

$$T_i(y)^{-1} M_{ij} T_i(y) = \sum\{T_i(y^{-1}xy): x \in K_j\} = M_{ij},$$

so $M_{ij} \in C(T_i)$. By Proposition 8.3 $M_{ij} = a1$, $a \in \mathbb{C}$. Taking traces we see

$$\begin{aligned}\operatorname{tr}(M_{ij}) = a\chi_i(1) &= \sum\{\operatorname{tr} T_i(x): x \in K_j\} \\ &= \sum\{\chi_i(x): x \in K_j\} = |K_j|\chi_i(K_j).\end{aligned}$$

Thus $a = |K_j|\chi_i(K_j)/\chi_i(1) = \omega_{ij}$.

Proposition 8.14. If $x \in K_s$ set $n_{ijs} = |\{(y, z) \in K_i \times K_j: yz = x\}|$. Then

(i) n_{ijs} depends only on i, j, and s, and not on the choice of x in K_s.
(ii) $|K_i||K_j|\chi_t(K_i)\chi_t(K_j) = \chi_t(1)\sum_s n_{ijs}|K_s|\chi_t(K_s)$, and
(iii) $\omega_{ti}\omega_{tj} = \sum_s n_{ijs}\omega_{ts}$ for all i, j, and t.

Proof. (i) If $y \in K_i$, $z \in K_j$, and $w \in G$, then $x = yz$ if and only if $w^{-1}xw = w^{-1}yw \cdot w^{-1}zw$.

(ii) With the notation of Proposition 8.13 we see that

$$\begin{aligned} M_{ti}M_{tj} &= \sum\{T_t(y): y \in K_i\}\sum\{T_t(z): z \in K_j\} \\ &= \sum\{T_t(yz):(y, z) \in K_i \times K_j\} \\ &= \sum_s \sum\{n_{ijs}T_t(x): x \in K_s\} = \sum_s n_{ijs}M_{ts}. \end{aligned}$$

Take traces.

(iii) Divide by $\chi_t(1)^2$ in (ii) and apply Proposition 8.13.

Theorem 8.15 (The Second Orthogonality Relation). If $\chi_1, \ldots, \chi_k$ are the distinct irreducible characters of G and K_i, K_j are conjugacy classes of G, then

$$\sum_{t=1}^{k} \chi_t(K_i)\chi_t(K_j^{-1}) = \delta_{ij}|G|/|K_j|.$$

Proof. Write K_m for the conjugacy class K_j^{-1}. By Proposition 8.14(ii) we see

$$\begin{aligned} |K_i|\,|K_m|\sum_t \chi_t(K_i)\chi_t(K_m) &= \sum_s n_{ims}|K_s|\sum_t \chi_t(1)\chi_t(K_s) \\ &= \sum_s n_{ims}|K_s|\rho(K_s) = n_{im1}|G| \end{aligned}$$

by Proposition 8.11. But

$$n_{im1} = |\{(y, z) \in K_i \times K_j^{-1}: yz = 1\}| = \delta_{ij}|K_i|.$$

Thus, since $|K_j^{-1}| = |K_j|$,

$$|K_i|\,|K_j|\sum_t \chi_t(K_i)\chi_t(K_j^{-1}) = \delta_{ij}|G|\,|K_i|,$$

and the stated result follows.

Theorem 8.16. The number k of irreducible characters of G is equal to the class number m of G.

Proof. Let G act on itself by conjugation, so the G-orbits are the m conjugacy classes. If θ is the character of the permutation action, then $\theta(x) = |C_G(x)|$ for each $x \in G$. By Burnside's Orbit Formula (8.12)

$$\begin{aligned} m = (\theta, 1_G) &= |G|^{-1}\sum\{\theta(x): x \in G\} = |G|^{-1}\sum\{|C_G(x)|: x \in G\} \\ &= |G|^{-1}\sum_{x \in G}\sum_{t=1}^{k} \chi_t(x)\chi_t(x^{-1}) \end{aligned}$$

by the Second Orthogonality Relation (8.15), since $|C_G(x)| = |G|/|\mathrm{cl}(x)|$. Thus

$$m = \sum_{t=1}^{k} |G|^{-1}\sum_{x \in G} \chi_t(x)\chi_t(x^{-1}) = \sum_{t=1}^{k} (\chi_t, \chi_t) = k.$$

Corollary 1. The set $\{\chi_1, \chi_2, \ldots, \chi_k\}$ of irreducible characters is a basis for the vector space cf(G) of complex-valued class functions on G.

Proof. The dimension of cf(G) is equal to the class number $m = k$, and the k irreducible characters are linearly independent class functions by Corollary 1 to Theorem 8.9.

Corollary 2. Every irreducible character χ_i is linear [i.e., $\chi_i(1) = 1$] if and only if G is abelian.

Proof. If $\chi_i(1) = 1$ for all i, then by the corollary to Proposition 8.11 we have $|G| = \sum_{i=1}^{k} |\chi_i(1)|^2 = k$, so every conjugacy class has just one element and $G = Z(G)$ is abelian. The converse was the corollary to Proposition 8.3.

The *character table* of G is the matrix whose rows are indexed by the irreducible characters $\chi_1, \ldots, \chi_k$ and whose columns are indexed by the conjugacy classes $K_1, \ldots, K_k$. By Theorem 8.15 distinct columns of the character table are orthogonal relative to the usual Hermitian inner product of complex column vectors.

For an easy example take $G = S_3$, with $K_1 = \{1\}$, $K_2 = \text{cl}(123)$, and $K_3 = \text{cl}(12)$. Besides the principal character $\chi_1 = 1_G$ there is a linear character χ_2, with $\chi_2(K_1) = \chi_2(K_2) = 1$, $\chi_2(K_3) = -1$ (see Example 3, page 260). It is immediate from the corollary to Proposition 8.11 and the Second Orthogonality Relation that the remaining character χ_3 has $\chi_3(K_1) = 2$, $\chi_3(K_2) = -1$, and $\chi_3(K_3) = 0$. Thus the character table is as indicated.

	K_1	K_2	K_3
χ_1	1	1	1
χ_2	1	1	-1
χ_3	2	-1	0

Exercise 8.6. Compute the character table of the quaternion group Q_2.

For the remainder of this section it will be convenient to write Irr(G) for the set $\{\chi_1, \chi_2, \ldots, \chi_k\}$ of irreducible characters of a group G.

If T is a representation of G with character χ define the *kernel* of χ to be $\ker \chi = \ker T$. It is an easy consequence of Exercise 8.5 that $\ker \chi$ is well defined, i.e., it depends only on χ and not on T.

Proposition 8.17. If χ is a character of G, then

$$\ker \chi = \{x \in G : \chi(x) = \chi(1)\}.$$

Proof. Let T be a representation with character χ. If $x \in \ker \chi$, then $T(x) = I$ so $\chi(x) = \text{tr}\, I = \chi(1)$. Suppose then that $\chi(x) = \chi(1)$. As in the proof of Proposition 8.8 we may assume that $\hat{T}(x)$ is a diagonal matrix whose entries

$a_1, \ldots, a_n$ are roots of unity in $\mathbb{C}$. Thus

$$\chi(x) = \sum_i a_i = \chi(1) = \sum_i |a_i|,$$

since each $|a_i| = 1$. Since equality holds in the triangle inequality we conclude that $a_i = b_i \cdot a_1$, with $0 < b_i \in \mathbb{R}$, for all i, and consequently $1 = |a_i| = b_i|a_1| = b_i$ for all i. But then $\chi(x) = \chi(1)a_1 = \chi(1)$, so $a_1 = 1$ and hence $\hat{T}(x) = I$ and $x \in \ker \chi$.

The ideas in the proof of Proposition 8.17 enable us to describe another useful subgroup of G associated with a character χ. Set $Z(\chi) = \{x \in G : |\chi(x)| = \chi(1)\}$. Thus $\ker \chi \subseteq Z(\chi)$. If T is a representation with character χ and if $x \in Z(\chi)$ we may assume, as in the proof above, that $\hat{T}(x)$ is diagonal, with entries $a_1, \ldots, a_n$ that are roots of unity in $\mathbb{C}$. As in the proof $a_i = b_i a_1$ for all i, with $0 < b_i \in \mathbb{R}$, and in fact each $b_i = 1$, so $\hat{T}(x) = a_1 1$. It follows easily that $Z(\chi)$ consists precisely of those $x \in G$ that are represented by scalar multiples of the identity in any representation with character χ. Two immediate consequences are recorded in the next proposition.

Proposition 8.18. If χ is a character of G, then $Z(\chi) \lhd G$. If χ is faithful, then $Z(\chi) \leq Z(G)$.

Suppose $K \lhd G$ and let T be a representation of G with character χ. If $K \leq \ker \chi$ we may without ambiguity define a representation $\tilde{T}$ of G/K by means of $\tilde{T}(xK) = T(x)$, all $xK \in G/K$. Thus the character $\tilde{\chi}$ of $\tilde{T}$ is given by $\tilde{\chi}(xK) = \chi(x)$, all $x \in G$. Conversely, if $\hat{T}$ is a representation of G/K with character $\tilde{\chi}$, and if $\eta: G \to G/K$ is the quotient map, then the composition $\hat{T}\eta$ is a representation T of G whose character $\chi = \tilde{\chi}\eta$, i.e., $\chi(x) = \tilde{\chi}(xK)$, and it is clear that $K \leq \ker \chi$. It is also clear in both cases that $\tilde{\chi}$ is irreducible if and only if χ is irreducible. It is customary to write $\tilde{\chi} = \chi$ and let the context determine whether χ is being viewed as a character of G or of G/K.

In particular, if $K \lhd G$, then $\mathrm{Irr}(G/K)$ consists of those $\chi \in \mathrm{Irr}(G)$ such that $K \leq \ker \chi$.

Proposition 8.19. The set of linear (i.e., degree 1) characters of G is $\mathrm{Irr}(G/G')$.

Proof. Since G/G' is abelian each $\chi \in \mathrm{Irr}(G/G')$ is linear by Corollary 2, Theorem 8.16. If $\chi \in \mathrm{Irr}(G)$ is linear, then $\chi: G \to \mathbb{C}^*$ is a homomorphism. Thus $G' \leq \ker \chi$, since $\mathbb{C}^*$ is abelian, and so $\chi \in \mathrm{Irr}(G/G')$.

Exercise 8.7. Compute the character tables of the dihedral groups D_4 and D_5. Note that in both cases the group elements are naturally represented as 2×2 matrices.

Recall from Chapter III that $\mathbb{A}$ denotes the field of algebraic numbers, i.e., all $a \in \mathbb{C}$ that are algebraic over $\mathbb{Q}$. Say that $a \in \mathbb{A}$ is an *algebraic integer* if its (monic) minimal polynomial $m_{\mathbb{Q}}(x)$ is in $\mathbb{Z}[x]$. Thus, for example, $\sqrt{2}$ is an algebraic integer but $\sqrt{2}/2$ is not.

Proposition 8.20. The following are equivalent:

(i) a is an algebraic integer;
(ii) $\mathbb{Z}[a]$ is a finitely generated $\mathbb{Z}$-module;
(iii) a is a root of a monic polynomial $f(x)$ in $\mathbb{Z}[x]$.

Proof. (i) $\Rightarrow$ (ii): If the minimal polynomial of a is

$$m(x) = a_0 + a_1 x + \cdots + x^k, \qquad a_i \in \mathbb{Z},$$

then

$$a^k = -a_0 \cdot 1 - a_1 \cdot a - \cdots - a_{k-1} \cdot a^{k-1} \in \mathbb{Z} \cdot 1 + \cdots + \mathbb{Z} \cdot a^{k-1},$$

and it follows easily that $\{1, a, \ldots, a^{k-1}\}$ is a generating set for the $\mathbb{Z}$-module $\mathbb{Z}[a]$.

(ii) $\Rightarrow$ (iii): Let $\{f_1(a), f_2(a), \ldots, f_m(a)\}$ be a generating set for $\mathbb{Z}[a]$, with $f_i(x) \in \mathbb{Z}[x]$, all i. Set

$$n = \max\{1 + \deg f_i(x) : 1 \leq i \leq m\}.$$

Since $a^n \in \mathbb{Z}[a]$ we may write $a^n = \sum_{i=1}^m b_i f_i(a)$, $b_i \in \mathbb{Z}$. Set

$$f(x) = x^n - \sum_{i=1}^m b_i f_i(x) \in \mathbb{Z}[x].$$

Then $f(x)$ is monic and $f(a) = 0$.

(iii) $\Rightarrow$ (i): Let $m(x) \in \mathbb{Q}[x]$ be the minimal polynomial of a. Then $f(x) = m(x)g(x)$ for some $g(x) \in \mathbb{Q}[x]$. Write $m(x) = (a/b)h(x)$ and $g(x) = (c/d)k(x)$, with $a, b, c, d \in \mathbb{Z}$ and $h(x)$, $k(x)$ primitive polynomials in $\mathbb{Z}[x]$. By Gauss's Lemma (Theorem II.5.13) $h(x)k(x)$ is also primitive. But then, since $bdf(x) = ach(x)k(x)$, we see that $bd = ac$, and hence $f(x) = h(x)k(x)$. It follows that $h(x)$ and $k(x)$ are monic, and hence $m(x) = h(x) \in \mathbb{Z}[x]$.

Corollary 1. If $a, b \in \mathbb{A}$ are algebraic integers, then $a + b$ and ab are algebraic integers.

Proof. Let $\{a_1, \ldots, a_m\}$ and $\{b_1, \ldots, b_n\}$ be sets of generators for $\mathbb{Z}[a]$ and $\mathbb{Z}[b]$ as $\mathbb{Z}$-modules. Then $\{a_i, b_j\}$ is a finite set of generators for $\mathbb{Z}[a, b]$, and $\mathbb{Z}[a + b]$, $\mathbb{Z}[ab]$ are submodules of $\mathbb{Z}[a, b]$ so they are finitely generated by the corollary to Theorem IV.2.9.

Corollary 2. If $a \in \mathbb{Q}$ and a is an algebraic integer, then $a \in \mathbb{Z}$.

Proof. The minimal polynomial over $\mathbb{Q}$ for a is $m(x) = x - a$, and $m(x) \in \mathbb{Z}[x]$ so $a \in \mathbb{Z}$.

If χ is a character of G, $|G| = n$, and $x \in G$, then $\chi(x) = a_1 + a_2 + \cdots + a_k$, where $k = \chi(1)$ and each a_i is an nth root of unity in $\mathbb{C}$ (see the proof of Proposition 8.8). But then each a_i is a root of $x^n - 1 \in \mathbb{Z}[x]$. By Proposition 8.20 and its first corollary we see that $\chi(x)$ is an algebraic integer. Note also that

$$|\chi(x)| \leq \sum_i |a_i| = k = \chi(1) \qquad \text{for all} \quad x \in G.$$

Recall that if $\mathrm{Irr}(G) = \{\chi_1, \ldots, \chi_k\}$ and $\{K_1, \ldots, K_k\}$ are the conjugacy classes in G, then $\omega_{ij} = |K_j|\chi_i(K_j)/\chi_i(1)$.

Proposition 8.21. For all i and j ω_{ij} is an algebraic integer.

Proof. By Proposition 8.14(iii) we have $\omega_{ti}\omega_{tj} = \sum_s n_{ijs}\omega_{ts}$ for all t, with $n_{ijs} \in \mathbb{Z}$. Fix i and t, let v be the column vector whose transpose is $v^t = (\omega_{t1}, \ldots, \omega_{tk})$, and let N be the $k \times k$ matrix whose j-s entry is n_{ijs}. Then $Nv = \omega_{ti}v$, so ω_{ti} is an eigenvalue of N and consequently ω_{ti} is a root of a monic polynomial in $\mathbb{Z}[x]$. Apply Proposition 8.20 and replace t, i by i, j.

Theorem 8.22. If $\mathrm{Irr}(G) = \{\chi_1, \ldots, \chi_k\}$, then each degree $\chi_i(1)$ is a divisor of $|G|$.

Proof. Since χ_i is a class function we have

$$1 = (\chi_i, \chi_i) = |G|^{-1} \sum_{j=1}^{k} |K_j|\chi_i(K_j)\chi_i(K_j^{-1}).$$

Thus

$$\begin{aligned} |G|/\chi_i(1) &= \big(|G|/\chi_i(1)\big)(\chi_i, \chi_i) \\ &= \big(1/\chi_i(1)\big) \sum_{j=1}^{k} |K_j|\chi_i(K_j)\chi_i(K_j^{-1}) = \sum_{j=1}^{k} \omega_{ij}\chi_i(K_j^{-1}), \end{aligned}$$

which is an algebraic integer by Proposition 8.21 and Corollary 1 to Proposition 8.20. But then $|G|/\chi_i(1)$ is both a rational number and an algebraic integer, so $|G|/\chi_i(1) \in \mathbb{Z}$.

Proposition 8.23. Suppose $\chi_i \in \mathrm{Irr}\, G$ and K_j is a conjugacy class of G, and that $\chi_i(1)$ and $|K_j|$ are relatively prime. Then either $\chi_i(K_j) = 0$ or else $|\chi_i(K_j)| = \chi_i(1)$, in which case $K_j \subseteq Z(\chi_i)$.

Proof. Choose $a, b \in \mathbb{Z}$ for which $a\chi_i(1) + b|K_j| = 1$. Multiply by $\chi_i(K_j)/\chi_i(1)$ to see that

$$a\chi_i(K_j) + b|K_j|\chi_i(K_j)/\chi_i(1) = \chi_i(K_j)/\chi_i(1)$$

or

$$a\chi_i(K_j) + b\omega_{ij} = \chi_i(K_j)/\chi_i(1).$$

Thus $\chi_i(K_j)/\chi_i(1)$ is an algebraic integer, by Proposition 8.21, and $|\chi_i(K_j)/\chi_i(1)| \leq 1$ as observed earlier. If the elements of K_j have order n let $\zeta \in \mathbb{C}$ be a primitive nth root of unity and let $\mathscr{G}$ be the Galois group $G(\mathbb{Q}(\zeta):\mathbb{Q})$. Since $\chi_i(K_j)$ is a sum of powers of ζ the same is true of $\sigma(\chi_i(K_j))$ for all $\sigma \in \mathscr{G}$. Furthermore, $\sigma(\chi_i(K_j)/\chi_i(1))$ is an algebraic integer, and $|\sigma(\chi_i(K_j)/\chi_i(1))| \leq 1$. Set

$$u = \prod\{\sigma(\chi_i(K_j)/\chi_i(1)) : \sigma \in \mathscr{G}\}.$$

Then u is an algebraic integer, $|u| \leq 1$, and, since u is fixed by all $\sigma \in \mathscr{G}$, $u \in \mathbb{Q}$. By Proposition 8.20, Corollary 2, $u \in \mathbb{Z}$, and hence $u = 0$ or $u = \pm 1$. If $u = 0$, then $\chi_i(K_j) = 0$. If $u = \pm 1$ take $\sigma = 1$ to see that $|\chi_i(K_j)| = \chi_i(1)$; hence $K_j \subseteq Z(\chi_i)$.

Theorem 8.24 (Burnside). If G is a nonabelian simple group, then the only conjugacy class having a prime power number of elements is $K_1 = \{1\}$.

Proof. Suppose K_j is a conjugacy class with $|K_j| = p^m$ for some prime p, and suppose $K_j \neq K_1$. Note that $m \geq 1$, since $Z(G) = 1$. Temporarily relabel the irreducible characters $\chi_2, \ldots, \chi_k$ so that $p \nmid \chi_i(1)$ if $2 \leq i \leq k_0$ but $p \mid \chi_i(1)$ if $k_0 + 1 \leq i \leq k$. Since G is simple $Z(\chi_i) = 1$ for all $i > 1$. By Proposition 8.23 it follows that $\chi_i(K_j) = 0$ for $2 \leq i \leq k_0$. Say $\chi_i(1) = pn_i$ for $k_0 + 1 \leq i \leq k$. Apply the Second Orthogonality relation to columns 1 and j of the character table to obtain

$$0 = 1 + p\sum\{n_i\chi_i(K_j) : k_0 + 1 \leq i \leq k\}.$$

But then

$$\alpha = \sum\{n_i\chi_i(K_j) : k_0 + 1 \leq i \leq k\}$$

is an algebraic integer, and $\alpha = -1/p \in \mathbb{Q}\backslash\mathbb{Z}$, contradicting Corollary 2 to Proposition 8.20.

The final theorem in this section is a first indication of the great power of character-theoretic methods for finite groups. It should be compared with the application of the Sylow theorems to groups of order pq in Chapter I.

Theorem 8.25 (Burnside). If p and q are distinct primes and G is a group of order $p^a q^b$, then G is solvable.

Proof. If G is not solvable it has a composition factor H that is simple and nonabelian, with $|H| = p^c q^d$. Note that $c \geq 1$ and $d \geq 1$ since p-groups are solvable. Let P be p-Sylow in H, choose $x \neq 1$ in $Z(P)$, and let K be the H-conjugacy class of x. Then $P \leq C_H(x)$, so $|K| = [H:C_H(x)]$ is a divisor of $[H:P] = q^d$, and $|K|$ is a prime power. By Theorem 8.24 $|K| = 1$, so $1 \neq x \in Z(H)$, contradicting the simplicity of H.

Corollary. A finite nonabelian simple group has at least three distinct primes dividing its order.

9. SOME GALOIS GROUPS

The determination of Galois groups for specific polynomials seems to be more of an art than a science. We present in this section some ad hoc methods that can be useful, and a number of examples.

Suppose F is a field, $f(x) \in F[x]$ is monic of degree n, and $K \supseteq F$ is a splitting field for $f(x)$. If $f(x)$ has roots $a_1, a_2, \ldots, a_n$ in K define the *discriminant* of $f(x)$ to be

$$D = D_f = \prod\{(a_i - a_j)^2 : 1 \le i < j \le n\}.$$

Note that $D = 0$ if and only if $f(x)$ has a repeated root in K. If $f(x)$ has no repeated roots, then K is a Galois extension of F. If $G = G(K:F)$, then each $\sigma \in G$ permutes the roots $\{a_i\}$ of $f(x)$, so clearly $\sigma(D) = D$, and hence $D \in \mathscr{F}G = F$. In particular D does not depend on the choice of the splitting field K.

In fact, D is left fixed by all possible permutations of $a_1, a_2, \ldots, a_n$ and consequently, by Theorem III.5.3, D is a polynomial over F in the elementary symmetric polynomials σ_j evaluated at $a_1, \ldots, a_n$, i.e., in the coefficients of $f(x)$.

In order to be able to evaluate some discriminants it will be useful to establish some classical results. For the next three propositions we take R to be a commutative ring with 1; $x_1, \ldots, x_n$ to be distinct indeterminates over R; and D to be the "generic discriminant" over R; i.e.,

$$D = \prod\{(x_i - x_j)^2 : 1 \le i < j \le n\}.$$

Proposition 9.1 (Vandermonde). Define $d = d(x_1, \ldots, x_n)$ to be

$$\prod\{x_i - x_j : 1 \le j < i \le n\},$$

so $d^2 = D$. Then

$$d = \begin{vmatrix} 1 & 1 & \cdots & 1 \\ x_1 & x_2 & \cdots & x_n \\ x_1^2 & x_2^2 & \cdots & x_n^2 \\ \vdots & \vdots & & \vdots \\ x_1^{n-1} & x_2^{n-1} & \cdots & x_n^{n-1} \end{vmatrix}.$$

Proof. The result is clear if $n = 2$, so suppose $n \ge 3$ and assume the result to hold for the $n - 1$ indeterminates $x_2, \ldots, x_n$. Multiply row $n - 1$ in the determinant by $-x_1$ and add to row n, then multiply row $n - 2$ by $-x_1$ and add to row $n - 1, \ldots,$ and finally multiply row 1 by $-x_1$ and add to row 2.

Then expand along the first column to see that the determinant is equal to

$$\begin{vmatrix} x_2 - x_1 & x_3 - x_1 & \cdots & x_n - x_1 \\ x_2^2 - x_2x_1 & x_3^2 - x_3x_1 & \cdots & x_n^2 - x_nx_1 \\ \vdots & \vdots & & \vdots \\ x_2^{n-1} - x_2^{n-2}x_1 & x_3^{n-1} - x_3^{n-2}x_1 & \cdots & x_n^{n-1} - x_n^{n-2}x_1 \end{vmatrix}$$

$$= \prod\{x_i - x_1 : 2 \le i \le n\} \begin{vmatrix} 1 & 1 & \cdots & 1 \\ x_2 & x_3 & \cdots & x_n \\ \vdots & \vdots & & \vdots \\ x_2^{n-2} & x_3^{n-2} & \cdots & x_n^{n-2} \end{vmatrix}.$$

The proposition follows by induction.

For each positive integer k the kth *power sum* of $x_1, x_2, \ldots, x_n$ is

$$s_k = x_1^k + x_2^k + \cdots + x_n^k \in R[x_1, x_2, \ldots, x_n].$$

Since each s_k is clearly a symmetric polynomial it can be expressed, by Theorem III.5.3, as a polynomial in the elementary symmetric polynomials. The next proposition provides a recursive scheme for expressing s_k in terms of $\sigma_1, \sigma_2, \ldots, \sigma_n$.

Proposition 9.2 (Newton's Identities). The kth power sum s_k satisfies

$$\begin{aligned} s_k &= s_{k-1}\sigma_1 - s_{k-2}\sigma_2 + \cdots + (-1)^k s_1\sigma_{k-1} + (-1)^{k+1}k\sigma_k \qquad &\text{if} \quad k \le n, \\ &= s_{k-1}\sigma_1 - s_{k-2}\sigma_2 + \cdots + (-1)^{n+1}s_{k-n}\sigma_n &\text{if} \quad k > n. \end{aligned}$$

Proof. We sketch a proof in terms of a generating function, a standard device in combinatorial theory. Let z be an indeterminate over $R[x_1, x_2, \ldots, x_n]$ and define

$$\sigma(z) = \prod\{1 - x_iz : 1 \le i \le n\}.$$

Observe that

$$\sigma(z) = 1 - \sigma_1 z + \sigma_2 z^2 - \cdots + (-1)^n\sigma_n z^n.$$

Define $s(z)$ to be the generating function for the sequence $\{s_k\}$, i.e., $s(z)$ is the formal power series $s(z) = \sum_{k=1}^{\infty} s_k z^k$. Then

$$\sigma'(z) = -\sum_{i=1}^{n} x_i \prod\{1 - x_jz : 1 \le j \le n, j \ne i\},$$

so

$$\frac{-z\sigma'(z)}{\sigma(z)} = \sum_{i=1}^{n} \frac{x_iz}{1 - x_iz} = \sum_{i=1}^{n}\sum_{k=1}^{\infty} x_i^k z^k$$

$$= \sum_{k=1}^{\infty}\left(\sum_{i=1}^{n} x_i^k\right)z^k = \sum_{k=1}^{\infty} s_k z^k = s(z),$$

i.e., $s(z)\sigma(z) = -z\sigma'(z)$. Thus (setting $\sigma_0 = 1$)

$$\left(\sum_{k=1}^{\infty} s_k z^k\right)\left(\sum_{i=0}^{n} (-1)^i \sigma_i z^i\right) = \sum_{j=1}^{n} (-1)^{j+1} j\sigma_j z^j.$$

Equating coefficients of z^k we obtain

$$s_k - s_{k-1}\sigma_1 + s_{k-2}\sigma_2 - \cdots + (-1)^{k+1} s_1 \sigma_{k-1} = (-1)^{k+1} k\sigma_k$$

if $k \leq n$ and

$$s_k - s_{k-1}\sigma_1 + s_{k-2}\sigma_2 - \cdots + (-1)^n s_{k-n}\sigma_n = 0$$

if $k > n$.

Proposition 9.3.

$$D = \begin{vmatrix} n & s_1 & \cdots & s_{n-1} \\ s_1 & s_2 & \cdots & s_n \\ s_2 & s_3 & \cdots & s_{n+1} \\ \vdots & \vdots & & \vdots \\ s_{n-1} & s_n & \cdots & s_{2n-2} \end{vmatrix}.$$

Proof. If A is the matrix with ij-entry x_j^i, $0 \leq i \leq n-1$, $1 \leq j \leq n-1$, then $\det A = d$ by Proposition 9.1, and so

$$D = d^2 = (\det A)(\det A^t) = \det(AA^t).$$

But AA^t is the matrix whose determinant is displayed in the statement of the proposition.

Corollary. The discriminant D_f of any polynomial $f(x)$ can be obtained as a polynomial in the coefficients of $f(x)$.

Proof. The coefficients of $f(x)$ are (to within sign) the elementary symmetric polynomials $\sigma_1, \sigma_2, \ldots, \sigma_n$ evaluated at the roots of $f(x)$. By the proposition D_f can be evaluated in terms of the power sums s_k, and then each s_k can be expressed in terms of the σ_i [i.e., the coefficients of $f(x)$] by means of Newton's identities.

It should be remarked that the discriminant D_f can also be obtained in terms of the *resultant* $R(f, f')$ of $f(x)$ and its derivative $f'(x)$ (see Van der Waerden [37, p. 87].) It is not necessary to compute the power sums s_k in order to compute $R(f, f')$. However, $R(f, f')$ is usually computed as the determinant of a $(2n-1) \times (2n-1)$ matrix, so there is no free lunch.

Some examples are in order.

EXAMPLES

1. If $f(x) = x^2 + bx + c \in F[x]$, then $\sigma_1 = -b$, $\sigma_2 = c$, $s_1 = \sigma_1$, and

$$s_2 = s_1\sigma_1 - 2\sigma_2 = \sigma_1^2 - 2\sigma_2 = b^2 - 2c.$$

Thus

$$D_f = \begin{vmatrix} 2 & -b \\ -b & b^2 - 2c \end{vmatrix} = b^2 - 4c.$$

Thus the familiar quadratic formula gives the roots of $f(x)$ as $\frac{1}{2}(-b \pm \sqrt{D_f})$ if char $F \neq 2$.

2. Suppose $f(x) = x^3 + ax^2 + bx + c \in F[x]$, with char $F \neq 3$. If we use $x = y - a/3$ to change variables then $f(x)$ takes the form $f(x) = g(y) = y^3 + py + q \in F[y]$. (Verify.) Since the roots of $g(y)$ are those of $f(x)$ each diminished by $a/3$ it is clear that $D_g = D_f$. For $g(y)$ we have $\sigma_1 = 0$, $\sigma_2 = p$, and $\sigma_3 = -q$. Newton's Identities yield $s_1 = 0$, $s_2 = -2p$, $s_3 = -3q$, and $s_4 = 2p^2$. Thus

$$D = \begin{vmatrix} 3 & 0 & -2p \\ 0 & -2p & -3q \\ -2p & -3q & 2p^2 \end{vmatrix} = -4p^3 - 27q^2.$$

It may be instructive to carry out the classical solution of the cubic equation $g(y) = 0$ in order to see the natural role played by the discriminant D_g. Continue to assume that char $F \neq 3$, and assume also that char $F \neq 2$.

Write $y = u + v$ with u and v satisfying $3uv = -p$. Then

$$y^3 = u^3 + v^3 + 3uv(u + v) = u^3 + v^3 + 3uvy,$$

and substituting into $g(y) = 0$ we obtain $u^3 + v^3 = -q$. If z is another (independent) indeterminate we have, since $u^3v^3 = -p^3/27$, that

$$(z - u^3)(z - v^3) = z^2 + qz - \frac{p^3}{27}.$$

If we set $h(z) = z^2 + qz - p^3/27$, then by Examples 1 and 2 above $D_h = q^2 + 4p^3/27 = -D_g/27$, and by the quadratic formula we may take $u^3 = -q/2 + \sqrt{-D_g/108}$ and $v^3 = -q/2 - \sqrt{-D_g/108}$. If we write $u = (-q/2 + \sqrt{-D_g/108})^{1/3}$ and $v = (-q/2 - \sqrt{-D_g/108})^{1/3}$ there are generally three values determined for each of u and v. The values must be paired, however, so that $3uv = -p$, giving three values (at most) for $y = u + v$. If $u = u_1$ and $v = v_1$ are one paired choice for cube roots, and if $\omega \neq 1$ is a cube root of unity, then the solutions to $g(y) = 0$ are

$$\begin{aligned}
y_1 &= u_1 + v_1 = (-q/2 + \sqrt{-D_g/108})^{1/3} + (-q/2 - \sqrt{-D_g/108})^{1/3},\\
y_2 &= \omega u_1 + \omega^2 v_1 = \omega(-q/2 + \sqrt{-D_g/108})^{1/3}\\
&\quad + \omega^2(-q/2 - \sqrt{-D_g/108})^{1/3},\\
y_3 &= \omega^2 u_1 + \omega v_1 = \omega^2(-q/2 + \sqrt{-D_g/108})^{1/3}\\
&\quad + \omega(-q/2 - \sqrt{-D_g/108})^{1/3}.
\end{aligned}$$

These equations are commonly called *Cardano's Formulas*, although they are the work of Scipio del Ferro and Nicolo Tartaglia.

Exercise 9.1. Take $F = \mathbb{Q}$ and use Cardano's Formulas to solve $f(x) = 0$ if

(1) $f(x) = x^3 - 9x - 28$,
(2) $f(x) = x^3 - 12x + 8$

[observe in (2) that $f(x)$ has 3 real roots].

If $f(x) \in F[x]$ has roots $a_1, a_2, \ldots, a_n$ in a splitting field K over F set

$$d_f = \prod\{a_i - a_j : 1 \le j < i \le n\},$$

a specialization of the Vandermonde determinant of Proposition 9.1. Thus $d_f^2 = D_f$.

Theorem 9.4. Suppose char $F \neq 2$, $f(x) \in F[x]$, and $f(x)$ has distinct roots in a splitting field K over F. View $G_f = G(K:F)$ as a subgroup of S_n and set $H = G_f \cap A_n$, the even subgroup. Then $\mathscr{F}H = F(d_f)$.

Proof. If $\sigma \in S_n$ is a transposition then $\sigma(d_f) = -d_f \neq d_f$. Thus if $\phi \in G_f$, then $\phi \in H$ if and only if $\phi(d_f) = d_f$, so $F(d_f) \subseteq \mathscr{F}H$ and $\mathscr{G}F(d_f) \le H$. Consequently $\mathscr{F}H \subseteq \mathscr{F}\mathscr{G}F(d_f) = F(d_f)$, so $F(d_f) = \mathscr{F}H$.

Corollary. The Galois group G_f of $f(x)$ is a subgroup of A_n if and only if $d_f \in F$, i.e., if and only if the discriminant D_f is a square in F.

The corollary above can be quite useful for the determination of Galois groups. Recall that if $f(x) \in F[x]$ is irreducible, then its Galois group $G = G_f$ acts transitively on its set of roots (see Exercise III.4.1).

Proposition 9.5. Suppose char $F \neq 2$ and $f(x) \in F[x]$ is irreducible, separable, and of degree 3. Then its Galois group $G = G_f$ is the alternating group A_3 if the discriminant D_f has a square root in F, otherwise $G = S_3$.

Proof. The only transitive subgroups of S_3 are A_3 and S_3. Thus the proposition is an immediate consequence of the corollary to Theorem 9.4.

Exercise 9.2. Determine the Galois groups of (1) $f(x) = x^3 - 9x - 28$ and (2) $f(x) = x^3 - 12x + 8$.

There is an analog of Proposition 9.5 for quartic (degree 4) polynomials, but a bit of preparation will be needed first.

Continue to assume that char $F \neq 2$. If

$$f(x) = x^4 + b_1x^3 + b_2x^2 + b_3x + b_4 \in F[x],$$

then the change of variables $y = x - b_1/4$ transforms $f(x)$ to a polynomial $g(y)$ for which the coefficient of y^3 is 0. Note that the discriminant is unchanged. Thus we may (and shall) assume that

$$f(x) = x^4 + bx^2 + cx + d \in F[x].$$

Suppose $f(x)$ has distinct roots a_1, a_2, a_3, a_4 in a splitting field K over F. Write $G = G_f$ for the Galois group $G(K:F)$ as usual.

The sole reason that the alternating group A_4 is not simple is the existence of the normal subgroup

$$V = \{1, (12)(34), (13)(24), (14)(23)\},$$

Klein's 4-group. In K set

$$\alpha_1 = a_1a_2 + a_3a_4, \qquad \alpha_2 = a_1a_3 + a_2a_4, \qquad \alpha_3 = a_1a_4 + a_2a_3$$

and observe that each α_i is fixed by the elements of V.

Proposition 9.6. In the above setting $\mathscr{F}(G \cap V) = F(\alpha_1, \alpha_2, \alpha_3)$.

Proof. We observed above that $F(\alpha_1, \alpha_2, \alpha_3) \subseteq \mathscr{F}(G \cap V)$. The element α_1 is fixed by the transposition (12) and hence by the subgroup $H_1 = \langle V \cup \{(12)\}\rangle$, a dihedral subgroup of order 8 in S_4. The transpositions (13) and (14) are representatives for the other two cosets of H_1 in S_4, and $(13)\alpha_1 = \alpha_3 \neq \alpha_1$, $(14)\alpha_1 = \alpha_2 \neq \alpha_1$ so H_1 is the subgroup of S_4 that fixes α_1. Similarly $H_2 = \langle V \cup \{(13)\}\rangle$ fixes α_2 and $H_3 = \langle V \cup \{(14)\}\rangle$ fixes α_3. Since $H_1 \cap H_2 \cap H_3 = V$ it follows that if $\sigma \in G$ fixes α_1, α_2, α_3 then $\sigma \in G \cap V$, so $\mathscr{G}F(\alpha_1, \alpha_2, \alpha_3) \subseteq G \cap V$. Thus

$$\mathscr{F}(G \cap V) \subseteq \mathscr{F}\mathscr{G}F(\alpha_1, \alpha_2, \alpha_3) = F(\alpha_1, \alpha_2, \alpha_3),$$

and the two are equal.

It is at least implicit in the remarks and proof above that α_1, α_2, and α_3 are permuted among themselves by all of S_4, and hence by G_f. It follows that $g(x) = (x - \alpha_1)(x - \alpha_2)(x - \alpha_3)$ is in $F[x]$. Let us determine its coefficients. Recall that $f(x) = x^4 + bx^2 + cx + d$, so $a_1 + a_2 + a_3 + a_4 = \sigma_1 = 0$, $\sigma_2 = b, \sigma_3 = -c$, and $\sigma_4 = a_1a_2a_3a_4 = d$. The coefficient of x^2 in $g(x)$ is

$$\begin{aligned} -(\alpha_1 + \alpha_2 + \alpha_3) &= -(a_1a_2 + a_3a_4 + a_1a_3 + a_2a_4 + a_1a_4 + a_2a_3) \\ &= -\sigma_2 = -b. \end{aligned}$$

Gathering terms judiciously we find that the coefficient of x in $g(x)$ is

$$\begin{aligned}\alpha_1\alpha_2 + \alpha_1\alpha_3 + \alpha_2\alpha_3 = a_1a_2a_3(a_1 + a_2 + a_3) + a_1a_2a_4(a_1 + a_2 + a_4)\\ + a_1a_3a_4(a_1 + a_3 + a_4)\\ + a_2a_3a_4(a_2 + a_3 + a_4) = -4a_1a_2a_3a_4\end{aligned}$$

(since $\sigma_1 = 0$), i.e., the coefficient is $-4\sigma_4 = -4d$.

Exercise 9.3. Show that the constant term of $g(x)$ is $-\alpha_1\alpha_2\alpha_3 = -\sigma_3^2 + 4\sigma_2\sigma_4 = -c^2 + 4bd$ (Newton's Identity for s_2 may be helpful).

Thus we have

$$\begin{aligned}g(x) &= (x - \alpha_1)(x - \alpha_2)(x - \alpha_3)\\ &= x^3 - bx^2 - 4dx - c^2 + 4bd \in F[x];\end{aligned}$$

$g(x)$ is called the *resolvent cubic* of $f(x)$. Set $L = F(\alpha_1, \alpha_2, \alpha_3)$, the splitting field for $g(x)$ over F, so $L = \mathscr{F}(G_f \cap V)$ by Proposition 9.6. By the Fundamental Theorem of Galois Theory

$$G_g = G(L:F) \cong G_f/\mathscr{G}L = G_f/(G_f \cap V).$$

Observe that

$$\alpha_1 - \alpha_2 = a_1a_2 + a_3a_4 - a_1a_3 - a_2a_4 = (a_1 - a_4)(a_2 - a_3),$$

and similarly

$$\alpha_1 - \alpha_3 = (a_1 - a_3)(a_2 - a_4), \alpha_2 - \alpha_3 = (a_1 - a_2)(a_3 - a_4).$$

It follows immediately that $D_g = D_f$.

Suppose now that $f(x)$ is irreducible over F, so G_f is a transitive subgroup of S_4. The possible orders for G_f are 24, 12, 8, and 4. Clearly S_4 and A_4 are transitive, and they are the unique subgroups of their orders. Subgroups of order 8 are 2-Sylow; there are three of them, all conjugate in S_4. Each is dihedral, one being $D = \langle(1234),(13)\rangle$, and they are transitive. There are two types of subgroups of order 4, both transitive. One is Klein's 4-group V, which is normal in S_4 (and unique); the other cyclic, e.g., $C = \langle(1234)\rangle$. The three cyclic subgroups of order 4 are all conjugate. Except for C and its conjugates every transitive subgroup contains V as a subgroup.

Set $m = [L:F]$, $L = F(\alpha_1, \alpha_2, \alpha_3)$ as above. Thus also $m = |G_g|$ since $G_g = G(L:F)$ and L is Galois over F. Since $G_g \cong G_f/G_f \cap V$ it follows from the discussion above that $G_f = S_4$ if and only if $m = 6$, $G_f = A_4$ if and only if $m = 3$, and $G_f = V$ if and only if $m = 1$. Since $V \subseteq D$ and $|C \cap V| = 2$ we have $m = 2$ both when $G_f = C$ and when $G_f = D$. If $G_f = C$, then $[K:F] = 4$ and $[L:F] = 2$, so $[K:L] = 2$, so $f(x)$ must be reducible over L in that case. On the other hand if $G_f = D$, then $[K:F] = 8$ and $[L:F] = 2$, so $[K:L] = 4$. In that case $G(K:L) = \mathscr{G}L = V$, which is transitive on $\{a_1, a_2, a_3, a_4\}$, so $f(x)$ is irreducible over L.

We summarize in the next proposition.

Proposition 9.7. Suppose char $F \neq 2$, $f(x) \in F[x]$ is separable and irreducible, deg $f(x) = 4$, and $g(x)$ is the resolvent cubic of $f(x)$. Let $L \subseteq K$ be the splitting field for $g(x)$ over F and set $m = [L:F]$. If $m = 6$, then $G_f = S_4$; if $m = 3$, then $G_f = A_4$; if $m = 2$ and $f(x)$ is irreducible over L, then $G_f = D$; if $m = 2$ and $f(x)$ is reducible over L, then $G_f = C$; and if $m = 1$, then $G_f = V$.

The restriction char $F \neq 2$ was not used in an essential way in the discussion above, only to change variables. In practice it is convenient to assume that char $F \neq 2$ and char $F \neq 3$ so that the resolvent cubic is more manageable.

Assume for the following examples that $F = \mathbb{Q}$. Thus separability is not in question.

EXAMPLES

1. If $f(x) = x^4 - x + 1 \in \mathbb{Q}[x]$, then reduction of coefficients mod 2 yields $\bar{f}(x) = x^4 + x + 1 \in \mathbb{Z}_2[x]$. Clearly $\bar{f}(x)$ has no roots in $\mathbb{Z}_2$, and the assumption that $\bar{f}(x)$ factors as a product of two quadratics in $\mathbb{Z}_2[x]$ leads quickly to a contradiction (verify), so $\bar{f}(x)$ is ireducible, so $f(x)$ is irreducible in $\mathbb{Z}[x]$, hence in $\mathbb{Q}[x]$ by Proposition II.5.15. The resolvent cubic of $f(x)$ is $g(x) = x^3 - 4x - 1$, which is also irreducible (e.g., since it has no roots in $\mathbb{Z}$, or since it has no roots mod 3). The discriminant is $D_f = D_g = 4^4 - 27 = 229$, which is not a square in $\mathbb{Q}$. Thus $G_g = S_3$ by Proposition 9.5, and $m = 6$. Thus $G_f = S_4$.

2. If $f(x) = x^4 - 6x^2 + 8x + 28 \in \mathbb{Q}[x]$, then $f(x+1) = x^4 + 4x^3 + 31$ which is irreducible in $\mathbb{Q}[x]$ by Exercise II.8.31, so $f(x)$ is also irreducible. The resolvent cubic is $g(x) = x^3 + 6x^2 - 112x - 736$. Change variables via $x = y - 2$ to transform $g(x)$ to $h(y) = y^3 - 124y - 496$. Then mod 3 we have $\bar{h}(y) = y^3 - y - 1 \in \mathbb{Z}_3[x]$, which is irreducible since it has no roots in $\mathbb{Z}_3$, and hence $h(y)$ is irreducible in $\mathbb{Q}[y]$. Since

$$D_g = D_h = 4 \cdot 124^3 - 27 \cdot 496^2 = 984064 = (31 \cdot 32)^2$$

we have $G_g = A_3$ and $m = 3$. Thus $G_f = A_4$.

3. If $f(x) = x^4 - 3x^2 + 6x - 3 \in \mathbb{Q}[x]$, then $f(x)$ is irreducible by Eisenstein's Criterion with $p = 3$. Its resolvent cubic is

$$g(x) = x^3 + 3x^2 + 12x = x(x^2 + 3x + 12),$$

so $m = 2$ and $L = \mathbb{Q}(\sqrt{-39})$. The ring of algebraic integers in $\mathbb{Q}(\sqrt{-39})$ is

$$R_{-7} = \{(a + b\sqrt{-39})/2 : a, b \in \mathbb{Z}, a \equiv b \pmod 2\}$$

(see Section II.5). It is easily verified that 3 is a prime in R_{-39}, which is a PID (Exercise II.5.7), so $f(x)$ is irreducible in $R_{-39}[x]$, again by Eisenstein's Criterion, and is irreducible in $L[x]$ by Proposition II.5.15. Thus $G_f = D$, dihedral of order 8, by Proposition 9.7.

4. If $f(x) = x^4 + 5x^2 + 5 \in \mathbb{Q}[x]$, then $f(x)$ is irreducible by Eisenstein's Criterion with $p = 5$. Thus

$$g(x) = x^3 - 5x^2 - 20x + 100 = (x - 5)(x^2 - 20),$$

and $L = \mathbb{Q}(\sqrt{5})$, $m = 2$. Since $f(x)$ is quadratic in x^2 it is easy to find the roots of $f(x)$ explicitly, and hence to factor $f(x)$ as

$$f(x) = [x^2 + (5 - \sqrt{5})/2][x^2 + (5 + \sqrt{5})/2]$$

in $L[x]$. Thus $G_f = C$, cyclic of order 4, by Proposition 9.7.

5. If $f(x) = x^4 + 1 \in \mathbb{Q}[x]$, then $f(x)$ is the cyclotomic polynomial $\Phi_8(x)$, so $f(x)$ is irreducible by Theorem III.4.2; alternatively $f(x + 1)$ is irreducible by Eisenstein's Criterion. Its resolvent cubic is

$$g(x) = x^3 - 4x = x(x - 2)(x + 2),$$

so $m = 1$ and $G_f = V$.

Exercise 9.4. Determine the Galois group over $\mathbb{Q}$ for each of the following quartics:

(a) $f(x) = x^4 - 2$;
(b) $f(x) = x^4 + 8x + 12$;
(c) $f(x) = 4x^4 + x^2 + 9$;
(d) $f(x) = \Phi_5(x) = x^4 + x^3 + x^2 + x + 1$ (see Proposition III.4.1);
(e) $f(x) = x^4 + px^2 + p$, p prime;
(f) $f(x) = x^4 + 4px + 3p$, p prime.

Exercise 9.5. (Solution of quartics). Suppose char $F \neq 2, 3$ and

$$f(x) = x^4 + bx^2 + cx + d \in F[x].$$

To solve $f(x) = 0$ add $(ux + v)^2$ to both sides and determine $u, v \in F$ so that $f(x) + (ux + v)^2 = (x^2 + w)^2$ for some $w \in F$. Show that u, v, and w satisfy $u^2 = 2w - b$, $2uv = -c$, $v^2 = w^2 - d$, and that $2w$ is a root of the resolvent cubic $g(x)$. Solve for w (by Cardano's formulas if necessary; any one of the three roots will do), then for u and v. Then the roots of $f(x)$ are the four roots of the two quadratics $x^2 + ux + w + v$ and $x^2 - ux + w - v$. Use this method to solve $x^4 - 7x^2 + 18x - 8 = 0$ (use $w = 1$).

Exercise 9.6. If $f(x) = x^4 - 4x + 2$ show that the Galois group G of $f(x)$ over $\mathbb{Q}$ is S_4. Show that $f(x)$ has a root $a_1 \in R$ with $0 < a_1 < 1$, and that if $L = \mathbb{Q}(a_1)$, then $[L:\mathbb{Q}] = 4$, $H = \mathcal{G}L \cong S_3$, and $H \nleq A_4$. Show that H is a maximal subgroup of S_4 (see Exercise I.12.22), and hence there are no fields properly between $\mathbb{Q}$ and L. Conclude that a_1 is not constructible, even though $[\mathbb{Q}(a_1):\mathbb{Q}] = 2^2$. Use Exercise 9.5 above to express a_1 in terms of radicals.

Suppose R is a UFD with field F of fractions, and $f(x) \in R[x]$ is monic of degree n with distinct roots $a_1, a_2, \ldots, a_n$ in a splitting field K over F. Let $G = G(K:F)$ be the Galois group of $f(x)$. As usual we may view G as a permutation group on $\{a_i\}$, hence as a subgroup of S_n, when it is convenient to do so. Let $x_1, x_2, \ldots, x_n$ be distinct indeterminates over $R[x]$ and set $R_1 = R[x_1, \ldots, x_n]$, $F_1 = F(x_1, \ldots, x_n)$, and $K_1 = K(x_1, \ldots, x_n)$. Note that K_1 is a splitting field for $f(x)$ over F_1, and that the Galois group $G_1 = G(K_1:F_1)$ is isomorphic with G. The isomorphism is effected simply by restricting each $\phi \in G_1$ to K; in fact, if G_1 and G are both viewed as permutation groups on $\{a_i\}$ they are identical, so we shall not usually distinguish between them.

If $\alpha = \alpha(x_1, x_2, \ldots, x_n) \in K_1$ and $\tau \in S_n$ define $\alpha_\tau = \alpha(x_{\tau^{-1}1}, \ldots, x_{\tau^{-1}n})$, and note that if also $\eta \in S_n$, then $(\alpha_\tau)_\eta = \alpha_{\tau\eta}$. This is essentially the permutation action discussed in Section III.5, with a slight change in notation.

Set $\theta = \sum_{i=1}^n a_i x_i \in K_1$ and note that if $\tau \in S_n$, then

$$\theta_\tau = \sum_i a_i x_{\tau^{-1}i} = \sum_i a_{\tau i} x_i.$$

Since the roots $\{a_i\}$ are distinct it is clear that if $\tau \neq \rho$ in S_n, then $\theta_\tau \neq \theta_\rho$. Define $g(x) = \prod\{x - \theta_\tau : \tau \in S_n\}$, a polynomial with $n!$ distinct roots in K_1. If we view $g(x)$ as a polynomial in x and in $x_1, \ldots, x_n$ with coefficients in K it is also clear that those coefficients are polynomials (say over R; actually over $\mathbb{Z}$) in the roots $\{a_i\}$, and that the coefficients are invariant under all permutations of $\{a_i\}$. It follows from Theorem III.5.3 that the coefficients are polynomials in the coefficients of $f(x)$; hence

$$g(x) \in R[x_1, \ldots, x_n, x] = R_1[x].$$

If $\phi \in G_1$ we have, of course, another action of ϕ on any $\alpha \in K_1$, denoted as usual by $\phi\alpha$, in which ϕ fixes each x_i and acts automorphically on the coefficients from K. It is not true in general that $\phi\alpha = \alpha_\phi$, although

$$\phi\theta = \sum_i a_{\phi i} x_i = \sum_i a_i x_{\phi^{-1}i} = \theta_\phi$$

as observed above. If $\tau \in S_n$ observe that

$$\phi(\theta_\tau) = (\phi\theta)_\tau = (\theta_\phi)_\tau = \theta_{\phi\tau}.$$

For fixed $\sigma \in S_n$ define

$$g_\sigma(x) = \prod\{x - \phi\theta_\sigma : \phi \in G\}.$$

If $\psi \in G$, then

$$\psi g_\sigma(x) = \prod\{x - \psi\phi\theta_\sigma : \phi \in G\} = g_\sigma(x),$$

so $g_\sigma(x) \in F_1[x]$, F_1 being the fixed field for G. In fact,

$$g_\sigma(x) \in F_1[x] \cap K[x_1, x_2, \ldots, x_n, x] = F[x_1, \ldots, x_n, x].$$

Observe that $[K_1 : F_1] = |G| = \deg g_\sigma(x)$, $g_\sigma(x)$ has θ_σ as a root and G is transitive on the set of roots of $g_\sigma(x)$ in K_1. It follows that $g_\sigma(x)$ is irreducible in $F_1[x]$ and that $K_1 = F_1(\theta_\sigma)$, with $g_\sigma(x)$ the minimal polynomial over F_1 for θ_σ.

Now let $\sigma_1 = 1, \sigma_2, \ldots, \sigma_k$ ($k = [S_n : G]$) be a set of (right) coset representatives for G in S_n and for each i write $g_i(x) = g_{\sigma_i}(x)$. Then

$$S_n = \{\phi\sigma_i : \phi \in G, 1 \le i \le k\}$$

and so

$$g(x) = \prod\{g_i(x) : 1 \le i \le k\},$$

representing $g(x)$ as a product of irreducible factors in $F[x_1, \ldots, x_n, x]$. Since $R_1[x] = R[x_1, \ldots, x_n, x]$ is also a UFD and $g(x) \in R_1[x]$ it follows from Proposition II.5.15 and the fact that $g(x)$ and all $g_i(x)$ are monic that $g(x) = \prod_i g_i(x)$ must be the prime factorization of $g(x)$ in $R_1[x]$.

For each $\tau \in S_n$ we have

$$g(x)_\tau = \left[\prod_i g_i(x)\right]_\tau = \prod_i g_{\sigma_i\tau}(x) = g(x),$$

with τ permuting the irreducible factors of $g(x)$, since $\{\sigma_i\tau : 1 \le i \le k\}$ is another set of coset representatives for G in S_n. Furthermore, since $\sigma_1 = 1$ we see that $g_{\sigma_1\tau}(x) = g_\tau(x) = g_1(x)$ if and only if $\tau \in G$. Thus in the action of S_n on the irreducible factors of $g(x)$ in $R_1[x]$ we have $\mathrm{Stab}_{S_n}[g_1(x)] = G$.

We use the ideas developed above to establish a theorem, due to Van der Waerden, that is often useful for determining Galois groups.

Theorem 9.8. Suppose R is a UFD with field F of fractions, P is a nonzero prime ideal in R, $\bar{R} = R/P$, and $\bar{F}$ is the field of fractions for $\bar{R}$. Suppose $f(x)$ is monic of degree n in $R[x]$ with $\bar{f}(x) \in \bar{R}[x]$ the result of reducing coefficients mod P, and suppose that $f(x)$ and $\bar{f}(x)$ have no repeated roots in respective splitting fields K and $\bar{K}$ over F and $\bar{F}$. If $G = G_f = G(K:F)$ and $\bar{G} = G_{\bar{f}} = G(\bar{K}:\bar{F})$, then $\bar{G}$ is (isomorphic with) a subgroup of G.

Proof. Take R_1, F_1, K_1 and $g(x)$ as above, and define $\bar{R}_1, \bar{F}_1, \bar{K}_1$ analogously. Since $g(x)$ factors as $\prod_i g_i(x)$ in $R_1[x]$ we may reduce all coefficients in R mod P and obtain $\bar{g}(x) = \prod_i \bar{g}_i(x)$. Each $\tau \in S_n$ permutes the factors $g_i(x)$ of $g(x)$ and hence also the factors $\bar{g}_i(x)$ of $\bar{g}(x)$, and G has been characterized as

Stab$[g_1(x)]$. The factors of $\bar{g}(x)$ are not necessarily irreducible in $R_1[x]$, they might factor further. The discussion preceding the theorem, as applied to $\bar{R}$, $\bar{G}$, etc., shows then that $\bar{G}$, as a subgroup of S_n, is the stabilizer of an irreducible factor of $\bar{g}_1(x)$. But then each $\psi \in \bar{G}$, viewed as an element of S_n, permutes the factors $\bar{g}_i(x)$, and $\bar{g}_1(x)_\psi$ can only be $\bar{g}_1(x)$ since ψ fixes an irreducible factor of $\bar{g}_1(x)$. Thus $\psi \in G$ and $\bar{G} \leq G$.

Corollary. Keep the hypotheses of the theorem but take $R = \mathbb{Z}$ and $P = (p)$ with p prime in $\mathbb{Z}$, so $\bar{R} = \bar{F} = \mathbb{Z}_p$. Suppose $\bar{f}(x)$ factors in $\mathbb{Z}_p[x]$ into irreducible factors $\bar{f}_1(x)\bar{f}_2(x)\ldots\bar{f}_m(x)$, with $\deg \bar{f}_i(x) = n_i \geq 1$. Then G contains a product $\sigma_1\sigma_2\cdots\sigma_m$ of disjoint cycles, σ_i being an n_i-cycle.

Proof. In this case the Galois group $\bar{G}$ is cyclic (Theorem III.3.11) say $\bar{G} = \langle\sigma\rangle$. Since $\bar{G}$ is transitive on the n_i roots in $\bar{K}$ of the irreducible factor $\bar{f}_i(x)$ for each i it is clear that σ has the indicated cycle structure, and $\sigma \in \bar{G} \leq G$.

The next two results also appear in Van der Waerden [37].

Proposition 9.9. If $G \leq S_n$ is transitive on $\{1, 2, \ldots, n\}$ and G contains a transposition σ and an $(n-1)$-cycle τ, then $G = S_n$.

Proof. By relabeling we may assume that $\tau = (23\ldots n)$ and then by transitivity take $\sigma = (1\ k)$ with $2 \leq k \leq n$. Then $\tau^i\sigma\tau^{-i} = (1\ j)$ where $2 \leq j \leq n$ and $j \equiv k + i (\text{mod } n-1)$. Thus G contains all $(1\ j)$, $2 \leq j \leq n$. An easy induction on n shows that these generate S_n.

Theorem 9.10. If $0 < n \in \mathbb{Z}$ there is some $f(x) \in \mathbb{Q}[x]$ with $G_f = S_n$.

Proof. In view of various examples above we may assume that $n \geq 4$. Choose $f_1(x)$, $f_2(x)$, $f_3(x) \in \mathbb{Z}[x]$, each monic of degree n, so that $\bar{f}_1(x)$ is irreducible in $\mathbb{Z}_2[x]$, $\bar{f}_2(x)$ splits in $\mathbb{Z}_3[x]$ as a product of a linear factor and an irreducible factor of degree $n-1$, and $\bar{f}_3(x)$ splits in $\mathbb{Z}_5[x]$ as a product of an irreducible quadratic factor and either one or two (distinct) irreducible factors of odd degree (see Exercise III.8.26). Then set

$$f(x) = -15f_1(x) + 10f_2(x) + 6f_3(x) \in \mathbb{Z}[x]$$

and set $G = G_f \leq S_n$. Since $\bar{f}(x) = \bar{f}_1(x)$ in $\mathbb{Z}_2[x]$ $f(x)$ is irreducible and G_f is transitive on the roots of $f(x)$. Since $\bar{f}(x) = \bar{f}_2(x)$ in $\mathbb{Z}_3[x]$ G contains an $(n-1)$-cycle by the corollary to Theorem 9.8. Since $\bar{f}(x) = \bar{f}_3(x)$ in $\mathbb{Z}_5[x]$ G contains a product ρ of a transposition σ and one or two cycles of odd length, also by the corollary. A suitable power of ρ is σ, and $G = S_n$ by Proposition 9.9.

For an example take $n = 6$. Then $f_1(x) = x^6 + x + 1$ is irreducible mod 2, $f_2(x) = x^6 + 2x^2 + x$ has irreducible factor $x^5 + 2x + 1$ mod 3, and $f_3(x) = x^6 + x^4 + x^3 + x$ factors as $(x^2 - 2)x(x^3 + 3x + 2)$, with $x^2 - 2$

and $x^3 + 3x + 2$ irreducible mod 5. Thus

$$f(x) = -15f_1(x) + 10f_2(x) + 6f_3(x) = x^6 + 6x^4 + 6x^3 + 20x^2 + x - 15$$

has Galois group $G_f = S_6$ over $\mathbb{Q}$.

The method of Theorem 9.10 is constructive only insofar as it is possible to exhibit the polynomials $f_1(x)$, $f_2(x)$, and $f_3(x)$. Unfortunately there seems to be no general constructive procedure available for producing irreducible polynomials of prescribed degrees over $\mathbb{Z}_p$. There are some tables; e.g., see Peterson and Weldon [30] for $p = 2$.

Another standard method for showing that every symmetric group S_n occurs as a Galois group over $\mathbb{Q}$ uses Hilbert's Irreducibility Theorem. See Hadlock [13] for an account. The same methods serve to show that every alternating group A_n occurs as a Galois group over $\mathbb{Q}$; a nice discussion appears in Paul Feit's senior thesis [10].

Exercise 9.7. Use the method of Theorem 9.10 to find polynomials in $\mathbb{Q}[x]$ with S_7 and S_8 as Galois groups.

We conclude this section with a final example. If

$$f(x) = x^5 - 821x + 3284 \in \mathbb{Z}[x],$$

then $f(x)$ is irreducible by Eisenstein's Criterion since 821 is a prime in $\mathbb{Z}$. Thus $5 \mid |G_f|$. An application of Proposition 9.3 yields $D_f = 821^4 \cdot 2^{16} \cdot 3^2$ (verify), so $G_f \leq A_5$ by the corollary to Theorem 9.4. In $\mathbb{Z}_7[x]$ we have

$$\bar{f}(x) = x^5 - 2x + 1 = (x - 1)(x - 3)(x^3 - 3x^2 - x - 2),$$

and $x^3 - 3x^2 - x - 2$ is irreducible mod 7. Thus there is a 3-cycle in G by the corollary to Theorem 9.8, so $15 \mid |G|$ and $|G| = 15, 30$, or 60. If $|G|$ were 15, then G would be cyclic (see Example 2, p. 21), but consideration of possible cycle types shows that A_5 has no elements of order 15. A subgroup of order 30 in A_5 would be normal, contradicting simplicity, and so $|G| = 60$ and $G = A_5$.

Exercise 9.8. Show that the following polynomials have Galois group A_5 over $\mathbb{Q}$.

(a) $x^5 + 20x + 16$,
(b) $x^5 + 719x - 2876$,
(c) $x^5 + 971x - 3884$.

Exercise 9.9 (D. Surowski). Choose $k \in \mathbb{Z}$ so that $3 \nmid k$ and $k \equiv 1 \pmod 7$, and set $m = 5k^2 - 1$. Show that $f(x) = x^5 + 5mx + 4m$ has Galois group A_5 over $\mathbb{Q}$. (The key steps are to compute D_f and to reduce coefficients mod 3 and mod 7.)

Appendix | Zorn's Lemma

A *partial ordering* on a set $\mathscr{S}$ is a relation $\leq$ that is (i) reflexive, (ii) antisymmetric, and (iii) transitive, i.e.,

(i) $A \leq A$,
(ii) if $A \leq B$ and $B \leq A$, then $A = B$, and
(iii) if $A \leq B$ and $B \leq C$, then $A \leq C$

for all A, B, and C in $\mathscr{S}$. If $\mathscr{S}$ has a partial ordering we say that $\mathscr{S}$ is a *partially ordered set*.

The prototypical example of a partially ordered set is a set $\mathscr{S}$ of subsets of some set S, where $A \leq B$ denotes set inclusion, i.e., $A \subseteq B$.

A *linear ordering* on a set $\mathscr{C}$ is a partial ordering $\leq$ in which any two elements A, B are comparable, i.e., $A \leq B$ or $B \leq A$. A *chain* in a partially ordered set $\mathscr{S}$ is a nonempty subset $\mathscr{C}$ of $\mathscr{S}$ that is linearly ordered by the ordering it inherits from $\mathscr{S}$ by restriction. An *upper bound* for a chain $\mathscr{C}$ in $\mathscr{S}$ is an element $B \in \mathscr{S}$ such that $A \leq B$ for all $A \in \mathscr{C}$ (it is *not* required that $B \in \mathscr{C}$). A *maximal* element in $\mathscr{S}$ is an element $M \in \mathscr{S}$ such that if $A \in \mathscr{S}$ and $M \leq A$, then $A = M$.

We take the point of view that the following statement, Zorn's Lemma, is an axiom of set theory.

Zorn's Lemma. If $\mathscr{S}$ is a nonempty partially ordered set in which every chain has an upper bound, then $\mathscr{S}$ has a maximal element.

There are many apparently different but logically equivalent formulations of Zorn's Lemma. Perhaps the best known is the Axiom of Choice, which asserts that the Cartesian product of any nonempty family of nonempty sets is not empty. An element of that Cartesian product is called a "choice function"; it can be thought of as simultaneously choosing an element from each of the sets in the family. A discussion of various formulations and their equivalences can be found in most set theory books. A brief and lucid account appears in Kelley [21].

Let us look briefly at an important application.

Suppose V is a (left) vector space over a division ring D. Recall that a subset X of V is *linearly dependent* if there are distinct elements $v_1, v_2, \ldots, v_n \in X$ and (not necessarily distinct) scalars $a_1, a_2, \ldots, a_n \in D$, with not all $a_i = 0$, such that

$$a_1v_1 + a_2v_2 + \cdots + a_nv_n = 0.$$

Otherwise X is *linearly independent.* The *span* of a subset X of V is the subspace of V generated by X, obtained by intersecting all subspaces that contain X as a subset. If $X \neq \varnothing$, then the span of X consists of all linear combinations $a_1v_1 + a_2v_2 + \cdots + a_nv_n$, where $1 \le n \in \mathbb{Z}$, $v_i \in X$, and $a_j \in D$. A *basis* for V is a linearly independent subset X of V such that the span of X is V.

Theorem A.1. If V is a vector space over a division ring D, then V has a basis.

Proof. Let $\mathscr{S}$ be the set of all linearly independent subsets of V. Then $\mathscr{S} \neq \varnothing$ since $\varnothing \in \mathscr{S}$ (if $V \neq 0$, then also each $\{v\} \in \mathscr{S}$ if $0 \neq v \in V$). Partially order $\mathscr{S}$ by ordinary set inclusion, i.e., $A \le B$ if and only if $A \subseteq B$. If $\mathscr{C}$ is a chain in $\mathscr{S}$ let $B = \bigcup\{A : A \in \mathscr{C}\}$. If $B \notin \mathscr{S}$, then there are distinct $v_1, \ldots, v_n \in B$ and scalars $a_1, \ldots, a_n \in D$, not all 0, such that $a_1v_1 + \cdots + a_nv_n = 0$. But each v_i is in some $A_i \in \mathscr{C}$, and $\mathscr{C}$ is a chain so one of $A_1, \ldots, A_n$ contains all the others. By relabeling we may assume that $A_i \subseteq A_n$ for all i. But then $v_1, \ldots, v_n \in A_n$, which is linearly independent, and we have a contradiction. Thus $B \in \mathscr{S}$, and B is clearly an upper bound for $\mathscr{C}$. By Zorn's Lemma we may conclude that there is a maximal element M in $\mathscr{S}$.

Let us show that M is a basis for V. To that end let W be the span of M. If $W \neq V$ choose $v \in V \backslash W$, and set $M_1 = M \bigcup \{v\}$. Then M_1 is linearly independent, since v is not a linear combination of elements of M, so $M_1 \in \mathscr{S}$, $M \le M_1$, and $M \neq M_1$, contradicting the maximality of M.

Theorem A.2. Any two bases for a vector space V over a division ring D have the same cardinality.

Proof. If V has a finite basis the result is a standard fact from elementary linear algebra. Suppose then that B_1 and B_2 are two infinite bases for V. Each $v \in B_1$ is (uniquely) a linear combination of a finite subset $B_2(v) \subseteq B_2$. If any $x \in B_2$ were in none of the sets $B_2(v)$, then the span of $B_2 \backslash \{x\}$ would include B_1, and hence would be all of V. But then B_2 would be linearly dependent, a contradiction, and hence $B_2 = \bigcup\{B_2(v) : v \in B_1\}$. Consequently $|B_2| \le \aleph_0 |B_1| = |B_1|$, where $\aleph_0$ is the first infinite cardinal. Similarly $|B_1| \le |B_2|$, and hence $|B_1| = |B_2|$ by the Schröder–Bernstein Theorem.

If B is any basis for a vector space V, then $|B|$ is called the *dimension* of V.

References

1. E. Artin, "Geometric Algebra," Wiley (Interscience), New York, 1957.
2. E. Artin, "Modern Higher Algebra, Galois Theory," Courant Institute of Mathematical Sciences Lecture Notes, New York Univ. Press, New York, 1957.
3. E. Artin, "Galois Theory," 2nd ed., Notre Dame Mathematical Lectures, No. 2, Univ. of Notre Dame Press, Notre Dame, Indiana, 1959.
4. E. Artin, "Algebraic Numbers and Algebraic Functions," Gordon and Breach, New York, 1967.
5. W. Bishop, How to construct a regular polygon, *Amer. Math. Monthly* **85**, 186–188 (1978).
6. Z. I. Borevich and I. R. Shafarevich, "Number Theory," Academic Press, New York, 1966.
7. H. S. M. Coxeter, "Introduction to Geometry," 2nd ed., Wiley, New York, 1969.
8. C. Curtis and I. Reiner, "Representation Theory of Finite Groups and Associative Algebras," Wiley (Interscience), New York, 1962.
9. N. G. DeBruijn, "Pólya's Theory of Counting," *in* "Applied Combinatorial Mathematics" (E. F. Beckenbach, ed.), p. 144. Wiley, New York, 1964.
10. P. Feit, The Hilbert irreducibility theorem and Galois groups over $\mathbb{Q}$, Senior Thesis, Harvard College, Cambridge, Massachusetts, 1981.
11. W. Feit, "Characters of Finite Groups," Benjamin, New York, 1967.
12. S. Haber and A. Rosenfeld, Groups as unions of proper subgroups, *Amer. Math. Monthly* **66**, 491–494 (1959).
13. C. Hadlock, "Field Theory and Its Classical Problems," Carus Math. Monographs, No. 19, Mathematical Association of America, Washington, D.C., 1978.
14. G. H. Hardy and E. M. Wright, "An Introduction to the Theory of Numbers," 4th ed., Oxford Univ. Press, London and New York, 1960.
15. I. Herstein, "Noncommutative Rings," Carus Math. Monographs, No. 15, Mathematical Association of America, Washington, D.C., 1968.
16. N. Jacobson, "Basic Algebra I," Freeman, San Francisco, California, 1974.
17. N. Jacobson, "Basic Algebra II," Freeman, San Francisco, California, 1980.
18. S. A. Jennings, The structure of the group ring of a p-group over a modular field, *Trans. Amer. Math. Soc.* **50**, 175–185 (1941).
19. E. Kamke, "The Theory of Sets," Dover, New York, 1950.
20. I. Kaplansky, "Fields and Rings," 2nd ed. Univ. of Chicago Press, Chicago, Illinois, 1972.

21. J. Kelley, "General Topology," Van Nostrand-Reinhold, Princeton, New Jersey, 1955.
22. A. G. Kurosh, "The Theory of Groups," Vol. 2, Chelsea, Bronx, New York, 1960.
23. S. Lang, "Algebra," Addison-Wesley, Reading, Massachusetts, 1965.
24. C. E. Linderholm, "Mathematics Made Difficult," World Publ., New York, 1972.
25. S. Lipka, Über die Irreduzibilität von Polynomen, *Math. Ann.* **118**, 235–245 (1941).
26. J. H. McKay, Another proof of Cauchy's group theorem, *Amer. Math. Monthly* **66**, 119 (1959).
27. N. C. Meyer, Jr., A new proof of the existence of the free group, *Amer. Math. Monthly* **77**, 870–873 (1970).
28. I. Niven, The transcendence of π, *Amer. Math. Monthly* **46**, 469–471 (1939).
29. R. S. Palais, The classification of real division algebras, *Amer. Math. Monthly* **75**, 366–368 (1968).
30. W. Peterson and E. Weldon, Jr., "Error Correcting Codes," 2nd ed., MIT Press, Cambridge, Massachusetts, 1972.
31. H. Pollard and H. G. Diamond, "The Theory of Algebraic Numbers," 2nd ed., Carus Math. Monographs, No. 9, Mathematical Association of America, Washington, D.C., 1978.
32. M. A. Rieffel, A general Wedderburn theorem, *Proc. Nat. Acad. Sci. U.S.A.* **54**, 1513 (1965).
33. E. Sasiada and P. M. Cohn, An example of a simple radical ring, *J. Algebra* **5**, 373–377 (1967).
34. I. R. Shafarevich, Constructions of fields of algebraic numbers with given solvable Galois group, *Izv. Akad. Nauk* SSSR *Ser. Mat.* **18**, 515–578 (1954).
35. E. Spitznagel, Jr., Note on the alternating group, *Amer. Math. Monthly* **75**, 68–69 (1968).
36. H. Stark, A complete determination of the complex quadratic fields of class-number one, *Michigan Math. J.* **14**, 1–27 (1967).
37. B. L. Van der Waerden, "Modern Algebra," Vol. 1, 2nd ed., Ungar, New York, 1953.
38. R. Walker, "Algebraic Curves," Dover, New York, 1950.
39. J. C. Wilson, A principal ideal ring that is not a Euclidean ring, *Math. Mag.* **46**, 34–38 (1973).
40. E. Witt, Über die Kommutativität Endlicher Schiefkörper, *Hamb. Abh.* **8**, 413 (1930).
41. O. Zariski and P. Samuel, "Commutative Algebra," Vol. 1, Van Nostrand-Reinhold, Princeton, New Jersey, 1958.

Index

A

B

C

N

O

P

Q

R

A CATALOG OF SELECTED

DOVER BOOKS

IN SCIENCE AND MATHEMATICS

Astronomy

BURNHAM'S CELESTIAL HANDBOOK, Robert Burnham, Jr. Thorough guide to the stars beyond our solar system. Exhaustive treatment. Alphabetical by constellation: Andromeda to Cetus in Vol. 1; Chamaeleon to Orion in Vol. 2; and Pavo to Vulpecula in Vol. 3. Hundreds of illustrations. Index in Vol. 3. 2,000pp. 6⅛ x 9¼.
Vol. I: 23567-X
Vol. II: 23568-8
Vol. III: 23673-0

EXPLORING THE MOON THROUGH BINOCULARS AND SMALL TELESCOPES, Ernest H. Cherrington, Jr. Informative, profusely illustrated guide to locating and identifying craters, rills, seas, mountains, other lunar features. Newly revised and updated with special section of new photos. Over 100 photos and diagrams. 240pp. 8¼ x 11. 24491-1

THE EXTRATERRESTRIAL LIFE DEBATE, 1750–1900, Michael J. Crowe. First detailed, scholarly study in English of the many ideas that developed from 1750 to 1900 regarding the existence of intelligent extraterrestrial life. Examines ideas of Kant, Herschel, Voltaire, Percival Lowell, many other scientists and thinkers. 16 illustrations. 704pp. 5⅜ x 8½. 40675-X

THEORIES OF THE WORLD FROM ANTIQUITY TO THE COPERNICAN REVOLUTION, Michael J. Crowe. Newly revised edition of an accessible, enlightening book recreates the change from an earth-centered to a sun-centered conception of the solar system. 242pp. 5⅜ x 8½. 41444-2

A HISTORY OF ASTRONOMY, A. Pannekoek. Well-balanced, carefully reasoned study covers such topics as Ptolemaic theory, work of Copernicus, Kepler, Newton, Eddington's work on stars, much more. Illustrated. References. 521pp. 5⅜ x 8½.
65994-1

A COMPLETE MANUAL OF AMATEUR ASTRONOMY: Tools and Techniques for Astronomical Observations, P. Clay Sherrod with Thomas L. Koed. Concise, highly readable book discusses: selecting, setting up and maintaining a telescope; amateur studies of the sun; lunar topography and occultations; observations of Mars, Jupiter, Saturn, the minor planets and the stars; an introduction to photoelectric photometry; more. 1981 ed. 124 figures. 26 halftones. 37 tables. 335pp. 6½ x 9¼.
42820-6

AMATEUR ASTRONOMER'S HANDBOOK, J. B. Sidgwick. Timeless, comprehensive coverage of telescopes, mirrors, lenses, mountings, telescope drives, micrometers, spectroscopes, more. 189 illustrations. 576pp. 5⅝ x 8¼. (Available in U.S. only.)
24034-7

STARS AND RELATIVITY, Ya. B. Zel'dovich and I. D. Novikov. Vol. 1 of *Relativistic Astrophysics* by famed Russian scientists. General relativity, properties of matter under astrophysical conditions, stars, and stellar systems. Deep physical insights, clear presentation. 1971 edition. References. 544pp. 5⅝ x 8¼. 69424-0

Chemistry

THE SCEPTICAL CHYMIST: The Classic 1661 Text, Robert Boyle. Boyle defines the term "element," asserting that all natural phenomena can be explained by the motion and organization of primary particles. 1911 ed. viii+232pp. 5⅜ x 8½. 42825-7

RADIOACTIVE SUBSTANCES, Marie Curie. Here is the celebrated scientist's doctoral thesis, the prelude to her receipt of the 1903 Nobel Prize. Curie discusses establishing atomic character of radioactivity found in compounds of uranium and thorium; extraction from pitchblende of polonium and radium; isolation of pure radium chloride; determination of atomic weight of radium; plus electric, photographic, luminous, heat, color effects of radioactivity. ii+94pp. 5⅜ x 8½. 42550-9

CHEMICAL MAGIC, Leonard A. Ford. Second Edition, Revised by E. Winston Grundmeier. Over 100 unusual stunts demonstrating cold fire, dust explosions, much more. Text explains scientific principles and stresses safety precautions. 128pp. 5⅜ x 8½. 67628-5

THE DEVELOPMENT OF MODERN CHEMISTRY, Aaron J. Ihde. Authoritative history of chemistry from ancient Greek theory to 20th-century innovation. Covers major chemists and their discoveries. 209 illustrations. 14 tables. Bibliographies. Indices. Appendices. 851pp. 5⅜ x 8½. 64235-6

CATALYSIS IN CHEMISTRY AND ENZYMOLOGY, William P. Jencks. Exceptionally clear coverage of mechanisms for catalysis, forces in aqueous solution, carbonyl- and acyl-group reactions, practical kinetics, more. 864pp. 5⅜ x 8½. 65460-5

ELEMENTS OF CHEMISTRY, Antoine Lavoisier. Monumental classic by founder of modern chemistry in remarkable reprint of rare 1790 Kerr translation. A must for every student of chemistry or the history of science. 539pp. 5⅜ x 8½. 64624-6

THE HISTORICAL BACKGROUND OF CHEMISTRY, Henry M. Leicester. Evolution of ideas, not individual biography. Concentrates on formulation of a coherent set of chemical laws. 260pp. 5⅜ x 8½. 61053-5

A SHORT HISTORY OF CHEMISTRY, J. R. Partington. Classic exposition explores origins of chemistry, alchemy, early medical chemistry, nature of atmosphere, theory of valency, laws and structure of atomic theory, much more. 428pp. 5⅜ x 8½. (Available in U.S. only.) 65977-1

GENERAL CHEMISTRY, Linus Pauling. Revised 3rd edition of classic first-year text by Nobel laureate. Atomic and molecular structure, quantum mechanics, statistical mechanics, thermodynamics correlated with descriptive chemistry. Problems. 992pp. 5⅜ x 8½. 65622-5

FROM ALCHEMY TO CHEMISTRY, John Read. Broad, humanistic treatment focuses on great figures of chemistry and ideas that revolutionized the science. 50 illustrations. 240pp. 5⅜ x 8½. 28690-8

Engineering

DE RE METALLICA, Georgius Agricola. The famous Hoover translation of greatest treatise on technological chemistry, engineering, geology, mining of early modern times (1556). All 289 original woodcuts. 638pp. 6¾ x 11. 60006-8

FUNDAMENTALS OF ASTRODYNAMICS, Roger Bate et al. Modern approach developed by U.S. Air Force Academy. Designed as a first course. Problems, exercises. Numerous illustrations. 455pp. 5⅜ x 8½. 60061-0

DYNAMICS OF FLUIDS IN POROUS MEDIA, Jacob Bear. For advanced students of ground water hydrology, soil mechanics and physics, drainage and irrigation engineering, and more. 335 illustrations. Exercises, with answers. 784pp. 6⅛ x 9¼. 65675-6

THEORY OF VISCOELASTICITY (Second Edition), Richard M. Christensen. Complete, consistent description of the linear theory of the viscoelastic behavior of materials. Problem-solving techniques discussed. 1982 edition. 29 figures. xiv+364pp. 6⅛ x 9¼. 42880-X

MECHANICS, J. P. Den Hartog. A classic introductory text or refresher. Hundreds of applications and design problems illuminate fundamentals of trusses, loaded beams and cables, etc. 334 answered problems. 462pp. 5⅜ x 8½. 60754-2

MECHANICAL VIBRATIONS, J. P. Den Hartog. Classic textbook offers lucid explanations and illustrative models, applying theories of vibrations to a variety of practical industrial engineering problems. Numerous figures. 233 problems, solutions. Appendix. Index. Preface. 436pp. 5⅜ x 8½. 64785-4

STRENGTH OF MATERIALS, J. P. Den Hartog. Full, clear treatment of basic material (tension, torsion, bending, etc.) plus advanced material on engineering methods, applications. 350 answered problems. 323pp. 5⅜ x 8½. 60755-0

A HISTORY OF MECHANICS, René Dugas. Monumental study of mechanical principles from antiquity to quantum mechanics. Contributions of ancient Greeks, Galileo, Leonardo, Kepler, Lagrange, many others. 671pp. 5⅜ x 8½. 65632-2

STABILITY THEORY AND ITS APPLICATIONS TO STRUCTURAL MECHANICS, Clive L. Dym. Self-contained text focuses on Koiter postbuckling analyses, with mathematical notions of stability of motion. Basing minimum energy principles for static stability upon dynamic concepts of stability of motion, it develops asymptotic buckling and postbuckling analyses from potential energy considerations, with applications to columns, plates, and arches. 1974 ed. 208pp. 5⅜ x 8½. 42541-X

METAL FATIGUE, N. E. Frost, K. J. Marsh, and L. P. Pook. Definitive, clearly written, and well-illustrated volume addresses all aspects of the subject, from the historical development of understanding metal fatigue to vital concepts of the cyclic stress that causes a crack to grow. Includes 7 appendixes. 544pp. 5⅜ x 8½. 40927-9

ROCKETS, Robert Goddard. Two of the most significant publications in the history of rocketry and jet propulsion: "A Method of Reaching Extreme Altitudes" (1919) and "Liquid Propellant Rocket Development" (1936). 128pp. 5⅜ x 8½. 42537-1

STATISTICAL MECHANICS: Principles and Applications, Terrell L. Hill. Standard text covers fundamentals of statistical mechanics, applications to fluctuation theory, imperfect gases, distribution functions, more. 448pp. 5⅜ x 8½. 65390-0

ENGINEERING AND TECHNOLOGY 1650–1750: Illustrations and Texts from Original Sources, Martin Jensen. Highly readable text with more than 200 contemporary drawings and detailed engravings of engineering projects dealing with surveying, leveling, materials, hand tools, lifting equipment, transport and erection, piling, bailing, water supply, hydraulic engineering, and more. Among the specific projects outlined–transporting a 50-ton stone to the Louvre, erecting an obelisk, building timber locks, and dredging canals. 207pp. 8⅜ x 11¼. 42232-1

THE VARIATIONAL PRINCIPLES OF MECHANICS, Cornelius Lanczos. Graduate level coverage of calculus of variations, equations of motion, relativistic mechanics, more. First inexpensive paperbound edition of classic treatise. Index. Bibliography. 418pp. 5⅜ x 8½. 65067-7

PROTECTION OF ELECTRONIC CIRCUITS FROM OVERVOLTAGES, Ronald B. Standler. Five-part treatment presents practical rules and strategies for circuits designed to protect electronic systems from damage by transient overvoltages. 1989 ed. xxiv+434pp. 6⅛ x 9¼. 42552-5

ROTARY WING AERODYNAMICS, W. Z. Stepniewski. Clear, concise text covers aerodynamic phenomena of the rotor and offers guidelines for helicopter performance evaluation. Originally prepared for NASA. 537 figures. 640pp. 6⅛ x 9¼. 64647-5

INTRODUCTION TO SPACE DYNAMICS, William Tyrrell Thomson. Comprehensive, classic introduction to space-flight engineering for advanced undergraduate and graduate students. Includes vector algebra, kinematics, transformation of coordinates. Bibliography. Index. 352pp. 5⅜ x 8½. 65113-4

HISTORY OF STRENGTH OF MATERIALS, Stephen P. Timoshenko. Excellent historical survey of the strength of materials with many references to the theories of elasticity and structure. 245 figures. 452pp. 5⅜ x 8½. 61187-6

ANALYTICAL FRACTURE MECHANICS, David J. Unger. Self-contained text supplements standard fracture mechanics texts by focusing on analytical methods for determining crack-tip stress and strain fields. 336pp. 6⅛ x 9¼. 41737-9

STATISTICAL MECHANICS OF ELASTICITY, J. H. Weiner. Advanced, self-contained treatment illustrates general principles and elastic behavior of solids. Part 1, based on classical mechanics, studies thermoelastic behavior of crystalline and polymeric solids. Part 2, based on quantum mechanics, focuses on interatomic force laws, behavior of solids, and thermally activated processes. For students of physics and chemistry and for polymer physicists. 1983 ed. 96 figures. 496pp. 5⅜ x 8½. 42260-7

Mathematics

FUNCTIONAL ANALYSIS (Second Corrected Edition), George Bachman and Lawrence Narici. Excellent treatment of subject geared toward students with background in linear algebra, advanced calculus, physics, and engineering. Text covers introduction to inner-product spaces, normed, metric spaces, and topological spaces; complete orthonormal sets, the Hahn-Banach Theorem and its consequences, and many other related subjects. 1966 ed. 544pp. 6⅛ x 9¼. 40251-7

ASYMPTOTIC EXPANSIONS OF INTEGRALS, Norman Bleistein & Richard A. Handelsman. Best introduction to important field with applications in a variety of scientific disciplines. New preface. Problems. Diagrams. Tables. Bibliography. Index. 448pp. 5⅜ x 8½. 65082-0

VECTOR AND TENSOR ANALYSIS WITH APPLICATIONS, A. I. Borisenko and I. E. Tarapov. Concise introduction. Worked-out problems, solutions, exercises. 257pp. 5⅝ x 8¼. 63833-2

THE ABSOLUTE DIFFERENTIAL CALCULUS (CALCULUS OF TENSORS), Tullio Levi-Civita. Great 20th-century mathematician's classic work on material necessary for mathematical grasp of theory of relativity. 452pp. 5⅝ x 8¼. 63401-9

AN INTRODUCTION TO ORDINARY DIFFERENTIAL EQUATIONS, Earl A. Coddington. A thorough and systematic first course in elementary differential equations for undergraduates in mathematics and science, with many exercises and problems (with answers). Index. 304pp. 5⅜ x 8½. 65942-9

FOURIER SERIES AND ORTHOGONAL FUNCTIONS, Harry F. Davis. An incisive text combining theory and practical example to introduce Fourier series, orthogonal functions and applications of the Fourier method to boundary-value problems. 570 exercises. Answers and notes. 416pp. 5⅜ x 8½. 65973-9

COMPUTABILITY AND UNSOLVABILITY, Martin Davis. Classic graduate-level introduction to theory of computability, usually referred to as theory of recurrent functions. New preface and appendix. 288pp. 5⅜ x 8½. 61471-9

ASYMPTOTIC METHODS IN ANALYSIS, N. G. de Bruijn. An inexpensive, comprehensive guide to asymptotic methods–the pioneering work that teaches by explaining worked examples in detail. Index. 224pp. 5⅜ x 8½ 64221-6

APPLIED COMPLEX VARIABLES, John W. Dettman. Step-by-step coverage of fundamentals of analytic function theory–plus lucid exposition of five important applications: Potential Theory; Ordinary Differential Equations; Fourier Transforms; Laplace Transforms; Asymptotic Expansions. 66 figures. Exercises at chapter ends. 512pp. 5⅜ x 8½. 64670-X

INTRODUCTION TO LINEAR ALGEBRA AND DIFFERENTIAL EQUATIONS, John W. Dettman. Excellent text covers complex numbers, determinants, orthonormal bases, Laplace transforms, much more. Exercises with solutions. Undergraduate level. 416pp. 5⅜ x 8½. 65191-6

CALCULUS OF VARIATIONS WITH APPLICATIONS, George M. Ewing. Applications-oriented introduction to variational theory develops insight and promotes understanding of specialized books, research papers. Suitable for advanced undergraduate/graduate students as primary, supplementary text. 352pp. 5⅜ x 8½. 64856-7

COMPLEX VARIABLES, Francis J. Flanigan. Unusual approach, delaying complex algebra till harmonic functions have been analyzed from real variable viewpoint. Includes problems with answers. 364pp. 5⅜ x 8½. 61388-7

AN INTRODUCTION TO THE CALCULUS OF VARIATIONS, Charles Fox. Graduate-level text covers variations of an integral, isoperimetrical problems, least action, special relativity, approximations, more. References. 279pp. 5⅜ x 8½. 65499-0

COUNTEREXAMPLES IN ANALYSIS, Bernard R. Gelbaum and John M. H. Olmsted. These counterexamples deal mostly with the part of analysis known as "real variables." The first half covers the real number system, and the second half encompasses higher dimensions. 1962 edition. xxiv+198pp. 5⅜ x 8½. 42875-3

CATASTROPHE THEORY FOR SCIENTISTS AND ENGINEERS, Robert Gilmore. Advanced-level treatment describes mathematics of theory grounded in the work of Poincaré, R. Thom, other mathematicians. Also important applications to problems in mathematics, physics, chemistry, and engineering. 1981 edition. References. 28 tables. 397 black-and-white illustrations. xvii+666pp. 6⅛ x 9¼. 67539-4

INTRODUCTION TO DIFFERENCE EQUATIONS, Samuel Goldberg. Exceptionally clear exposition of important discipline with applications to sociology, psychology, economics. Many illustrative examples; over 250 problems. 260pp. 5⅜ x 8½. 65084-7

NUMERICAL METHODS FOR SCIENTISTS AND ENGINEERS, Richard Hamming. Classic text stresses frequency approach in coverage of algorithms, polynomial approximation, Fourier approximation, exponential approximation, other topics. Revised and enlarged 2nd edition. 721pp. 5⅜ x 8½. 65241-6

INTRODUCTION TO NUMERICAL ANALYSIS (2nd Edition), F. B. Hildebrand. Classic, fundamental treatment covers computation, approximation, interpolation, numerical differentiation and integration, other topics. 150 new problems. 669pp. 5⅜ x 8½. 65363-3

THREE PEARLS OF NUMBER THEORY, A. Y. Khinchin. Three compelling puzzles require proof of a basic law governing the world of numbers. Challenges concern van der Waerden's theorem, the Landau-Schnirelmann hypothesis and Mann's theorem, and a solution to Waring's problem. Solutions included. 64pp. 5⅜ x 8½. 40026-3

THE PHILOSOPHY OF MATHEMATICS: An Introductory Essay, Stephan Körner. Surveys the views of Plato, Aristotle, Leibniz & Kant concerning propositions and theories of applied and pure mathematics. Introduction. Two appendices. Index. 198pp. 5⅜ x 8½. 25048-2

TENSOR CALCULUS, J.L. Synge and A. Schild. Widely used introductory text covers spaces and tensors, basic operations in Riemannian space, non-Riemannian spaces, etc. 324pp. 5⅝ x 8¼. 63612-7

ORDINARY DIFFERENTIAL EQUATIONS, Morris Tenenbaum and Harry Pollard. Exhaustive survey of ordinary differential equations for undergraduates in mathematics, engineering, science. Thorough analysis of theorems. Diagrams. Bibliography. Index. 818pp. 5⅜ x 8½. 64940-7

INTEGRAL EQUATIONS, F. G. Tricomi. Authoritative, well-written treatment of extremely useful mathematical tool with wide applications. Volterra Equations, Fredholm Equations, much more. Advanced undergraduate to graduate level. Exercises. Bibliography. 238pp. 5⅜ x 8½. 64828-1

FOURIER SERIES, Georgi P. Tolstov. Translated by Richard A. Silverman. A valuable addition to the literature on the subject, moving clearly from subject to subject and theorem to theorem. 107 problems, answers. 336pp. 5⅜ x 8½. 63317-9

INTRODUCTION TO MATHEMATICAL THINKING, Friedrich Waismann. Examinations of arithmetic, geometry, and theory of integers; rational and natural numbers; complete induction; limit and point of accumulation; remarkable curves; complex and hypercomplex numbers, more. 1959 ed. 27 figures. xii+260pp. 5⅜ x 8½. 42804-4

POPULAR LECTURES ON MATHEMATICAL LOGIC, Hao Wang. Noted logician's lucid treatment of historical developments, set theory, model theory, recursion theory and constructivism, proof theory, more. 3 appendixes. Bibliography. 1981 ed. ix+283pp. 5⅜ x 8½. 67632-3

CALCULUS OF VARIATIONS, Robert Weinstock. Basic introduction covering isoperimetric problems, theory of elasticity, quantum mechanics, electrostatics, etc. Exercises throughout. 326pp. 5⅜ x 8½. 63069-2

THE CONTINUUM: A Critical Examination of the Foundation of Analysis, Hermann Weyl. Classic of 20th-century foundational research deals with the conceptual problem posed by the continuum. 156pp. 5⅜ x 8½. 67982-9

CHALLENGING MATHEMATICAL PROBLEMS WITH ELEMENTARY SOLUTIONS, A. M. Yaglom and I. M. Yaglom. Over 170 challenging problems on probability theory, combinatorial analysis, points and lines, topology, convex polygons, many other topics. Solutions. Total of 445pp. 5⅜ x 8½. Two-vol. set.
Vol. I: 65536-9 Vol. II: 65537-7

INTRODUCTION TO PARTIAL DIFFERENTIAL EQUATIONS WITH APPLICATIONS, E. C. Zachmanoglou and Dale W. Thoe. Essentials of partial differential equations applied to common problems in engineering and the physical sciences. Problems and answers. 416pp. 5⅜ x 8½. 65251-3

THE THEORY OF GROUPS, Hans J. Zassenhaus. Well-written graduate-level text acquaints reader with group-theoretic methods and demonstrates their usefulness in mathematics. Axioms, the calculus of complexes, homomorphic mapping, *p*-group theory, more. 276pp. 5⅜ x 8½. 40922-8

Math–Decision Theory, Statistics, Probability

ELEMENTARY DECISION THEORY, Herman Chernoff and Lincoln E. Moses. Clear introduction to statistics and statistical theory covers data processing, probability and random variables, testing hypotheses, much more. Exercises. 364pp. 5⅜ x 8½. 65218-1

STATISTICS MANUAL, Edwin L. Crow et al. Comprehensive, practical collection of classical and modern methods prepared by U.S. Naval Ordnance Test Station. Stress on use. Basics of statistics assumed. 288pp. 5⅜ x 8½. 60599-X

SOME THEORY OF SAMPLING, William Edwards Deming. Analysis of the problems, theory, and design of sampling techniques for social scientists, industrial managers, and others who find statistics important at work. 61 tables. 90 figures. xvii +602pp. 5⅜ x 8½. 64684-X

LINEAR PROGRAMMING AND ECONOMIC ANALYSIS, Robert Dorfman, Paul A. Samuelson and Robert M. Solow. First comprehensive treatment of linear programming in standard economic analysis. Game theory, modern welfare economics, Leontief input-output, more. 525pp. 5⅜ x 8½. 65491-5

PROBABILITY: An Introduction, Samuel Goldberg. Excellent basic text covers set theory, probability theory for finite sample spaces, binomial theorem, much more. 360 problems. Bibliographies. 322pp. 5⅜ x 8½. 65252-1

GAMES AND DECISIONS: Introduction and Critical Survey, R. Duncan Luce and Howard Raiffa. Superb nontechnical introduction to game theory, primarily applied to social sciences. Utility theory, zero-sum games, n-person games, decision-making, much more. Bibliography. 509pp. 5⅜ x 8½. 65943-7

INTRODUCTION TO THE THEORY OF GAMES, J. C. C. McKinsey. This comprehensive overview of the mathematical theory of games illustrates applications to situations involving conflicts of interest, including economic, social, political, and military contexts. Appropriate for advanced undergraduate and graduate courses; advanced calculus a prerequisite. 1952 ed. x+372pp. 5⅜ x 8½. 42811-7

FIFTY CHALLENGING PROBLEMS IN PROBABILITY WITH SOLUTIONS, Frederick Mosteller. Remarkable puzzlers, graded in difficulty, illustrate elementary and advanced aspects of probability. Detailed solutions. 88pp. 5⅜ x 8½. 65355-2

PROBABILITY THEORY: A Concise Course, Y. A. Rozanov. Highly readable, self-contained introduction covers combination of events, dependent events, Bernoulli trials, etc. 148pp. 5⅜ x 8¼. 63544-9

STATISTICAL METHOD FROM THE VIEWPOINT OF QUALITY CONTROL, Walter A. Shewhart. Important text explains regulation of variables, uses of statistical control to achieve quality control in industry, agriculture, other areas. 192pp. 5⅜ x 8½. 65232-7

Math–Geometry and Topology

ELEMENTARY CONCEPTS OF TOPOLOGY, Paul Alexandroff. Elegant, intuitive approach to topology from set-theoretic topology to Betti groups; how concepts of topology are useful in math and physics. 25 figures. 57pp. 5⅜ x 8½. 60747-X

COMBINATORIAL TOPOLOGY, P. S. Alexandrov. Clearly written, well-organized, three-part text begins by dealing with certain classic problems without using the formal techniques of homology theory and advances to the central concept, the Betti groups. Numerous detailed examples. 654pp. 5⅜ x 8½. 40179-0

EXPERIMENTS IN TOPOLOGY, Stephen Barr. Classic, lively explanation of one of the byways of mathematics. Klein bottles, Moebius strips, projective planes, map coloring, problem of the Koenigsberg bridges, much more, described with clarity and wit. 43 figures. 210pp. 5⅜ x 8½. 25933-1

CONFORMAL MAPPING ON RIEMANN SURFACES, Harvey Cohn. Lucid, insightful book presents ideal coverage of subject. 334 exercises make book perfect for self-study. 55 figures. 352pp. 5⅝ x 8¼. 64025-6

THE GEOMETRY OF RENÉ DESCARTES, René Descartes. The great work founded analytical geometry. Original French text, Descartes's own diagrams, together with definitive Smith-Latham translation. 244pp. 5⅜ x 8½. 60068-8

PRACTICAL CONIC SECTIONS: The Geometric Properties of Ellipses, Parabolas and Hyperbolas, J. W. Downs. This text shows how to create ellipses, parabolas, and hyperbolas. It also presents historical background on their ancient origins and describes the reflective properties and roles of curves in design applications. 1993 ed. 98 figures. xii+100pp. 6½ x 9¼. 42876-1

THE THIRTEEN BOOKS OF EUCLID'S ELEMENTS, translated with introduction and commentary by Thomas L. Heath. Definitive edition. Textual and linguistic notes, mathematical analysis. 2,500 years of critical commentary. Unabridged. 1,414pp. 5⅜ x 8½. Three-vol. set. Vol. I: 60088-2 Vol. II: 60089-0 Vol. III: 60090-4

GEOMETRY OF COMPLEX NUMBERS, Hans Schwerdtfeger. Illuminating, widely praised book on analytic geometry of circles, the Moebius transformation, and two-dimensional non-Euclidean geometries. 200pp. 5⅝ x 8¼. 63830-8

DIFFERENTIAL GEOMETRY, Heinrich W. Guggenheimer. Local differential geometry as an application of advanced calculus and linear algebra. Curvature, transformation groups, surfaces, more. Exercises. 62 figures. 378pp. 5⅜ x 8½. 63433-7

CURVATURE AND HOMOLOGY: Enlarged Edition, Samuel I. Goldberg. Revised edition examines topology of differentiable manifolds; curvature, homology of Riemannian manifolds; compact Lie groups; complex manifolds; curvature, homology of Kaehler manifolds. New Preface. Four new appendixes. 416pp. 5⅜ x 8½. 40207-X

History of Math

THE WORKS OF ARCHIMEDES, Archimedes (T. L. Heath, ed.). Topics include the famous problems of the ratio of the areas of a cylinder and an inscribed sphere; the measurement of a circle; the properties of conoids, spheroids, and spirals; and the quadrature of the parabola. Informative introduction. clxxxvi+326pp; supplement, 52pp. 5⅜ x 8½. 42084-1

A SHORT ACCOUNT OF THE HISTORY OF MATHEMATICS, W. W. Rouse Ball. One of clearest, most authoritative surveys from the Egyptians and Phoenicians through 19th-century figures such as Grassman, Galois, Riemann. Fourth edition. 522pp. 5⅜ x 8½. 20630-0

THE HISTORY OF THE CALCULUS AND ITS CONCEPTUAL DEVELOPMENT, Carl B. Boyer. Origins in antiquity, medieval contributions, work of Newton, Leibniz, rigorous formulation. Treatment is verbal. 346pp. 5⅜ x 8½. 60509-4

THE HISTORICAL ROOTS OF ELEMENTARY MATHEMATICS, Lucas N. H. Bunt, Phillip S. Jones, and Jack D. Bedient. Fundamental underpinnings of modern arithmetic, algebra, geometry, and number systems derived from ancient civilizations. 320pp. 5⅜ x 8½. 25563-8

A HISTORY OF MATHEMATICAL NOTATIONS, Florian Cajori. This classic study notes the first appearance of a mathematical symbol and its origin, the competition it encountered, its spread among writers in different countries, its rise to popularity, its eventual decline or ultimate survival. Original 1929 two-volume edition presented here in one volume. xxviii+820pp. 5⅜ x 8½. 67766-4

GAMES, GODS & GAMBLING: A History of Probability and Statistical Ideas, F. N. David. Episodes from the lives of Galileo, Fermat, Pascal, and others illustrate this fascinating account of the roots of mathematics. Features thought-provoking references to classics, archaeology, biography, poetry. 1962 edition. 304pp. 5⅜ x 8½. (Available in U.S. only.) 40023-9

OF MEN AND NUMBERS: The Story of the Great Mathematicians, Jane Muir. Fascinating accounts of the lives and accomplishments of history's greatest mathematical minds–Pythagoras, Descartes, Euler, Pascal, Cantor, many more. Anecdotal, illuminating. 30 diagrams. Bibliography. 256pp. 5⅜ x 8½. 28973-7

HISTORY OF MATHEMATICS, David E. Smith. Nontechnical survey from ancient Greece and Orient to late 19th century; evolution of arithmetic, geometry, trigonometry, calculating devices, algebra, the calculus. 362 illustrations. 1,355pp. 5⅜ x 8½. Two-vol. set. Vol. I: 20429-4 Vol. II: 20430-8

A CONCISE HISTORY OF MATHEMATICS, Dirk J. Struik. The best brief history of mathematics. Stresses origins and covers every major figure from ancient Near East to 19th century. 41 illustrations. 195pp. 5⅜ x 8½. 60255-9

Physics

OPTICAL RESONANCE AND TWO-LEVEL ATOMS, L. Allen and J. H. Eberly. Clear, comprehensive introduction to basic principles behind all quantum optical resonance phenomena. 53 illustrations. Preface. Index. 256pp. 5⅜ x 8½. 65533-4

QUANTUM THEORY, David Bohm. This advanced undergraduate-level text presents the quantum theory in terms of qualitative and imaginative concepts, followed by specific applications worked out in mathematical detail. Preface. Index. 655pp. 5⅜ x 8½. 65969-0

ATOMIC PHYSICS: 8th edition, Max Born. Nobel laureate's lucid treatment of kinetic theory of gases, elementary particles, nuclear atom, wave-corpuscles, atomic structure and spectral lines, much more. Over 40 appendices, bibliography. 495pp. 5⅜ x 8½. 65984-4

A SOPHISTICATE'S PRIMER OF RELATIVITY, P. W. Bridgman. Geared toward readers already acquainted with special relativity, this book transcends the view of theory as a working tool to answer natural questions: What is a frame of reference? What is a "law of nature"? What is the role of the "observer"? Extensive treatment, written in terms accessible to those without a scientific background. 1983 ed. xlviii+172pp. 5⅜ x 8½. 42549-5

AN INTRODUCTION TO HAMILTONIAN OPTICS, H. A. Buchdahl. Detailed account of the Hamiltonian treatment of aberration theory in geometrical optics. Many classes of optical systems defined in terms of the symmetries they possess. Problems with detailed solutions. 1970 edition. xv+360pp. 5⅜ x 8½. 67597-1

PRIMER OF QUANTUM MECHANICS, Marvin Chester. Introductory text examines the classical quantum bead on a track: its state and representations; operator eigenvalues; harmonic oscillator and bound bead in a symmetric force field; and bead in a spherical shell. Other topics include spin, matrices, and the structure of quantum mechanics; the simplest atom; indistinguishable particles; and stationary-state perturbation theory. 1992 ed. xiv+314pp. 6⅛ x 9¼. 42878-8

LECTURES ON QUANTUM MECHANICS, Paul A. M. Dirac. Four concise, brilliant lectures on mathematical methods in quantum mechanics from Nobel Prize–winning quantum pioneer build on idea of visualizing quantum theory through the use of classical mechanics. 96pp. 5⅜ x 8½. 41713-1

THIRTY YEARS THAT SHOOK PHYSICS: The Story of Quantum Theory, George Gamow. Lucid, accessible introduction to influential theory of energy and matter. Careful explanations of Dirac's anti-particles, Bohr's model of the atom, much more. 12 plates. Numerous drawings. 240pp. 5⅜ x 8½. 24895-X

ELECTRONIC STRUCTURE AND THE PROPERTIES OF SOLIDS: The Physics of the Chemical Bond, Walter A. Harrison. Innovative text offers basic understanding of the electronic structure of covalent and ionic solids, simple metals, transition metals and their compounds. Problems. 1980 edition. 582pp. 6⅛ x 9¼. 66021-4

QUANTUM MECHANICS: Principles and Formalism, Roy McWeeny. Graduate student-oriented volume develops subject as fundamental discipline, opening with review of origins of Schrödinger's equations and vector spaces. Focusing on main principles of quantum mechanics and their immediate consequences, it concludes with final generalizations covering alternative "languages" or representations. 1972 ed. 15 figures. xi+155pp. 5⅜ x 8½. 42829-X

INTRODUCTION TO QUANTUM MECHANICS WITH APPLICATIONS TO CHEMISTRY, Linus Pauling & E. Bright Wilson, Jr. Classic undergraduate text by Nobel Prize winner applies quantum mechanics to chemical and physical problems. Numerous tables and figures enhance the text. Chapter bibliographies. Appendices. Index. 468pp. 5⅜ x 8½. 64871-0

METHODS OF THERMODYNAMICS, Howard Reiss. Outstanding text focuses on physical technique of thermodynamics, typical problem areas of understanding, and significance and use of thermodynamic potential. 1965 edition. 238pp. 5⅜ x 8½. 69445-3

TENSOR ANALYSIS FOR PHYSICISTS, J. A. Schouten. Concise exposition of the mathematical basis of tensor analysis, integrated with well-chosen physical examples of the theory. Exercises. Index. Bibliography. 289pp. 5⅜ x 8½. 65582-2

THE ELECTROMAGNETIC FIELD, Albert Shadowitz. Comprehensive undergraduate text covers basics of electric and magnetic fields, builds up to electromagnetic theory. Also related topics, including relativity. Over 900 problems. 768pp. 5⅝ x 8¼. 65660-8

GREAT EXPERIMENTS IN PHYSICS: Firsthand Accounts from Galileo to Einstein, Morris H. Shamos (ed.). 25 crucial discoveries: Newton's laws of motion, Chadwick's study of the neutron, Hertz on electromagnetic waves, more. Original accounts clearly annotated. 370pp. 5⅜ x 8½. 25346-5

RELATIVITY, THERMODYNAMICS AND COSMOLOGY, Richard C. Tolman. Landmark study extends thermodynamics to special, general relativity; also applications of relativistic mechanics, thermodynamics to cosmological models. 501pp. 5⅜ x 8½. 65383-8

STATISTICAL PHYSICS, Gregory H. Wannier. Classic text combines thermodynamics, statistical mechanics, and kinetic theory in one unified presentation of thermal physics. Problems with solutions. Bibliography. 532pp. 5⅜ x 8½. 65401-X

Paperbound unless otherwise indicated. Available at your book dealer, online at **www.doverpublications.com**, or by writing to Dept. GI, Dover Publications, Inc., 31 East 2nd Street, Mineola, NY 11501. For current price information or for free catalogs (please indicate field of interest), write to Dover Publications or log on to **www.doverpublications.com** and see every Dover book in print. Dover publishes more than 500 books each year on science, elementary and advanced mathematics, biology, music, art, literary history, social sciences, and other areas.